Student Solutions Manual for
Barnett, Ziegler, and Byleen's
ANALYTIC TRIGONOMETRY
WITH APPLICATIONS

Seventh Edition

FRED SAFIER
City College of San Francisco

Brooks/Cole Publishing Company

I(T)P® *An International Thomson Publishing Company*

Pacific Grove • Albany • Belmont • Bonn • Boston • Cincinnati • Detroit
Johannesburg • London • Madrid • Melbourne • Mexico City • New York
Paris • Singapore • Tokyo • Toronto • Washington

Sponsoring Editor: *Melissa Henderson*
Marketing Team: *Caroline Croley, Debra Johnston*
Editorial Assistants: *Shelley Gesicki, Joanne Von
    Zastrow*

Production: *Dorothy Bell*
Cover Design: *Roger Knox*
Cover Art: *Istvan Bodoczky*
Printing and Binding: *Webcom Limited*

For more information, contact:

BROOKS/COLE PUBLISHING COMPANY
511 Forest Lodge Road
Pacific Grove, CA 93950
USA

International Thomson Publishing Europe
Berkshire House 168-173
High Holborn
London WC1V 7AA
England

Thomas Nelson Australia
102 Dodds Street
South Melbourne, 3205
Victoria, Australia

Nelson Canada
1120 Birchmount Road
Scarborough, Ontario
Canada M1K 5G4

International Thomson Editores
Seneca 53
Col. Polanco
11560 México, D. F., México

International Thomson Publishing GmbH
Königswinterer Strasse 418
53227 Bonn
Germany

International Thomson Publishing Asia
60 Albert Street
#15-01 Albert Complex
Singapore 189969

International Thomson Publishing Japan
Hirakawacho Kyowa Building, 3F
2-2-1 Hirakawacho
Chiyoda-ku, Tokyo 102
Japan

Printed in Canada

10   9   8   7   6   5   4   3

ISBN 0-534-35839-X

CONTENTS

Chapter 1   Right Triangle Ratios

EXERCISE 1.1   Angles, Degrees, and Arcs

1. Since one complete revolution has measure $360°$, $\frac{1}{2}$ revolution has measure $\frac{1}{2}(360°) = 180°$.

3. Since one complete revolution has measure $360°$, $\frac{1}{8}$ revolution has measure $\frac{1}{8}(360°) = 45°$.

5. Since one complete revolution has measure $360°$, $\frac{2}{3}$ revolution has measure $\frac{2}{3}(360°) = 240°$.

7. Since $123°$ is between $90°$ and $180°$, this is an obtuse angle.

9. A $180°$ angle is called a straight angle.

11. Since $45°$ is between $0°$ and $90°$, this is an acute angle.

13. None of these.

15. An angle of one degree measure is formed by rotating the terminal side of the angle $\frac{1}{360}$ of a complete revolution in a counterclockwise direction from the initial side.

17. Since $21' = \frac{21°}{60}$ and $4" = \frac{4°}{3,600}$, then $43°21'4" = \left(43 + \frac{21}{60} + \frac{4}{3,600}\right)° \approx 43.351°$ to three decimal places

19. Since $12' = \frac{12°}{60}$ and $47" = \frac{47°}{3,600}$, then $2°12'47" = \left(2 + \frac{12}{60} + \frac{47}{3,600}\right)° \approx 2.213°$ to three decimal places

21. Since $17' = \frac{17°}{60}$ and $41" = \frac{41°}{3,600}$, then $103°17'41" = \left(103 + \frac{17}{60} + \frac{41}{3,600}\right)° \approx 103.295°$ to three decimal places

23. $13.633° = 13°(0.633 \times 60)' = 13°37.98' = 13°37'(0.98 \times 60)" \approx 13°37'59"$

25. $83.017° = 83°(0.017 \times 60)' = 83°1.02' = 83°1'(0.02 \times 60)" \approx 83°1'1"$

27. $187.204° = 187°(0.204 \times 60)' = 187°12.24' = 187°12'(0.24 \times 60)' \approx 187°12'14"$

29. There are two methods.

    a. Convert the first to decimal degree form and compare with the second.
    $$47°33'41" = \left(47 + \frac{33}{60} + \frac{41}{3,600}\right)° \approx 47.561° \text{ to three decimal places}$$
    Since $47.561° < 47.572°$, $47.572°$ is larger.

    b. Convert the second to DMS form and compare with the first.
    $47.572° = 47°(0.572 \times 60)' = 47°34.32' = 47°34'(0.32 \times 60)" = 47°34'19.2"$
    Since $47°34'19.2" > 47°33'41"$, the $47°34'19.2°$ angle, or $47.572°$, is larger.

31. To compare α and β, we convert β to decimal form. Since $9' = \dfrac{9°}{60}$ and $17'' = \dfrac{17°}{3,600}$,

    then $27°9'17'' = \left(27 + \dfrac{9}{60} + \dfrac{17}{3,600}\right)° \approx 27.155°$. Since $27.155° < 27.163°$, we conclude that $\alpha < \beta$.

33. To compare α and β, we convert β to decimal form. Since $47' = \dfrac{47°}{60}$ and $13'' = \dfrac{13°}{3,600}$, then

    $12°47'13'' = \left(12 + \dfrac{47}{60} + \dfrac{13}{3,600}\right)° \approx 12.787°$. Since $12.807° > 12.787°$, we conclude that $\alpha > \beta$.

35. $47°37'49'' + 62°40'15'' \rightarrow$ DMS
    $110°18'4''$

37. $90° - 67°37'29'' \rightarrow$ DMS
    $22°22'31''$

39. Since $\dfrac{s}{C} = \dfrac{\theta}{360°}$, then

    $\dfrac{s}{1000 \text{ cm}} = \dfrac{36°}{360°}$

    $s = \dfrac{36}{360} (1000 \text{ cm}) = 100 \text{ cm}$

41. Since $\dfrac{s}{C} = \dfrac{\theta}{360°}$, then

    $\dfrac{25 \text{ km}}{C} = \dfrac{20°}{360°}$

    $C = \dfrac{360}{20} (25 \text{ km}) = 450 \text{ km}$

43. Since $\dfrac{s}{C} = \dfrac{\theta}{360°}$ and $C = 2\pi r$, then $\dfrac{s}{2\pi r} = \dfrac{\theta}{360°}$

    $\dfrac{s}{2(\pi)(5,400,000 \text{ mi})} \approx \dfrac{2.6°}{360°}$

    $s \approx \dfrac{2(\pi)(5,400,000 \text{ mi})(2.6)}{360}$

    $\approx 240,000 \text{ mi}$ (to nearest 10,000 mi)

45. Since $\dfrac{s}{C} = \dfrac{\theta}{360°}$ and $\theta = 12°31'4'' = \left(12 + \dfrac{31}{60} + \dfrac{4}{3,600}\right)° = 12.517°$, then

    $\dfrac{50.2 \text{ cm}}{C} \approx \dfrac{12.517°}{360°}$

    $C \approx \dfrac{360}{12.517} (50.2 \text{ cm}) \approx 1440 \text{ cm}$ (to nearest 10 cm)

47. Since $\dfrac{A}{\pi r^2} = \dfrac{\theta}{360°}$, then $\dfrac{A}{\pi(25.2 \text{ cm})^2} = \dfrac{47.3°}{360°}$

    $A \approx \dfrac{47.3}{360} (\pi)(25.2 \text{ cm})^2 \approx 262 \text{ cm}^2$

49. Since $\dfrac{A}{\pi r^2} = \dfrac{\theta}{360°}$, then $\dfrac{98.4 \text{ m}^2}{\pi(12.6 \text{ m})^2} = \dfrac{\theta}{360°}$

    $\theta = \dfrac{98.4}{(\pi)(12.6)^2} \cdot 360° = 71.0°$

51. Since $\dfrac{s}{C} = \dfrac{\theta}{360°}$ and $C = 2\pi r$, then $\dfrac{s}{2\pi r} = \dfrac{\theta}{360°}$

$$r = \dfrac{s}{2\pi} \cdot \dfrac{360°}{\theta}$$

$$\approx \dfrac{11.5 \text{ mm}}{2(\pi)} \cdot \dfrac{360}{118.2} \approx 5.58 \text{ mm}$$

53. Since $\dfrac{s}{C} = \dfrac{\theta}{360°}$ and $C = 2\pi r$, then $\dfrac{s}{2\pi r} = \dfrac{\theta}{360°}$

$$s = 2\pi r \cdot \dfrac{\theta}{360°}$$

$$\approx 2(\pi)(5.49 \text{ mm}) \cdot \dfrac{119.7}{360} \approx 11.5 \text{ mm}$$

In Problems 55–61 we use the diagram and reason as follows: Since the cities have the same longitude, θ is given by their difference in latitude.

55. Since $\dfrac{s}{C} = \dfrac{\theta}{360°}$ and $C = 2\pi r$, then $\dfrac{s}{2\pi r} = \dfrac{\theta}{360°}$; $\theta = 47°40' - 37°50' = 9°50' = \left(9 + \dfrac{50}{60}\right)°$

$$s = 2\pi r \cdot \dfrac{\theta}{360°} \approx 2(\pi)(3960 \text{ mi}) \cdot \dfrac{9 + \dfrac{50}{60}}{360} \approx 679 \text{ mi}$$

57. Since $\dfrac{s}{C} = \dfrac{\theta}{360°}$ and $C = 2\pi r$, then $\dfrac{s}{2\pi r} = \dfrac{\theta}{360°}$; $\theta = 40°50' - 32°50' = 8°$

$$s = 2\pi r \cdot \dfrac{\theta}{360°} \approx 2(\pi)(3960) \cdot \dfrac{8}{360} \approx 553 \text{ mi}$$

59. To find the length of s in nautical miles, since 1 nautical mile is the length of 1' on the circle shown in the diagram, we need only find how many minutes are in the angle θ. Since
$$\theta = 47°40' - 37°50' = 9°50' = (9 \times 60 + 50)', \ \theta = 590'$$
Therefore, $s = 590$ nautical miles.

61. To find the length of s in nautical miles, since 1 nautical mile is the length of 1' on the circle shown in the diagram, we need only find how many minutes are in the angle θ. Since
$$\theta = 40°50' - 32°50' = 8° = (8 \times 60)', \ \theta = 480'.$$
Therefore, $s = 480$ nautical miles.

63. (A) Since $\dfrac{s}{C} = \dfrac{\theta}{360°}$ and $C = 2\pi r$, then $\dfrac{s}{2\pi r} = \dfrac{\theta}{360°}$; $\theta = 8°$ and $r = 500$ ft.

Hence
$$s = 2\pi r \cdot \dfrac{\theta}{360°} = 2\pi \cdot 500 \ \dfrac{8°}{360°} \approx 70 \text{ ft.}$$

(B) The arc length of a circular sector is very close to the chord length if the central angle of the sector is small and the radius of the sector is large, which is the case in this problem.

EXERCISE 1.2   Similar Triangles

1. The measures of the third angle in each triangle are the same, since the sum of the measures of the three angles in any triangle is 180°. Thus, if $A + B + C = 180°$ and $A' + B' + C' = 180°$ and $A = A'$ and $B = B'$, then $C = 180° - (A + B) = 180° - (A' + B') = C'$.

3. Since $\dfrac{a}{a'} = \dfrac{b}{b'}$, by Euclid's Theorem, then $\dfrac{5}{2} = \dfrac{15}{b'}$, $b' = \dfrac{2(15)}{5} = 6$

5. Since $\dfrac{a}{a'} = \dfrac{c}{c'}$, by Euclid's Theorem, then $\dfrac{12}{2.0} = \dfrac{c}{18}$, $c = \dfrac{(18)(12)}{2.0} = 110$

7. Since $\dfrac{b}{b'} = \dfrac{c}{c'}$, by Euclid's Theorem, then $\dfrac{52,000}{8.0} = \dfrac{18,000}{c'}$, $c' = \dfrac{(8.0)(18,000)}{52,000} = 2.8$

9. Two similar triangles can have equal sides if, and only if, they are congruent, that is, if the two triangles would coincide when one is moved on top of the other.

11. Since the triangles are similar, the sides are proportional and we can write

$$\dfrac{a}{0.47} = \dfrac{51 \text{ in}}{1.0} \qquad\qquad \dfrac{c}{1.1} = \dfrac{51 \text{ in}}{1.0}$$

$$a = \dfrac{(0.47)(51 \text{ in})}{1.0} = 24 \text{ in} \qquad c = \dfrac{(1.1)(51 \text{ in})}{1.0} = 56 \text{ in}$$

13. Since the triangles are similar, the sides are proportional and we can write

$$\dfrac{b}{1.0} = \dfrac{23.4 \text{ m}}{0.47} \qquad\qquad \dfrac{c}{1.1} = \dfrac{23.4 \text{ m}}{0.47}$$

$$b = \dfrac{23.4 \text{ m}}{0.47} = 50 \text{ m} \qquad c = \dfrac{(1.1)(23.4 \text{ m})}{0.47} = 55 \text{ m}$$

15. Since the triangles are similar, the sides are propotional and we can write

$$\dfrac{a}{0.47} = \dfrac{2.478 \times 10^9 \text{ yd}}{1.0} \qquad\qquad \dfrac{c}{1.1} = \dfrac{2.489 \times 10^9 \text{ yd}}{1.0}$$

$$a = (0.47)(2.489 \times 10^9 \text{ yd}) \qquad c = (1.1)(2.489 \times 10^9 \text{ yd})$$
$$= 1.2 \times 10^9 \text{ yd} \qquad\qquad = 2.7 \times 10^9 \text{ yd}$$

17. Since the triangles are similar, the sides are proportional and we can write

$$\dfrac{a}{0.47} = \dfrac{8.39 \times 10^{-5} \text{ mm}}{1.1} \qquad\qquad \dfrac{b}{1.0} = \dfrac{8.39 \times 10^{-5} \text{ mm}}{1.1}$$

$$a = \dfrac{(0.47)(8.39 \times 10^{-5} \text{ mm})}{1.1} \qquad b = \dfrac{8.39 \times 10^{-5} \text{ mm}}{1.1}$$
$$= 3.6 \times 10^{-5} \text{ mm} \qquad\qquad = 7.6 \times 10^{-5} \text{ mm}$$

19. We make a scale drawing of the triangle, choosing a' to be 2.00 in, $\angle A' = 70°$, $\angle C' = 90°$. Now measure c' (approximately 2.13 in) and set up a proportion. Thus,

$$\dfrac{c}{2.13 \text{ in}} = \dfrac{101 \text{ ft}}{2.00 \text{ in}}$$

$$c \approx \dfrac{2.13}{2.00}(101 \text{ ft}) \approx 108 \text{ ft}$$

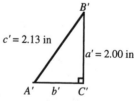

Drawing not to scale

21. In the drawing, we note that triangles *LBT* and *LNM* are similar.
$MN = 9$ ft.

$NB = \dfrac{1}{2}$ (length of court) $= \dfrac{1}{2}$ (78 ft) $= 39$ ft. $TB = 3$ ft. *BL* is to be found. Let

$BL = x$. Then, $\dfrac{BL}{NL} = \dfrac{TB}{MN}$.

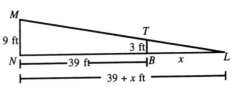

$NL = NB + BL$. Thus, $\dfrac{x}{39 + x} = \dfrac{3}{9}$.

$9(39 + x)\dfrac{x}{39 + x} = 9(39 + x)\dfrac{3}{9}$ (clear of

fractions)

$$9x = (39 + x)3 = 117 + 3x$$
$$6x = 117$$
$$x = 19.5 \text{ ft}$$

23. Since the triangles *ABC* and *DEC* in the figure are similar, we can write $\dfrac{AB}{DE} = \dfrac{AC}{CD}$. Then,

$$\frac{AB}{5.5 \text{ ft}} = \frac{24 \text{ ft}}{2.1 \text{ ft}}$$
$$AB = \frac{5.5}{2.1}(24 \text{ ft}) = 63 \text{ ft}$$

25. Let us make a scale drawing of the figure in the
text as follows: pick any convenient length, say
2 in, for $A'C'$. Copy the 15° angle *CAB* and 90°
angle *ACB* using a protractor. Now, measure
$B'C'$ (approximately 0.55 in) and set up a
proportion. Thus,

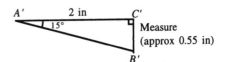

Measure (approx 0.55 in)

$$\frac{x}{0.55 \text{ in}} = \frac{4.0 \text{ km}}{2 \text{ in}}$$
$$x \approx \frac{0.55}{2}(4.0 \text{ km}) \approx 1.1 \text{ km}$$

27. (A) Triangles *PAC*, *FBC*, *ACP'*, and *ABF'* are all right triangles. Angles *APC* and *BFC* are equal,
and angles *CP'A* and *BF'A* are equal--alternate interior angles of parallel lines cut by a transversal
are equal (see Text, Appendix C.1). Thus, triangles *PAC* and *FBC* are similar, and triangles
ACP' and *ABF'* are similar, since, in each case, the angles of each triangle are equal.

(B) Since the triangles *PAC* and *FBC* are similar, corresponding sides are proportional. Hence,

$\dfrac{AC}{PA} = \dfrac{BC}{FB}$ or $\dfrac{AB + BC}{PA} = \dfrac{BC}{FB}$ or $\dfrac{h + h'}{u} = \dfrac{h'}{f}$

Since the triangles *ACP'* and *ABF'* are also similar, corresponding sides are also proportional.
Hence,

$\dfrac{AC}{CP'} = \dfrac{AB}{BF'}$ or $\dfrac{AB + BC}{CP'} = \dfrac{AB}{FB}$ or $\dfrac{h + h'}{v} = \dfrac{h}{f}$

(C) Adding the results in part (B) yields

$$\frac{h + h'}{u} + \frac{h + h'}{v} = \frac{h'}{f} + \frac{h}{f}$$

or

$$\frac{h + h'}{u} + \frac{h + h'}{v} = \frac{h + h'}{f}$$

Dividing both sides by $h + h'$, we obtain

$$\frac{1}{u} + \frac{1}{v} = \frac{1}{f}$$

(D) Use the formula just derived with $u = 3$ m = 3000 mm and $f = 50$ mm. Then

$$\frac{1}{3000} + \frac{1}{v} = \frac{1}{50}$$

$$\frac{1}{v} = \frac{1}{50} - \frac{1}{3000}$$

$$\frac{1}{v} = \frac{59}{3000}$$

$$v = \frac{3000}{59} \approx 50.847 \text{ mm}$$

EXERCISE 1.3   Trigonometric Ratios and Right Triangles

1. $\dfrac{a}{c}$ 3. $\dfrac{b}{a}$ 5. $\dfrac{c}{a}$ 7. sin θ 9. tan θ 11. csc θ

13.   Set calculator in degree mode and use sin key.
        sin 25.6° = 0.432.

15.   Set calculator in degree mode, convert to decimal degrees, and use tan key.

$$35°20' = \left(35 + \frac{20}{60}\right)^{\circ} = (35.3333333\ldots)^{\circ}$$

        tan 35°20' = tan(35.333333...)° = 0.709

17.   Use the reciprocal relationship sec θ = 1/cos θ. Set calculator in degree mode, use cos key, then take reciprocal.
        sec 44.8° = 1.41

19.   Set calculator in degree mode and use cos key.
        cos 72.9° = 0.294

21.   Use the reciprocal relationship cot θ = 1/tan θ. Set calculator in degree mode, use tan key, then take reciprocal.
        cot 54.9° = 0.703

23.   Use the reciprocal relationship csc θ = 1/sin θ. Set calculator in degree mode, convert to decimal degrees, use sin key, then take reciprocal.

$$67°30' = \left(67 + \frac{30}{60}\right)^{\circ} = 67.5°$$

        csc 67°30' = csc 67.5° = 1.08

25. If $\sin \theta = 0.8032$, then
$\theta = \sin^{-1} 0.8032 = 53.44°$

27. $\theta = \arccos 0.7153$
$= 44.332° = 44°20'$

29. $\theta = \tan^{-1} 1.948$
$= 62.826° = 62°50'$

31. The triangle is uniquely determined. Angle α can be found using $\tan \alpha = \dfrac{a}{b}$; angle $\beta = 90° - \alpha$; The hypotenuse c can be found using the Pythagorean Theorem or by using $\sin \alpha = \dfrac{b}{c}$.

33. The triangle is not uniquely determined. In fact, there are infinitely many different size triangles with the same acute angles--all are similar to each other.

35. *Solve for the complementary angle:* $90° - \theta = 90° - 58°40' = 31°20'$

Solve for b: Since $\theta = 58°40' = \left(58 + \dfrac{40}{60}\right)° = (58.666...)°$ and $c = 15.0$ mm, we look for a trigonometric ratio that involves θ and c (the known quantities) and b (the unknown quantity). We choose the sine.

$\sin \theta = \dfrac{b}{c}$

$b = c \sin \theta = (15.0 \text{ mm})(\sin 58.666...°) = 12.8$ mm

Solve for a: We choose the cosine to find a. Thus,

$\cos \theta = \dfrac{a}{c}$

$a = c \cos \theta = (15.0)(\cos 58.666...°) = 7.80$ mm

37. *Solve for the complementary angle:* $90° - \theta = 90° - 83.7° = 6.3°$
Solve for a: Since $\theta = 83.7°$ and $b = 3.21$ km, we look for a trigonometric ratio that involves θ and b (the known quantities) and a (the unknown quantity). We choose the tangent.

$\tan \theta = \dfrac{b}{a}$

$a = \dfrac{b}{\tan \theta} = \dfrac{3.21 \text{ km}}{\tan 83.7°} = 0.354$ km

Solve for c: We choose the sine to find c. Thus,

$\sin \theta = \dfrac{b}{c}$

$c = \dfrac{b}{\sin \theta} = \dfrac{3.21 \text{ km}}{\sin 83.7°} = 3.23$ km

39. *Solve for the complementary angle:* $90° - \theta = 90° - 71.5° = 18.5°$
Solve for a: Since $\theta = 71.5°$ and $b = 12.8$ in, we look for a trigonometric ratio that involves θ and b (the known quantities) and a (the unknown quantity). We choose the tangent.

$\tan \theta = \dfrac{b}{a}$

$a = \dfrac{b}{\tan \theta} = \dfrac{12.8 \text{ in}}{\tan 71.5°} = 4.28$ in

Solve for c: We choose the sine to find c. Thus,

$\sin \theta = \dfrac{b}{c}$

$c = \dfrac{b}{\sin \theta} = \dfrac{12.8 \text{ in}}{\sin 71.5°} = 13.5$ in

41. *Solve for θ:* $\sin\theta = \dfrac{b}{c} = \dfrac{63.8\text{ ft}}{134\text{ ft}} = 0.476$

$\theta = \sin^{-1} 0.476 = 28.4° = 28°30'$ to nearest 10'

Solve for the complementary angle: $90° - \theta = 90° - 28°30' = 61°30'$

Solve for a: We will use the tangent. Thus, $\tan\theta = \dfrac{b}{a}$

$$a = \frac{b}{\tan\theta} = \frac{63.8\text{ ft}}{\tan 28°30'} = 118\text{ ft}$$

43. *Solve for θ:* $\tan\theta = \dfrac{b}{a} = \dfrac{132\text{ mi}}{108\text{ mi}} = 1.22$

$\theta = \tan^{-1} 1.22 = 50.7°$ to nearest 0.1°

Solve for the complementary angle: $90° - \theta = 90° - 50.7° = 39.3°$

Solve for c: We will use the sine. Thus, $\sin\theta = \dfrac{b}{c}$

$$c = \frac{b}{\sin\theta} = \frac{132\text{ mi}}{\sin 50.7°} = 171\text{ mi}$$

45. The calculator was accidentally set in radian mode. Changing the mode to degree, $a = 235\sin(14.1) = 57.2$ m and $b = 235\cos(14.1) = 228$ m.

47. (A) If $\theta = 11°$, $(\sin\theta)^2 + (\cos\theta)^2 = (\sin 11°)^2 + (\cos 11°)^2 = (0.1908...)^2 + (0.9816...)^2 = 1$

(B) If $\theta = 6.09°$, $(\sin\theta)^2 + (\cos\theta)^2 = (\sin 6.09°)^2 + (\cos 6.09°)^2 = (0.106...)^2 + (0.994...)^2 = 1$

(C) If $\theta = 43°24'47''$, $(\sin\theta)^2 + (\cos\theta)^2 = (\sin 43°24'47'')^2 + (\cos 43°24'47'')^2$
$= (0.687...)^2 + (0.726...)^2 = 1$

49. (A) If $\theta = 19°$, $\sin\theta - \cos(90° - \theta) = \sin 19° - \cos(90° - 19°)$
$= \sin 19° - \cos 71° = 0.3256 - 0.3256 = 0$

(B) If $\theta = 49.06°$, $\sin\theta - \cos(90° - \theta) = \sin 49.06° - \cos(90° - 49.06°)$
$= \sin 49.06° - \cos 40.94° = 0.7554 - 0.7554 = 0$

(C) If $\theta = 72°51'12''$, $\sin\theta - \cos(90° - \theta) = \sin 72°51'12'' - \cos(90° - 72°51'12'')$
$= \sin 72°51'12'' - \cos(17°8'48'')$
$= 0.9556 - 0.9556 = 0$

51. *Solve for the complementary angle:* $90° - \theta = 90° - 83°12' = 6°48'$

Solve for b: We choose the tangent to find b. Thus, $\tan\theta = \dfrac{b}{a}$

$$\begin{aligned}
b &= a\tan\theta \\
&= (23.82\text{ mi})(\tan 83°12') \\
&= 199.8\text{ mi}
\end{aligned}$$

Solve for c: We choose the cosine to find c. Thus, $\cos\theta = \dfrac{a}{c}$

$$c = \frac{a}{\cos\theta} = \frac{23.82\text{ mi}}{(\cos 83°12')} = 201.2\text{ mi}$$

53.   *Solve for θ:* $\tan\theta = \dfrac{b}{a} = \dfrac{42.39 \text{ cm}}{56.04 \text{ cm}}$; $\tan\theta = 0.7564$; $\theta = \tan^{-1} 0.7564 = 37.105° = 37°6'$

Solve for the complementary angle: $90° - \theta = 90° - 37°6' = 52°54'$

Solve for c: We will use the sine. Thus, $\sin\theta = \dfrac{b}{c}$

$$c = \dfrac{b}{\sin\theta} = \dfrac{42.39 \text{ cm}}{\sin 37°6'} = 70.27 \text{ cm}$$

55.   *Solve for θ:* $\sin\theta = \dfrac{b}{c} = \dfrac{35.06 \text{ cm}}{50.37 \text{ cm}} = 0.6960$

$$\theta = \sin^{-1}(0.6960) = 44.11°$$

Solve for the complementary angle: $90° - \theta = 90° - 44.11° = 45.89°$

Solve for a: We choose the cosine to find a. Thus, $\cos\theta = \dfrac{a}{c}$

$$a = c\cos\theta$$
$$= (50.37 \text{ cm})(\cos 44.11°) = 36.17 \text{ cm}$$

57.   According to the Pythagorean theorem, $a^2 + b^2 = c^2$. Then, using Definition 1, we have

$$(\sin\theta)^2 + (\cos\theta)^2 = \left(\dfrac{b}{c}\right)^2 + \left(\dfrac{a}{c}\right)^2 = \dfrac{b^2 + a^2}{c^2} = \dfrac{a^2 + b^2}{c^2} = \dfrac{c^2}{c^2} = 1$$

59.   (A) According to Definition 1, $\cot\theta = \dfrac{a}{b} = \dfrac{1}{b/a} = \dfrac{1}{\tan\theta}$

(B) According to Definition 1, $\csc(90° - \theta) = \dfrac{c}{a} = \sec\theta$

61.   (A) In right triangle OAD, $\sin\theta = \dfrac{\text{Opp}}{\text{Hyp}} = \dfrac{AD}{OC} = \dfrac{AD}{1} = AD$

(B) In right triangle OCD, $\tan\theta = \dfrac{\text{Opp}}{\text{Adj}} = \dfrac{DC}{OD} = \dfrac{DC}{1} = DC$

(C) In right triangle ODE, $\csc\theta = \csc OED = \dfrac{\text{Hyp}}{\text{Opp}} = \dfrac{OE}{OD} = \dfrac{OE}{1} = OE$

63.   (A) As θ approaches $90°$, AD approaches OD, which has measure 1. Thus, $\sin\theta\,(= AD)$ approaches 1.

(B) As θ approaches $90°$, EC approaches being parallel to the x axis. Thus, DC increases without bound, so $\tan\theta\,(= DC)$ increases without bound.

(C) As θ approaches $90°$, OE approaches OF, which has measure 1. Thus, $\csc\theta\,(= OE)$ approaches 1.

65.   (A) As θ approaches $0°$, OA approaches OB, which has measure 1. Thus, $\cos\theta\,(= OA)$ approaches 1.

(B) As θ approaches $0°$, EC approaches being parallel to the y axis. Thus, ED increases without bound, so $\cot\theta\,(= ED)$ increases without bound.

(C) As θ approaches $0°$, OC approaches OB, which has measure 1. Thus, $\sec\theta\,(= OC)$ approaches 1.

EXERCISE 1.4   Right Triangle Applications

1. $\sin 61° = \dfrac{\text{Opp}}{\text{Hyp}} = \dfrac{x}{8.0 \text{ m}}$

$x = (8.0 \text{ m})(\sin 61°) = 7.0 \text{ m}$

3.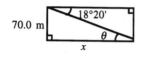

$\cot \theta = \dfrac{\text{Adj}}{\text{Opp}}$

$\cot 18°20' = \dfrac{x}{70.0 \text{ m}}$

$x = (70.0 \text{ m})(\cot 18°20') = 211 \text{ m}$

(If line p crosses parallel lines m and n then angles α and β have the same measure. Thus, $\theta = 18°20'$.)

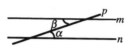

5.   We first sketch a figure and label the known parts.

$\tan 15° = \dfrac{\text{Opp}}{\text{Adj}}$

$= \dfrac{x}{4.0 \text{ km}}$

$x = (4.0 \text{ km})(\tan 15°)$

$= 1.1 \text{ km}$

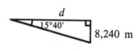

7.   We first sketch a figure and label the known parts.

$\tan 15°40' = \dfrac{\text{Opp}}{\text{Adj}} = \dfrac{8{,}240 \text{ m}}{d}$

$d = \dfrac{8{,}240 \text{ m}}{\tan 15°40'} = 29{,}400 \text{ m or } 29.4 \text{ km}$

9.   We first sketch a figure and label the known parts.

$\tan \theta = \dfrac{\text{Opp}}{\text{Adj}} = \dfrac{3{,}300 \text{ ft}}{8{,}200 \text{ ft}} = 0.40\ldots$

$\theta = \tan^{-1}(0.40\ldots) = 22°$

11. (A) In triangle ABC, $\angle \theta$ is complementary to 75°, thus $\theta = 15°$.

$BC = x = \text{roof overhang}$

$AC = 19 \text{ ft}$

$\tan \theta = \dfrac{\text{Opp}}{\text{Adj}} = \dfrac{x}{19 \text{ ft}}$

$x = (19 \text{ ft})(\tan 15°) = 5.1 \text{ ft}$

(B) In triangle $A'BC$, $\theta' = 27°$

$A'C = y = \text{how far down shadow will reach}$

$\tan 27° = \dfrac{\text{Opp}}{\text{Adj}} = \dfrac{y}{5.1 \text{ ft}}$

$y = (5.1 \text{ ft})(\tan 27°) = 2.6 \text{ ft}$

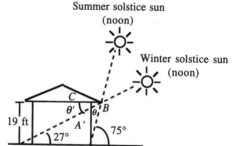

13. We first sketch a figure and label the known parts. From geometry we know that each angle of an equilateral triangle has measure 60°.

$$\sin 60° = \frac{h}{4.0 \text{ m}}$$
$$h = (4.0 \text{ m})(\sin 60°) = 3.5 \text{ m}$$

15. We note that since AB is a side of a nine-sided regular polygon,

$$\angle BCA = \frac{1}{9} \text{ (circumference)} = \frac{1}{9}(360°) = 40°.$$

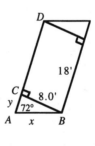

Since ABC is an isosceles triangle,

$$\angle FCA = \frac{1}{2}\angle BCA = \frac{1}{2}(40°) = 20°,$$

and also,

$$AF = \frac{1}{2}AB = \frac{1}{2}x.$$

Therefore, in right triangle AFC, $\sin 20° = \dfrac{(1/2)x}{8.32 \text{ cm}}$

$$x = 2(8.32 \text{ cm})(\sin 20°) = 5.69 \text{ cm}$$

17. We note: by the symmetry of the cone, ATC is an isosceles right triangle, hence
   $$\angle TAC = \angle TCA = 45°$$
   Since the mast is perpendicular to the deck, TBA and TBC are also right triangles, and since each has a 45° angle, these are also isosceles right triangles. Then it follows that $TB = AB$ and $TB = AC$. Since the length TB is given as 67.0 feet, the diameter

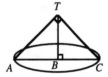

$$AC = AB + BC = 67.0 \text{ feet} + 67.0 \text{ feet}$$
$$= 134.0 \text{ ft.}$$

19. Label the reguired sides AD and AB. Then
   $$AD = AC + CD = AC + 18.$$

In right triangle ABC,

$$\sin A = \frac{BC}{AB} \qquad\qquad \tan A = \frac{BC}{AC}$$
$$\sin 72° = \frac{8.0 \text{ feet}}{x} \qquad \tan 72° = \frac{8.0 \text{ feet}}{y}$$
$$x = \frac{8.0 \text{ feet}}{\sin 72°} \qquad\qquad y = \frac{8.0 \text{ feet}}{\tan 72°}$$
$$= 8.4 \text{ feet} \qquad\qquad = 2.6 \text{ feet}$$

Thus the sides of the parallelogram are $AB = 8.4$ feet and $AD = 18$ feet + 2.6 feet ≈ 21 feet.

21. (A) We note that $\angle TSC = 90° - \alpha$, hence $\angle C = \alpha$.

   Thus, in triangle CST,
   $$\cos \alpha = \frac{\text{Adj}}{\text{Hyp}} = \frac{r}{r+h}$$

   (B) $(r+h)\cos \alpha = (r+h) \cdot \dfrac{r}{r+h}$
   $$r\cos \alpha + h\cos \alpha = r$$
   $$h\cos \alpha = r - r\cos \alpha$$
   $$h\cos \alpha = r(1 - \cos \alpha)$$
   $$r = \frac{h\cos \alpha}{1 - \cos \alpha}$$

Earth

(C) $r = \dfrac{(335 \text{ miles}) \cos 22°47'}{1 - \cos 22°47'} = 3960 \text{ miles}$

23.   We note that since ABC is an isosceles triangle,

$\angle FCA = \dfrac{1}{2} \angle BCA = \dfrac{1}{2}(2\theta) = \theta,$

and also,

$AF = \dfrac{1}{2} AB = \dfrac{1}{2}(6.0 \text{ mi}) = 3.0 \text{ mi}.$

We are to find AC, the radius of the circle.

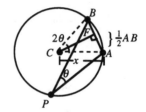

In right triangle AFC,

$\sin \angle FCA = \dfrac{AF}{AC} \qquad \sin 21° = \dfrac{3.0 \text{ mi}}{x}$

$\sin \theta = \dfrac{3.0 \text{ mi}}{x} \qquad x = \dfrac{3.0 \text{ mi}}{\sin 21°} = 8.4 \text{ mi}$

25.   In the figure, r denotes the radius of the parallel of latitude, R the radius of the
earth, i.e., the radius of the equator. Clearly, $\cos \theta = \dfrac{\text{Adj}}{\text{Hyp}} = \dfrac{r}{R}$.

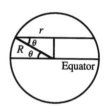

Since $L = 2\pi r$ and $E = 2\pi R$, we have

$r = R \cos \theta$

$2\pi r = 2\pi R \cos \theta$

$L = E \cos \theta$

E is given as 24,900 miles, In the particular case of San Francisco, $\theta = 38°$.
Hence,

$L = (24,900 \text{ mi}) \cos 38° = 19,600 \text{ mi}$

27.   (A) A lifeguard can run faster than he can swim, so it would seem that he should run along the beach
first before entering the water. [Parts (D) and (E) suggest how far the lifeguard should run before
swimming to get to the swimmer in the least time].

(B) Let $PB = y$ and $PS = x$. In the right triangle SPB,

$\csc \theta = \dfrac{x}{c} \qquad \cot \theta = \dfrac{y}{c}$

$x = c \csc \theta \qquad y = c \cot \theta$

Thus the lifeguard runs a distance $d - y = d - c \cot \theta$
at speed p.

This requires a time $t_1 = \dfrac{\text{distance run}}{\text{rate run}} = \dfrac{d - c \cot \theta}{p}$

The lifeguard then swims a distance $x = c \csc \theta$ at speed q.

This requires a time $t_2 = \dfrac{\text{distance swum}}{\text{rate swum}} = \dfrac{c \csc \theta}{q}$

Hence total time $T = t_1 + t_2 = \dfrac{d - c \cot \theta}{p} + \dfrac{c \csc \theta}{q}$

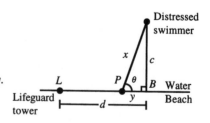

Distressed
swimmer

Lifeguard
tower

Water
Beach

(C) $T = \dfrac{(380 \text{ m}) - (76 \text{ m})\cot 51°}{5.1 \text{ m/sec}} + \dfrac{(76 \text{ m})\csc 51°}{1.7 \text{ m/sec}} = 119.97 \text{ sec}$

(D) T decreases, then increases as θ goes from 55° to 85°. T has a minimum value of 116.66 sec when $\theta = 70°$.

(E) We have already found that the distance run, $LP = d - c \cot \theta$.
Hence, $LP = 380 \text{ m} - (76 \text{ m}) \cot 70° = 352 \text{ m}$.

29. (A) Since the pipeline costs twice as much in the water as on the land it appears that the total cost of the pipeline would depend on θ.

(B) We note that the pipeline consists of ocean section TP, and shore section $PW = 10 \text{ mi} - SP$.

Let $x = TP$ and $y = SP$. In right triangle SPT, $\quad \cos \theta = \dfrac{4 \text{ mi}}{x}, \qquad \tan \theta = \dfrac{y}{4 \text{ mi}}$

$x = (4 \text{ mi})\sec \theta \qquad y = (4 \text{ mi}) \tan \theta$

Thus,
C = the cost of the pipeline

$= \left(\begin{array}{c}\text{cost of ocean}\\\text{section per mile}\end{array}\right)\left(\begin{array}{c}\text{number of}\\\text{ocean miles} = x\end{array}\right) + \left(\begin{array}{c}\text{cost of shore}\\\text{section per mile}\end{array}\right)\left(\begin{array}{c}\text{number of shore}\\\text{miles} = 10 - y\end{array}\right)$

$= \left(40,000 \,\dfrac{\$}{\text{mi}}\right)(4 \text{ mi})\sec \theta + \left(20,000 \,\dfrac{\$}{\text{mi}}\right)[10 - (4 \text{ mi})\tan \theta]$

$C = 160,000 \sec \theta + 20,000(10 - 4 \tan \theta)$

(C) $C = 160,000 \sec 15° + 20,000(10 - 4 \tan 15°) = \$344,200$

(D) As θ increases from 15° to 45°, C decreases and then increases. C has the minimum value of \$338,600 when $\theta = 30°$.

(E) From part (A), the shore section $= 10 - y = 10 - (4 \text{ mi})\tan \theta = 10 - 4 \tan 30° = 7.69$ miles. The ocean section $= x = (4 \text{ mi})\sec \theta = 4 \sec 30° = 4.62$ miles

31. A simple way to solve the system of equations

$\tan 42° = \dfrac{y}{x} \qquad\qquad \tan 25° = \dfrac{y}{1.0 + x}$ for y is to clear of fractions,

then eliminate x from the resulting equivalent system of equations.
$x \tan 42° = y \qquad (1.0 + x)(\tan 25°) = y$

$x = \dfrac{y}{\tan 42°}$

Therefore, $\left(1.0 + \dfrac{y}{\tan 42°}\right)(\tan 25°) = y$

$\tan 25° + \dfrac{\tan 25°}{\tan 42°} y = y \qquad$ (Distributive property)

$\tan 25° = y - \dfrac{\tan 25°}{\tan 42°} y = \left(1 - \dfrac{\tan 25°}{\tan 42°}\right)y$

$y = \dfrac{\tan 25°}{1 - \dfrac{\tan 25°}{\tan 42°}} = 0.97 \text{ km}$

33.   Labeling the diagram with the information given in the
      problem we note: We are asked to find $d =$ how far apart the
      two buildings are, and $h =$ the height of the apartment
      building. Note also that $x + h = 847$, so that $x = 847 - h$.

      In right triangle SLT, $\tan 43.2° = \dfrac{x}{d} = \dfrac{847 - h}{d}$.

      In right triangle SLB, $\tan 51.4° = \dfrac{h}{d}$.

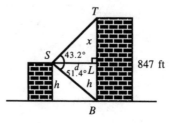

      We solve the system of equations $\tan 51.4° = \dfrac{h}{d}$ and $\tan 43.2° = \dfrac{847 - h}{d}$ by clearing of fractions,
      then eliminating h.

      (1) $d \tan 51.4° = h$ and $d \tan 43.2° = 847 - h$
          Adding, $d \tan 51.4° + d \tan 43.2° = 847$
          $$d(\tan 51.4° + \tan 43.2°) = 847$$
          $$d = \frac{847}{\tan 51.4° + \tan 43.2°} = 386 \text{ ft apart}$$
      Substituting in (1), $h = d \tan 51.4° = (386 \text{ ft}) \tan 51.4° = 484$ ft high.

35.   We are given $t = 2$ sec, $v = 11.1$ ft/sec, $\theta = 10.0°$. Thus,
      $$g = \frac{v}{(\sin \theta)t} = \frac{11.1 \text{ ft/sec}}{(\sin 10.0°)(2 \text{ sec})} = 32.0 \text{ ft/sec}^2$$

37.   From the Pythagorean theorem:

Since $AB = s$ and $MB = \dfrac{s}{2}$ Since $CN = CM = \dfrac{s}{2}$

$\quad A M^2 = AB^2 + MB^2$ $\quad N M^2 = CN^2 + CM^2$

$\qquad = s^2 + \left(\dfrac{s}{2}\right)^2 = \dfrac{5s^2}{4}$ $\qquad = \left(\dfrac{s}{2}\right)^2 + \left(\dfrac{s}{2}\right)^2 = \dfrac{2s^2}{4}$

$\quad AM = s\dfrac{\sqrt{5}}{2}$ $\quad NM = s\dfrac{\sqrt{2}}{2}$

From the fact that $NA = MA$, thus triangle, AMN is isosceles: AE bisects NM, hence

$$ME = \frac{1}{2}NM = \frac{1}{2} \cdot s\frac{\sqrt{2}}{2} = s\frac{\sqrt{2}}{4}$$

From the Pythagorean theorem, once again:

$$AE^2 + EM^2 = MA^2$$

$$AE^2 + \left(s\frac{\sqrt{2}}{4}\right)^2 = \left(s\frac{\sqrt{5}}{2}\right)^2$$

$$AE^2 + \frac{2s^2}{16} = \frac{5s^2}{4}$$

$$AE^2 = \frac{18s^2}{16}$$

$$AE = s\frac{3\sqrt{2}}{4}$$

From the fact that triangle AMN is isosceles, once again:

$$\angle NMA = \angle MNA = \frac{1}{2}(180° - \theta)$$

Since $\angle NFM = \angle AEN = 90°$ and $\angle NMF = \angle ENA = \frac{1}{2}(180° - \theta)$, triangles NMF and ANE are similar. Hence,

$$\frac{NF}{NM} = \frac{AE}{AN} = \frac{AE}{MA} = s \frac{3\sqrt{2}}{4} \div s \frac{\sqrt{5}}{2} = \frac{3\sqrt{2}}{2\sqrt{5}}.$$

Finally, $\sin \theta = \frac{NF}{NA} = \frac{NF}{NM} \cdot NM \div NA = \frac{3\sqrt{2}}{2\sqrt{5}} \cdot s \frac{\sqrt{2}}{2} \div s \frac{\sqrt{5}}{2} = \frac{3\sqrt{2}\sqrt{2}}{4\sqrt{5}} \cdot \frac{2}{\sqrt{5}} = \frac{3}{5}$

CHAPTER 1   REVIEW EXERCISE

1. $2°1'20" = (2 \cdot 60 \cdot 60 + 1 \cdot 60 + 20)" = 7,280"$

2.   Since a circumference has degree measure 360, $\frac{1}{6}$ a circumference has degree measure $\frac{1}{6}$ of 360;

     that is, $\frac{1}{6}(360) = 60$ degrees, written 60°.

3.   Since $\frac{a}{a'} = \frac{c}{c'}$ by Euclid's Theorem, then $\frac{a}{2} = \frac{20,000}{5}$; $a = \frac{(2)(20,000)}{5} = 8,000$

4.   Since $23' = \frac{23°}{60}$, then $36°23' = \left(36 + \frac{23}{60}\right)° \approx 36.38°$ to two decimal places

5.   An angle of degree measure 1 is an angle formed by rotating the terminal side of the angle 1/360th of a complete revolution in a counterclockwise direction from the initial side.

6.   First find the third angle in each triangle. In the first, $A + B + C = 180°$, thus $A = 180° - B - C = 180° - 120° - 39° = 21°$. In the second, $A + B + C = 180°$, thus $B = 180° - A - C = 180° - 21° - 120° = 39°$. In the third, $A + B + C = 180°$, thus $C = 180° - A - B = 180° - 21° - 39° = 120°$. Hence all three angles are equal in all three triangles. Hence, all three triangles are similar.

7.   No. Similar triangles have equal angles.

8.   The sum of all of the angles in a triangle is 180°. Two obtuse angles would add up to more than 180°.

9.   We draw a figure (the one shown is not drawn to scale) and label the known parts. Since triangles BTA and LMY are similar, we have

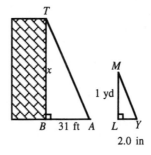

$$\frac{BT}{LM} = \frac{BA}{LY}$$

$$\frac{x}{1 \text{ yd}} = \frac{31 \text{ ft}}{2.0 \text{ in}}$$

$$\frac{x}{36 \text{ in}} = \frac{31 \text{ ft}}{2.0 \text{ in}}$$

$$x = \frac{(31 \text{ ft})(36 \text{ in})}{2.0 \text{ in}}$$

$$= 560 \text{ ft (to two significant digits)}$$

10. (A) $\dfrac{b}{c}$ (B) $\dfrac{c}{a}$ (C) $\dfrac{b}{a}$ (D) $\dfrac{c}{b}$ (E) $\dfrac{a}{c}$ (F) $\dfrac{a}{b}$

11. *Solve for the complementary angle:* $90° - \theta = 90° - 35.2° = 54.8°$

    Solve for a: We choose the cosine to find a. Thus, $\cos \theta = \dfrac{a}{c}$

    $$a = c \cos \theta$$
    $$= (20.2 \text{ cm})(\cos 35.2°) = 16.5 \text{ cm}$$

    Solve for b: We choose the sine to solve for b. Thus, $\sin \theta = \dfrac{b}{c}$

    $$b = c \sin \theta$$
    $$= (20.2 \text{ cm})(\sin 35.2°) = 11.6 \text{ cm}$$

12. Since $\dfrac{s}{C} = \dfrac{\theta}{360°}$, then $\dfrac{8.00 \text{ cm}}{20.0 \text{ cm}} = \dfrac{\theta}{360°}$ and $\theta = \dfrac{8.00}{20.0}(360°) = 144°$

13. In 20 minutes the tip of the hand travels through 1/3 of the circumference of a circle of radius 2 in (since it travels through an entire circumference in 60 minutes). Thus, $s = \dfrac{1}{3} C$ and $C = 2\pi r$. That is,

    $$s = \dfrac{1}{3}(2\pi r) = \dfrac{1}{3}(2\pi)(2 \text{ in}) = \dfrac{4}{3}\pi \text{ in} \approx 4.19 \text{ in.}$$

14. There are two methods.

    a. Convert the first to decimal degree form and compare with the second.

    $$27°14' = \left(27 + \dfrac{14}{60}\right)° \approx 27.23°$$

    Since $27.23° > 27.19°$, $27°14'$ is larger.

    b. Convert the second to DMS form and compare with the first.
    $27.19° = 27°(0.19 \times 60°) \approx 27°11.4'$
    Since $27.14' > 27°11.4'$, $27°14'$ is larger.

15. (A) 72°55'49"   (B) 113.837° → DMS
        72.930°   113°50'13"

16. (A) 90° + −33°27'51" → DMS   (B) 12(28°32'14" − 13°40'22") → DMS
        56°32'9"   178°22'24"

17. Since $\dfrac{a}{a'} = \dfrac{b}{b'}$, by Euclid's Theorem, then

    $$\dfrac{4.1 \times 10^{-6} \text{ mm}}{1.5 \times 10^{-4} \text{ mm}} = \dfrac{b}{2.6 \times 10^{-4} \text{ mm}}$$

    $$b = (2.6 \times 10^{-4} \text{ mm})\dfrac{4.1 \times 10^{-6}}{1.5 \times 10^{-4}} = 7.1 \times 10^{-6} \text{ mm}$$

18. (A) $\cos \theta$ (B) $\tan \theta$ (C) $\sin \theta$ (D) $\sec \theta$ (E) $\csc \theta$ (F) $\cot \theta$

19. Two right triangles having an acute angle of one equal to an acute angle of the other are similar, and corresponding sides of similar triangles are proportional. Thus, in the similar triangles,
    $\sin \theta = \dfrac{\text{Opp}}{\text{Hyp}} = \dfrac{\text{Opp}'}{\text{Hyp}'}$ are the same quantity.

20. *Solve for the complementary angle:* $90° - \theta = 90° - 62°20' = 27°40'$

   Solve for b: We choose the tangent to find b. Thus, $\tan \theta = \dfrac{b}{a}$

$$b = a \tan \theta$$
$$= (4.00 \times 10^{-8} \text{ m})(\tan 62°20')$$
$$= 7.63 \times 10^{-8} \text{ m}$$

   Solve for c: We choose the cosine to find c. Thus, $\cos \theta = \dfrac{a}{c}$

$$c = \frac{a}{\cos \theta}$$
$$= \frac{4.00 \times 10^{-8} \text{ m}}{\cos 62°20'} = 8.61 \times 10^{-8} \text{ m}$$

21. (A) If $\tan \theta = 1.662$, then $\theta = \tan^{-1} 1.662 = 58.97°$.

   (B) $\theta = \arccos 0.5607 = 55.896° = 55°50'$

   (C) $\theta = \sin^{-1} 0.0138 = 0.7907° = 0°47'27''$

22. If a calculator is set in degree mode, then it yields $\cos(5.47) = 0.9954...$ Therefore, display (b) is in degree mode and display (a) must be in radian mode.

23. *Solve for θ:* We choose the tangent to solve for θ. Thus, $\tan \theta = \dfrac{b}{a} = \dfrac{13.3 \text{ mm}}{15.7 \text{ mm}} = 0.8471$
$$\theta = \tan^{-1} 0.8471 = 40.3°$$

   Solve for the complementary angle: $90° - \theta = 90° - 40.3° = 49.7°$

   Solve for c: We choose the sine to solve for c. Thus, $\sin \theta = \dfrac{b}{c}$

$$c = \frac{b}{\sin \theta} = \frac{13.3 \text{ mm}}{\sin 40.3} = 20.6 \text{ mm}$$

24. $40.3° = 40°(0.3 \times 60)' = 40°20'$ (to nearest 10'); $90° - \theta = 90° - 40°20' = 49°40'$

25. We first sketch a figure and label the known parts. From geometry we know that each angle of an equilateral triangle has measure 60°.
$$\sin 60° = \frac{h}{10 \text{ ft}}$$
$$h = (10 \text{ ft})(\sin 60°) = 8.7 \text{ ft}$$

26. Since $\dfrac{s}{C} = \dfrac{\theta}{360°}$ and $C = 2\pi r$, then
$$\frac{s}{2\pi r} = \frac{\theta}{360°}$$
$$\frac{s}{2(\pi)(1500 \text{ ft})} \approx \frac{36°}{360°}$$
$$s \approx \frac{2(\pi)(1500 \text{ ft})(36)}{360} \approx 940 \text{ ft}$$

27. Since $\dfrac{A}{\pi r^2} = \dfrac{\theta}{360°}$, then $\dfrac{A}{\pi(18.3 \text{ ft})^2} = \dfrac{36.5°}{360°}$
$$A = \frac{36.5}{360} \pi (18.3 \text{ ft})^2 = 107 \text{ ft}^2$$

28. *Solve for θ:* $\theta = 90° - (90° - \theta) = 90° - 23°43' = 66°17'$

Solve for a: We choose the cosine to find a. Thus, $\cos\theta = \dfrac{a}{c}$

$$a = c\cos\theta$$
$$= (232.6\text{ km})(\cos 66°17') = 93.56\text{ km}$$

Solve for b: We choose the sine to find b. Thus, $\sin\theta = \dfrac{b}{c}$

$$b = c\sin\theta$$
$$= (232.6\text{ km})(\sin 66°17') = 213.0\text{ km}$$

29. *Solve for θ:* We choose the cosine to find θ. Thus, $\cos\theta = \dfrac{a}{c} = \dfrac{2{,}421\text{ m}}{4{,}883\text{ m}} = 60.28°$

Solve for the complementary angle: $90° - \theta = 90° - 60.28° = 29.72°$

Solve for b: We choose the sine to find b. Thus, $\sin\theta = \dfrac{b}{c}$

$$b = c\sin\theta$$
$$= (4{,}883\text{ m})(\sin 60.28°) = 4{,}241\text{ m}$$

30. Use the reciprocal relationship $\csc\theta = 1/\sin\theta$. Set calculator in degree mode, use sin key, then take reciprocal.
$\csc 72.3142° = 1.0496$

31. We note: triangles *PFH* and *PBT* are similar, hence
$$\frac{PF}{FH} = \frac{PB}{BT}$$
$$\frac{s}{5.5\text{ ft}} = \frac{20\text{ ft} + s}{18\text{ ft}}$$
$$5.5(18\text{ ft})\cdot\frac{s}{5.5\text{ ft}} = 5.5(18\text{ ft})\cdot\frac{20\text{ ft} + s}{18\text{ ft}}$$
$$18s = 5.5(20\text{ ft} + s) = 110\text{ ft} + 5.5s$$
$$12.5s = 110\text{ ft}$$
$$s = \frac{110\text{ ft}}{12.5} = 8.8\text{ ft}$$

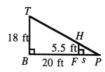

32. In right triangle, *ABC*, the length of the ramp = *AB*.
$$\sin A = \frac{BC}{AB}$$
$$AB = \frac{BC}{\sin A}$$
$$= \frac{4.25\text{ ft}}{\sin 10.0°} = 24.5\text{ ft}$$

The distance of the end of the ramp from the porch = *AC*.
$$\tan A = \frac{BC}{AC}$$
$$AC = \frac{BC}{\tan A}$$
$$= \frac{4.25\text{ ft}}{\tan 10.0°} = 24.1\text{ ft}$$

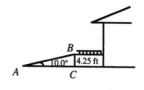

33. From the figure it is clear that $\tan \theta = \dfrac{a}{b}$. Given: percentage of inclination $\dfrac{a}{b} = 4\% = 0.04$,

   then $\tan \theta = 0.04$; $\theta = 2.3°$

   Given: angle of inclination $\theta = 4°$, then $\dfrac{a}{b} = \tan 4° = 0.07$ or 7%

34. Since $\dfrac{s}{c} = \dfrac{\theta}{360°}$ and $C = 2\pi r$, then $\dfrac{s}{2\pi r} = \dfrac{\theta}{360°}$; $s = 2\pi r \cdot \dfrac{\theta}{360°}$. Since the cities have the same

   longitude, θ is given by their difference in latitude.

   $$\theta = 44°31' - 30°42' = 13°49' = \left(13 + \dfrac{49}{60}\right)°.$$

   Thus, we have

   $$s \approx 2(\pi)(3960)\,\dfrac{13 + \dfrac{49}{60}}{360} \approx 954 \text{ miles}$$

35. We first sketch a figure and label the known parts. From the figure we note:

   $$h + 1400 \text{ ft} = 2800 \text{ ft}$$
   $$h = 1400 \text{ ft}$$
   $$\tan 64° = \dfrac{2800 \text{ ft}}{g}$$
   $$\tan \theta = \dfrac{h}{g} = \dfrac{1400 \text{ ft}}{g}$$

   Thus, $\quad g = \dfrac{2800 \text{ ft}}{\tan 64°}$

   and $\quad \tan \theta = \dfrac{1400 \text{ ft}}{\left(\dfrac{2800 \text{ ft}}{\tan 64°}\right)}$.

   Then, $\quad \tan \theta = \dfrac{1400 \tan 64°}{2800}$

   $$= \dfrac{1}{2}\tan 64° = 1.025$$
   $$\theta = 46°$$

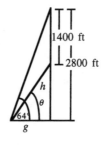

1400 ft

2800 ft

36.

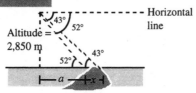

   We note: $\cot 52° = \dfrac{a}{2{,}850 \text{ m}}$, $\cot 43° = \dfrac{a + x}{2{,}850 \text{ m}}$.

   Then $\cot 43° - \cot 52° = \dfrac{a + x}{2{,}850 \text{ m}} - \dfrac{a}{2{,}850 \text{ m}} = \dfrac{a + x - a}{2{,}850 \text{ m}} = \dfrac{x}{2{,}850 \text{ m}}$

   $x = 2{,}850 \text{ m}(\cot 43° - \cot 52°) = 830 \text{ m}$

37. We note:

In right triangle BCP_1, $\cot 73.5° = \dfrac{x}{h}$

In right triangle BCP_2, $\cot 54.2° = \dfrac{x + 525 \text{ m}}{h}$

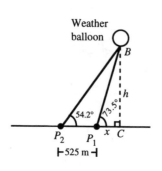

Weather
balloon

Then,
$$\cot 54.2° - \cot 73.5° = \frac{x + 525 \text{ m}}{h} - \frac{x}{h}$$
$$= \frac{x + 525 \text{ m} - x}{h}$$
$$= \frac{525 \text{ m}}{h}$$

$$h = \frac{525 \text{ m}}{\cot 54.2° - \cot 73.5°} = 1{,}240 \text{ m}$$

38. We use the figure and the notation of Problem 25, Ex. 1-4.
r = radius of the parallel of latitude
R = radius of the earth
θ = latitude

$\cos \theta = \dfrac{r}{R}$, $r = R \cos \theta$

L = length of the parallel of latitude
$L = 2\pi r$
$L = 2\pi R \cos \theta$

To keep the sun in the same position, the plane must fly at a rate v sufficient to fly a distance L in 24 hours, thus
$$v = \frac{L}{24} = \frac{2\pi R \cos \theta}{24}$$
$$v \approx \frac{2\pi(3960 \text{ mi})\cos 42°50'}{24 \text{ hr}} \approx 760 \text{ mi/hr}$$

39. (A) Since α and β are complementary (see diagram) $\beta = 90° - \alpha$.

(B) In the right triangle containing r, h, and α, $\tan \alpha = \dfrac{r}{h}$. Thus, $r = h \tan \alpha$.

(C) Using similar triangles, we can write $\dfrac{H}{h} = \dfrac{R}{r}$. Then
$$H = \frac{R}{r} h,$$
$$H - h = \frac{R}{r} h - h = h\left(\frac{R}{r} - 1\right) = h\left(\frac{R - r}{r}\right) = \frac{h}{r}(R - r)$$
Since $\dfrac{r}{h} = \tan \alpha$, $\dfrac{h}{r} = 1 + \dfrac{r}{h} = 1 + \tan \alpha = \cot \alpha$. Hence, we can write $H - h = (R - r)\cot \alpha$.

40.

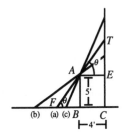

(A) See the figure. If the shortest ladder is represented in position (a), then as its foot moves away from the fence (position (b)) the ladder must get longer. And, as its foot moves toward the fence (position (c)) the ladder must also get longer.

(B) The length of the ladder is given by $FA + AT$. In triangle FBA, $\sin \theta = \dfrac{5}{FA}$, thus $FA = \dfrac{5}{\sin \theta} = 5 \csc \theta$. In triangle AET, AE is drawn parallel to BC, thus, $AE = BC = 4$.

Then, since triangle AET is similar to triangle FBA, $\cos \theta = \dfrac{4}{AT}$, thus $AT = \dfrac{4}{\cos \theta} = 4 \sec \theta$.

Thus $\ell = FA + AT = 5 \csc \theta + 4 \sec \theta$.

(C)

θ	25	35	45	55	65	75	85
L	16.25	13.60	12.73	13.08	14.98	20.63	50.91

(D) L decreases and then increases. L in Table 1 has a minimum value of 12.73 when $\theta = 45°$.

(E) Make up another table for values of θ close to 45° and on either side of 45°.

21

Chapter 2   Trigonometric Functions

EXERCISE 2.1   Degrees and Radians

1. The central angle of a circle has radian measure 1 if it intercepts an arc of length equal to the radius of the circle.

3. Since 30° corresponds to a radian measure of $\pi/6$ rad, we have:

$60° = 2 \cdot 30°$ corresponds to $2 \cdot \pi/6$ or $\pi/3$ rad.	$210° = 7 \cdot 30°$ corresponds to $7 \cdot \pi/6$ or $7\pi/6$ rad.
$90° = 3 \cdot 30°$ corresponds to $3 \cdot \pi/6$ or $\pi/2$ rad.	$240° = 8 \cdot 30°$ corresponds to $8 \cdot \pi/6$ or $4\pi/3$ rad.
$120° = 4 \cdot 30°$ corresponds to $4 \cdot \pi/6$ or $2\pi/3$ rad.	$270° = 9 \cdot 30°$ corresponds to $9 \cdot \pi/6$ or $3\pi/2$ rad.
$150° = 5 \cdot 30°$ corresponds to $5 \cdot \pi/6$ or $5\pi/6$ rad.	$300° = 10 \cdot 30°$ corresponds to $10 \cdot \pi/6$ or $5\pi/3$ rad.
$180° = 6 \cdot 30°$ corresponds to $6 \cdot \pi/6$ or π rad.	$330° = 11 \cdot 30°$ corresponds to $11 \cdot \pi/6$ or $11\pi/6$ rad.
	$360° = 12 \cdot 30°$ corresponds to $12 \cdot \pi/6$ or 2π rad.

5. The angle of radian measure 1 is larger, since 1 radian corresponds to a degree measure of $180°/\pi$, or approximately 57.3°.

7.

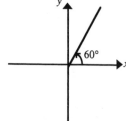

Coterminal angles have measures
$60° + 360° = 420°$ and
$60° - 360° = -300°$

9.

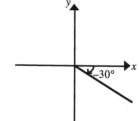

Coterminal angles have measures
$-30° + 360° = 330°$ and
$-30° - 360° = -390°$

11. $\theta_r = \dfrac{\pi \text{ rad}}{180°} \theta_d$

    $= \dfrac{\pi}{180°} 18°$

    $= \dfrac{\pi}{10}$ Exact form

    ≈ 0.3142 to four significant digits

13. $\theta_r = \dfrac{\pi \text{ rad}}{180°} \theta_d$

    $= \dfrac{\pi}{180°} 130°$

    $= \dfrac{13\pi}{18}$ Exact form

    ≈ 2.269 to four significant digits

15. $\theta_d = \dfrac{180°}{\pi \text{ rad}} \theta_r$

    $= \dfrac{180°}{\pi} (1.6)$

    $= \dfrac{288°}{\pi}$ Exact form

    $\approx 91.67°$ to four significant digits

17. $\theta_d = \dfrac{180°}{\pi \text{ rad}} \theta_r$

    $= \dfrac{180°}{\pi} \cdot \dfrac{\pi}{60}$

    $= 3°$ exactly

19.

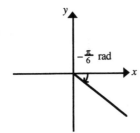

21.

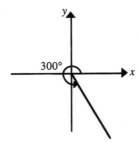

23.

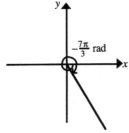

25. (A) $(8.30)^r$ $(-11.6)^r$
        $475.555°$ $-664.631°$

    (B) Set calculator in radian mode
            $563°$ $-1230°$
            9.826 rad -21.468 rad

27. Since $\theta = \dfrac{s}{r}$, we have (A) $\theta = \dfrac{12 \text{ m}}{4.0 \text{ m}} = 3$ rad (B) $\theta = \dfrac{18 \text{ m}}{4.0 \text{ m}} = 4.5$ rad

29. (A) $s = r\theta$ (B) $s = \dfrac{\pi}{180} r\theta$

           $= (25.0)(2.33) \approx 58.3$ m $= \dfrac{\pi}{180}(25.0)(19.0) \approx 8.29$ m

    (C) $s = r\theta$ (D) $s = \dfrac{\pi}{180} r\theta$

           $= (25.0)(0.821) \approx 20.5$ m $= \dfrac{\pi}{180}(25.0)(108) \approx 47.1$ m

31. Since $\theta_d = \dfrac{180}{\pi}\theta_r$, if both sides of this equation are doubled, then

        $2\theta_d = \dfrac{180}{\pi}2\theta_r$

    Thus if θ_r is doubled, θ_d is also doubled.

33. Since $\theta_r = \dfrac{s}{r}$, if s is held constant while r is doubled,

        $\theta_r = \dfrac{s}{2r} = \dfrac{1}{2}\dfrac{s}{r}$

    Thus θ_r is cut in half.

35. (A) $A = \frac{1}{2}r^2\theta$

$= \frac{1}{2}(14.0)^2 (0.473) \approx 46.4$ cm^2

(B) $A = \frac{\pi}{360}r^2\theta$

$= \frac{\pi}{360}(14.0)^2 (25.0) = 42.8$ cm^2

(C) $A = \frac{1}{2}r^2\theta$

$= \frac{1}{2}(14.0)^2 (1.02) \approx 1.0 \times 10^2$ cm^2

(D) $A = \frac{\pi}{360}r^2\theta$

$= \frac{\pi}{360}(14.0)^2 (112) \approx 192$ cm^2

37. Since $s = r\theta$ with θ in radian measure, if $r = 1$ and $\theta = m$, then $s = 1 \cdot m = m$.

39. 432° is coterminal with (432 – 360)° or 72°. Since 72° is between 0° and 90°, its terminal side lies in quadrant I.

41. $-\frac{14\pi}{3}$ is coterminal with $-\frac{14\pi}{3} + 2\pi = -\frac{8\pi}{3}$, and thus with $-\frac{8\pi}{3} + 2\pi = -\frac{2\pi}{3}$. Since $-\frac{2\pi}{3}$ is between $-\frac{\pi}{2}$ and $-\pi$, its terminal side lies in quadrant III.

43. 1,243° is coterminal with (1,243 – 360)° or 883°, and thus with (883 – 360)° or 523°, and thus with (523 – 360)° = 163°. Since 163° is between 90° and 180°, its terminal side lies in quadrant II.

45. $\theta_r = \frac{\pi \text{ rad}}{180°}(\theta_d) = \frac{\pi}{180}(57.3421) = 1.0008$ rad

47. $\theta_d = \frac{180°}{\pi \text{ rad}}(\theta_r) = \frac{180}{\pi}(0.3184) = 18.2430°$

49. $\theta_r = \frac{\pi \text{ rad}}{180°}(\theta_d) = \frac{\pi}{180}\left(26 + \frac{23}{60} + \frac{14}{3,600}\right) = 0.4605$ rad

51. At 2:30, the minute hand has moved $\frac{1}{2}$ of a circumference from its position at the top of the clock. The hour hand has moved $2\frac{1}{2}$ twelfths of a circumference from the same position. Therefore, they form an angle of

$$\frac{1}{2}C - \frac{2\frac{1}{2}}{12}C \text{ radians,}$$

where C = a total circumference or 2π radians. Thus, the desired angle is

$$\frac{1}{2}(2\pi) - \frac{2\frac{1}{2}}{12}(2\pi) = \pi - \frac{5\pi}{12} = \frac{7\pi}{12} \text{ rad} \approx 1.83 \text{ rad}$$

53. We are to find the arc length subtended by a central angle of 32°, in a circle of radius 22 cm.

$$s = \frac{\pi}{180}r\theta = \frac{\pi}{180}(22)(32) = 12 \text{ cm}$$

55. We are to find the angle, in degrees, that subtends an arc of 24 inches, in a circle of radius 72 inches.

$$s = \frac{\pi}{180}r\theta; \quad \theta = \frac{180s}{\pi r} = \frac{180(24)}{\pi(72)} = 19°$$

57. Since $s = r\theta$ we have

$$s = (54.3 \text{ cm})\frac{3\pi}{2} \approx 256 \text{ cm}$$

59. Since $s = r\theta$, we have

$$\text{diameter} \approx s = (1.5 \times 10^8 \text{ km})(9.3 \times 10^{-3} \text{ rad})$$
$$\approx 1.4 \times 10^6 \text{ km}$$

61. Since $s = \frac{\pi}{180} r\theta$, we have

$$\text{width of field} \approx s = \frac{\pi}{180}(1{,}250 \text{ ft})(8) = 175 \text{ ft}$$

63.

Since $s = r\theta$, we have

$$\text{width of object} \approx s = (250 \text{ mi})(5 \times 10^{-7} \text{ rad}) = 1.25 \times 10^{-4} \text{ mi}$$
$$1.25 \times 10^{-4} \text{ mi} = (1.25 \times 10^{-4} \text{ mi})(1{,}609 \text{ m/mi}) \approx 0.2 \text{ m}$$
$$1.25 \times 10^{-4} \text{ mi} = (0.2 \text{ m})(39.37 \text{ in}) \approx 7.9 \text{ in}$$

65. Assuming that an angle corresponding to an entire circumference is swept out in 1 year (52 weeks), and that the amount swept out in one week is proportional to the time, we can write

$$\frac{\text{angle}}{\text{time}} = \frac{\text{angle}}{\text{time}}$$
$$\frac{\theta}{1} = \frac{2\pi}{52}$$
$$\theta = \frac{2\pi}{52} = \frac{\pi}{26} \text{ rad} \approx 0.12 \text{ rad}$$

67. We use the proportion $\dfrac{\text{error in distance}}{\text{error in time}} = \dfrac{\text{actual distance}}{\text{actual time}}$. Let x = error in distance. The actual time,

$$1 \text{ year} = (365 \text{ days})\left(24 \frac{\text{hours}}{\text{day}}\right)\left(3{,}600 \frac{\text{seconds}}{\text{hour}}\right).$$

Thus,

$$\frac{x}{365 \text{ seconds}} = \frac{2\pi r}{365 \cdot 24 \cdot 3{,}600 \text{ seconds}}$$
$$x = 365 \cdot \frac{2\pi(9.3 \times 10^7 \text{ miles})}{365 \cdot 24 \cdot 3{,}600}$$
$$= \frac{2\pi(9.3 \times 10^7 \text{ miles})}{24 \cdot 3{,}600}$$
$$= 6{,}800 \text{ miles}$$

69. Since $A = \frac{1}{2}r^2\theta$ and $P = s + 2r = r\theta + 2r$, we can eliminate θ between the two equations and write

$$2A = r^2\theta, \ \theta = \frac{2A}{r^2}.$$

$$P = r\left(\frac{2A}{r^2}\right) + 2r = \frac{2A}{r} + 2r.$$

Thus, $P = \dfrac{2(52.39)}{10.5} + 2(10.5) \approx 31 \text{ ft}$

71. (A) Since one revolution corresponds to 2π radians, n revolutions corresponds to $2\pi n$ radians.

(B) No. Radian measure is independent of the size of the circle used; hence, the size of the wheel used.

(C) Since there are $2\pi n$ radians in n revolutions, in 5 revolutions there are $2\pi(5) = 10\pi \approx 31.42$ rad, and in 3.6 revolutions there are $2\pi(3.6) = 7.2\pi \approx 22.62$ rad.

73. Since the two wheels are coupled together, the distance (arc length) that the drive wheel turns is equal to the distance that the shaft turns. Thus,

$$s = r_1\theta_1 \qquad s = r_2\theta_2$$
$$r_1\theta_1 = r_2\theta_2$$
$$\theta_1 = \frac{r_2}{r_1}\theta_2 = \frac{26}{12} \text{ (3 revolutions)}$$
$$= 6.5 \text{ revolutions}$$

In Problem 71, we noted that there are $2\pi n$ rad in n revolutions, hence there are $2\pi(6.5) \approx 40.8$ rad in 6.5 revolutions.

75. Since $s = \frac{\pi}{180}r\theta$, we have $\frac{s}{r} = \frac{\pi}{180}\theta$, $\theta = \frac{180}{\pi}\left(\frac{s}{r}\right)$. Here,

$$r = \frac{1}{2}(32) = 16 \text{ in and } s = 20 \text{ ft} = 20 \text{ ft}\left(12 \ \frac{in}{ft}\right) = 240 \text{ in}$$

Thus, $\theta = \frac{180°}{\pi}\left(\frac{240}{16}\right) = 859°$.

EXERCISE 2.2 Linear and Angular Velocity

1. $V = r\omega = 6(0.5) = 3$ mm/sec

3. $\omega = \frac{V}{r} = \frac{102}{6.0} = 17$ rad/sec

5. A radial line from the center to a point on the circumference of a rotating circle sweeps out an angle at a uniform rate called *angular velocity*. This uniform rate is usually given in radians per unit time.

7. $\omega = \frac{\theta}{t} = \frac{2\pi}{1.7} = 3.7$ rad/hr

9. $\omega = \frac{\theta}{t} = \frac{8.07}{13.6} = 0.593$ rad/sec

11. 1,500 revolutions per second = $1{,}500 \cdot 2\pi$ rad/sec = $3{,}000\pi$ rad/sec

$$V = r\omega = \left(\frac{1}{2} \cdot 16\right)(3{,}000\pi) \text{ mm/sec} = 24{,}000\pi \text{ mm/sec}$$

$$= 24{,}000\pi \ \frac{mm}{sec} \cdot \frac{1}{1000} \frac{m}{mm} = 75 \ \frac{m}{sec}$$

13. $\omega = \frac{V}{r} = \frac{20{,}000 \text{ mph}}{4{,}300 \text{ mi}} = 4.65$ rad/hr

15. $\omega = \frac{V}{r} = \frac{335.3 \text{ m/sec}}{\frac{1}{2}(3.000 \text{ m})} = 223.5$ rad/sec $= (223.5 \text{ rad/sec})\left(\frac{1}{2\pi}\frac{\text{revolution}}{\text{radian}}\right) = 35.6$ rev/sec

17. The earth travels 1 revolution, or 2π radian, in 1 year, or $24 \cdot 365$ hours. Thus,

$$\omega = \frac{\theta}{t} = \frac{2\pi}{24 \cdot 365} = \frac{\pi}{4,380} \frac{\text{rad}}{\text{hr}}$$

$$V = r\omega = \left(\frac{\pi}{4,380} \frac{\text{rad}}{\text{hr}}\right)(9.3 \times 10^7 \text{ mi})$$

$$= 6.67 \times 10^4 \text{ mi/hr or } 66,700 \text{ mi/hr}$$

19. (A) Jupiter travels 1 revolution, or 2π radian, in 9 hr 55 min, or $\left(9 + \frac{55}{60}\right)$ hr. Thus,

$$\omega = \frac{\theta}{t} = \frac{2\pi}{9 + \frac{55}{60}} = 0.633 \frac{\text{rad}}{\text{hr}}$$

(B) Note that $r = \frac{1}{2} \times$ diameter. Thus,

$$V = r\omega = \left(\frac{1}{2} \times 88,700 \text{ miles}\right)\left(0.633 \frac{\text{rad}}{\text{hr}}\right) = 28,100 \text{ mi/hr}$$

21. The satellite travels 1 revolution, or 2π radian, in 23.93 hr. So,

$$\omega = \frac{\theta}{t} = \frac{2\pi}{23.93} = 0.2626 \frac{\text{rad}}{\text{hr}}$$

The radius, r, of the satellite's orbit is given by adding the satellite's distance above the earth's surface to the radius of the earth (see sketch). Thus,

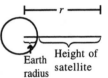

$$V = r\omega = (22,300 + 3,964 \text{ miles})\left(0.2626 \frac{\text{rad}}{\text{hr}}\right) = 6,900 \text{ mi/hr}$$

23. Using subscript E to denote quantities associated with the earth, and subscript S to denote quantities associated with the satellite, we can write:

$$\omega_E = \frac{\theta_E}{t} \qquad \omega_S = \frac{\theta_S}{t} \qquad \theta_E = \omega_E t \qquad \theta_S = \omega_S t$$

Using the hint, $2\pi = \theta_S - \theta_E$, hence,

$$2\pi = \omega_S t - \omega_E t = (\omega_S - \omega_E)t$$

$$t = \frac{2\pi}{\omega_S - \omega_E}$$

Thus, $\quad t = \dfrac{2\pi}{\dfrac{2\pi}{1.51} - \dfrac{2\pi}{23.93}} = 1.61 \text{ hr}$

25. (A) 1 rps corresponds to an angular velocity of 2π rad/sec, so that at the end of t sec, $\theta = 2\pi t$.

(B) In triangle ABC, we can write $\tan \theta = \dfrac{a}{b} = \dfrac{a}{15}$. Thus, $a = 15 \tan \theta$. But $\theta = \omega t$, where ω is given by 1 revolution per second $= 2\pi \dfrac{\text{rad}}{\text{sec}}$. Thus, $\theta = 2\pi t$, and $a = 15 \tan 2\pi t$.

(C) The speed of the light spot on the wall increases as t increases from 0.00 to 0.24. When $t = 0.25$, $a = 15 \tan(\pi/2)$, which is not defined. The light has made one quarter turn and the spot is no longer on the wall.

t sec	0.00	0.04	0.08	0.12	0.16	0.20	0.24
a ft	0.00	3.85	8.25	14.09	23.64	46.17	238.42

EXERCISE 2.3   Trigonometric Fucntions

1. $P(a, b) = (3, 4)$, $r = \sqrt{a^2 + b^2} = \sqrt{3^2 + 4^2} = 5$

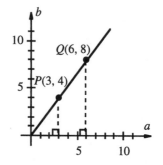

$$\sin\theta = \frac{b}{r} = \frac{4}{5} \qquad \csc\theta = \frac{r}{b} = \frac{5}{4}$$

$$\cos\theta = \frac{a}{r} = \frac{3}{5} \qquad \sec\theta = \frac{r}{a} = \frac{5}{3}$$

$$\tan\theta = \frac{b}{a} = \frac{4}{3} \qquad \cot\theta = \frac{a}{b} = \frac{3}{4}$$

$Q(a, b) = (6, 8)$ $r = \sqrt{a^2 + b^2} = \sqrt{6^2 + 8^2} = 10$

$$\sin\theta = \frac{b}{r} = \frac{8}{10} = \frac{4}{5} \qquad \csc\theta = \frac{r}{b} = \frac{10}{8} = \frac{5}{4}$$

$$\cos\theta = \frac{a}{r} = \frac{6}{10} = \frac{3}{5} \qquad \sec\theta = \frac{r}{a} = \frac{10}{6} = \frac{5}{3}$$

$$\tan\theta = \frac{b}{a} = \frac{8}{6} = \frac{4}{3} \qquad \cot\theta = \frac{a}{b} = \frac{6}{8} = \frac{3}{4}$$

3. $P(a, b) = (4, -3)$, $r = \sqrt{a^2 + b^2} = \sqrt{4^2 + (-3)^2} = 5$

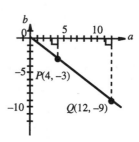

$$\sin\theta = \frac{b}{r} = \frac{-3}{5} = -\frac{3}{5} \qquad \csc\theta = \frac{r}{b} = \frac{5}{-3} = -\frac{5}{3}$$

$$\cos\theta = \frac{a}{r} = \frac{4}{5} \qquad \sec\theta = \frac{r}{a} = \frac{5}{4}$$

$$\tan\theta = \frac{b}{a} = \frac{-3}{4} = -\frac{3}{4} \qquad \cot\theta = \frac{a}{b} = \frac{4}{-3} = -\frac{4}{3}$$

$Q(a, b) = (12, -9)$ $r = \sqrt{a^2 + b^2} = \sqrt{12^2 + (-9)^2} = 15$

$$\sin\theta = \frac{b}{r} = \frac{-9}{15} = -\frac{3}{5} \qquad \csc\theta = \frac{r}{b} = \frac{15}{-9} = -\frac{5}{3}$$

$$\cos\theta = \frac{a}{r} = \frac{12}{15} = \frac{4}{5} \qquad \sec\theta = \frac{r}{a} = \frac{15}{12} = \frac{5}{4}$$

$$\tan\theta = \frac{b}{a} = \frac{-9}{12} = -\frac{3}{4} \qquad \cot\theta = \frac{a}{b} = \frac{12}{-9} = -\frac{4}{3}$$

5. We sketch a reference triangle and label what we know.

Since $\cos \theta = \dfrac{a}{r} = \dfrac{3}{5}$, we know that $a = 3$ and $r = 5$.

Use the Pythagorean theorem to find b:
$$3^2 + b^2 = 5^2$$
$$b^2 = 25 - 9 = 16$$
$$b = 4$$
b is positive since $P(a, b)$ is in quadrant I.

We can now find the other five functions using Definition 1:

$$\sin \theta = \frac{b}{r} = \frac{4}{5} \qquad \csc \theta = \frac{r}{b} = \frac{5}{4}$$

$$\tan \theta = \frac{b}{a} = \frac{4}{3} \qquad \cot \theta = \frac{a}{b} = \frac{3}{4}$$

$$\sec \theta = \frac{r}{a} = \frac{5}{3}$$

7. We sketch a reference triangle and label what we know.

Since $\cos \theta = \dfrac{a}{r} = \dfrac{3}{5}$, we know that $a = 3$ and $r = 5$.

Use the Pythagorean theorem to find b:
$$3^2 + b^2 = 5^2$$
$$b^2 = 25 - 9 = 16$$
$$b = -4$$
b is negative since $P(a, b)$ is in quadrant IV.

We can now find the other five functions using Definition 1:

$$\sin \theta = \frac{b}{r} = \frac{-4}{5} = -\frac{4}{5} \qquad \csc \theta = \frac{r}{b} = \frac{5}{-4} = -\frac{5}{4}$$

$$\tan \theta = \frac{b}{a} = \frac{-4}{3} = -\frac{4}{3} \qquad \cot \theta = \frac{a}{b} = \frac{3}{-4} = -\frac{3}{4}$$

$$\sec \theta = \frac{r}{a} = \frac{5}{3}$$

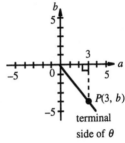

9. We sketch a reference triangle and label what we know.

Since $\csc \theta = \dfrac{r}{b} = -\dfrac{5}{4} = \dfrac{5}{-4}$, we kow that $b = -4$ and $r = 5$
(r is never negative). Use the Pythagorean theorem to find a:
$$a^2 + (-4)^2 = 5^2$$
$$a^2 = 25 - 16 = 9$$
$$a = -3$$
a is negative since $P(a, b)$ is in quadrant III.

We can now find the other five functions using Definition 1:

$$\sin \theta = \frac{b}{r} = \frac{-4}{5} = -\frac{4}{5} \qquad \sec \theta = \frac{r}{a} = \frac{5}{-3} = -\frac{5}{3}$$

$$\cos \theta = \frac{a}{r} = \frac{-3}{5} = -\frac{3}{5} \qquad \cot \theta = \frac{a}{b} = \frac{-3}{-4} = \frac{3}{4}$$

$$\tan \theta = \frac{b}{a} = \frac{-4}{-3} = \frac{4}{3}$$

11. No. For all those values of x for which both are defined, $\cos x = 1/(\sec x)$, hence both are either positive or both are negative.

13. Degree mode: $\tan 89° = 57.29$

15. Radian mode: $\cos (3 \text{ rad}) = -0.9900$

17. Use the reciprocal relationship $\csc \theta = \dfrac{1}{\sin \theta}$. Degree mode: $\csc 162° = \dfrac{1}{\sin 162°} = 3.236$

19. Use the reciprocal relationship $\cot \theta = \dfrac{1}{\tan \theta}$. Degree mode: $\cot 341° = \dfrac{1}{\tan 341°} = -2.904$

21. Radian mode: $\sin 13 = 0.4202$

23. Use the reciprocal relationship $\cot \theta = \dfrac{1}{\tan \theta}$. Radian mode: $\cot 2 = \dfrac{1}{\tan 2} = -0.4577$

25. Use the reciprocal relationship $\sec \theta = \dfrac{1}{\cos \theta}$. Radian mode: $\sec 74 = \dfrac{1}{\cos 74} = 5.824$

27. Degree mode: $\sin 428° = 0.9272$

29. Radian mode: $\cos (-12) = 0.8439$

31. Use the reciprocal relationship $\cot \theta = \dfrac{1}{\tan \theta}$. Degree mode: $\cot (-167°) = \dfrac{1}{\tan (-167°)} = 4.331$

33. $(a, b) = (\sqrt{3}, 1)$, $r = \sqrt{a^2 + b^2} = \sqrt{(\sqrt{3})^2 + 1^2} = \sqrt{4} = 2$

$\sin \theta = \dfrac{b}{r} = \dfrac{1}{2}$ $\qquad$ $\cos \theta = \dfrac{a}{r} = \dfrac{\sqrt{3}}{2}$ $\qquad$ $\tan \theta = \dfrac{b}{a} = \dfrac{1}{\sqrt{3}}$

$\csc \theta = \dfrac{r}{b} = \dfrac{2}{1} = 2$ $\qquad$ $\sec \theta = \dfrac{r}{a} = \dfrac{2}{\sqrt{3}}$ $\qquad$ $\cot \theta = \dfrac{a}{b} = \dfrac{\sqrt{3}}{1} = \sqrt{3}$

35. $(a, b) = (1, -\sqrt{3})$, $r = \sqrt{a^2 + b^2} = \sqrt{1^2 + (\sqrt{3})^2} = \sqrt{4} = 2$

$\sin \theta = \dfrac{b}{r} = \dfrac{-\sqrt{3}}{2} = -\dfrac{\sqrt{3}}{2}$ $\qquad$ $\cos \theta = \dfrac{a}{r} = \dfrac{1}{2}$ $\qquad$ $\tan \theta = \dfrac{b}{a} = \dfrac{-\sqrt{3}}{1} = -\sqrt{3}$

$\csc \theta = \dfrac{r}{b} = \dfrac{2}{-\sqrt{3}} = -\dfrac{2}{\sqrt{3}}$ $\qquad$ $\sec \theta = \dfrac{r}{a} = \dfrac{2}{1} = 2$ $\qquad$ $\cot \theta = \dfrac{a}{b} = \dfrac{1}{-\sqrt{3}} = -\dfrac{1}{\sqrt{3}}$

37. I, IV $\qquad$ 39. I, III $\qquad$ 41. I, IV $\qquad$ 43. III, IV $\qquad$ 45. II, IV

47. III, IV

49. Since $\sin \theta < 0$ and $\cot \theta > 0$, the terminal side of θ lies in quadrant III. We sketch a reference triangle and label what we know. Since $\sin \theta = \dfrac{b}{r} = -\dfrac{2}{3} = \dfrac{-2}{3}$, we know that $b = -2$ and $r = 3$ (r is never negative). Use the Pythagorean theorem to find a:

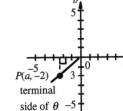

$$a^2 + (-2)^2 = 3^2$$
$$a^2 = 9 - 4 = 5$$
$$a = -\sqrt{5}$$

a is negative since the terminal side of θ lies in quadrant III.

We can now find the other five functions using Definition 1:

$$\csc \theta = \frac{r}{b} = \frac{3}{-2} = -\frac{3}{2} \qquad \sec \theta = \frac{r}{a} = \frac{3}{-\sqrt{5}} = -\frac{3}{\sqrt{5}}$$

$$\cos \theta = \frac{a}{r} = \frac{-\sqrt{5}}{3} = -\frac{\sqrt{5}}{3} \qquad \cot \theta = \frac{a}{b} = \frac{-\sqrt{5}}{-2} = \frac{\sqrt{5}}{2}$$

$$\tan \theta = \frac{b}{a} = \frac{-2}{-\sqrt{5}} = \frac{2}{\sqrt{5}}$$

51. Since $\sec \theta > 0$ and $\sin \theta < 0$, the terminal side of θ lies in quadrant IV. We sketch a reference triangle and label what we know. Since $\sec \theta = \dfrac{r}{a} = \sqrt{3} = \dfrac{\sqrt{3}}{1}$, we know that $r = \sqrt{3}$ and $a = 1$. Use the Pythagorean theorem to find b:

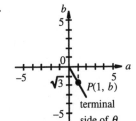

$$1^2 + b^2 = (\sqrt{3})^2$$
$$b^2 = 3 - 1 = 2$$
$$b = -\sqrt{2}$$

b is negative since the terminal side of θ lies in quadrant IV.

We can now find the other five functions using Definition 1:

$$\sin \theta = \frac{b}{r} = \frac{-\sqrt{2}}{\sqrt{3}} = -\frac{\sqrt{2}}{\sqrt{3}} \qquad \csc \theta = \frac{r}{b} = \frac{\sqrt{3}}{-\sqrt{2}} = -\frac{\sqrt{3}}{\sqrt{2}}$$

$$\cos \theta = \frac{a}{r} = \frac{1}{\sqrt{3}} \qquad \cot \theta = \frac{a}{b} = \frac{1}{-\sqrt{2}} = -\frac{1}{\sqrt{2}}$$

$$\tan \theta = \frac{b}{a} = \frac{-\sqrt{2}}{1} = -\sqrt{2}$$

53. $\cos \alpha = \cos \beta$, since the terminal sides of each angle will coincide and therefore the same point $P(a, b) \neq (0, 0)$ can be chosen on the terminal side s. Thus, $\cos \alpha = \dfrac{a}{r} = \cos \beta$, where $r = \sqrt{a^2 + b^2} \neq 0$.

55. Use the reciprocal identity $\cot x = \dfrac{1}{\tan x}$. $\cot x = \dfrac{1}{-2.18504} = -0.45766$

57. Degree mode: $\cos 308.25° = 0.6191$

59. Radian mode: $\tan 1.371 = 4.938$

61. Use the reciprocal relationship $\cot \theta = \dfrac{1}{\tan \theta}$.

Degree mode: $\cot(-265.33°) = \dfrac{1}{\tan(-265.33°)} = -0.08169$

63. Use the reciprocal relationship $\sec \theta = \dfrac{1}{\cos \theta}$

Radian mode: $\sec(-4.013) = \dfrac{1}{\cos(-4.013)} = -1.553$

65. Degree mode: $\cos 208°12'55" = \cos(208.21528...°)$ (Convert to decimal degrees, if necessary.)
$= -0.8812$

67. Use the reciprocal relationsip $\csc \theta = \dfrac{1}{\sin \theta}$

Degree mode: $\csc 112°5'38" = \csc(112.09389...°)$ (Covert to decimal degrees, if necessary.)
$= \dfrac{1}{\sin(112.09389...°)} = 1.079$

69. Use the reciprocal relationship $\sec \theta = \dfrac{1}{\cos \theta}$

Radian mode: $\sec(-1,000) = \dfrac{1}{\cos(-1,000)} = 1.778$

71. Degree mode: $\sin(405.33°) = 0.7112$

73. Degree mode: $\cos(-168°32'5") = \cos(-168.53472...°) = -0.9800$

75. When the terminal side of an angle lies along the vertical axis, the coordinates of any point on the terminal side have the form $(0, b)$, that is, $a = 0$. Therefore, $\tan \theta = \dfrac{b}{a}$ and $\sec \theta = \dfrac{r}{a}$ are not defined.

77. (A) Since $\theta = \dfrac{s}{r}$, and $s = 6$, and r, the measure of CA, is 5, we have $\theta = \dfrac{6}{5} = 1.2$ rad

(B) Since $\cos \theta = \dfrac{a}{r}$ and $\sin \theta = \dfrac{b}{r}$, we have
$a = r \cos \theta = 5 \cos 1.2$ $b = r \sin \theta = 5 \sin 1.2$
Thus, $(a, b) = (5 \cos 1.2, 5 \sin 1.2) = (1.81, 4.66)$

79. (A) Since $\theta = \dfrac{s}{r}$, and $s = 2$, and r, the measure of CA, is 1, we have $\theta = \dfrac{2}{1} = 2$ rad

(B) Since $\cos \theta = \dfrac{a}{r}$ and $\sin \theta = \dfrac{b}{r}$, we have
$a = r \cos \theta = 1 \cos 2$ $b = r \sin \theta = 1 \sin 2$
Thus, $(a, b) = (1 \cos 2, 1 \sin 2) = (-0.416, 0.909)$

81. From the figure, we note: $s = r\theta$

$$r = \sqrt{a^2 + b^2} = \sqrt{4^2 + 3^2} = \sqrt{25} = 5$$

$$\tan\theta = \frac{b}{a} = \frac{3}{4}$$

$$\theta = \tan^{-1}\frac{3}{4}$$

Thus, $s = 5 \tan^{-1}\frac{3}{4} \approx 3.22$ units (calculator in radian mode)

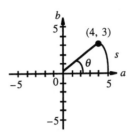

83. If $\theta = 0°$, $I = k \cos 0° = k \cdot 1 = k$

If $\theta = 20°$, $I = k \cos 20° = k\,(0.94) = 0.94k$

If $\theta = 40°$, $I = k \cos 40° = k\,(0.77) = 0.77k$

If $\theta = 60°$, $I = k \cos 60° = k\,(0.50) = 0.50k$

If $\theta = 80°$, $I = k \cos 80° = k\,(0.17) = 0.17k$

85. If $\theta = 15°$ (summer solstice), $I = k \cos 15° = k(0.97) = 0.97k$

If $\theta = 63°$ (winter solstice), $I = k \cos 63° = k(0.45) = 0.45k$

87. (A) If $n = 6$, $A = \frac{n}{2} \sin\left(\frac{360}{n}\right)^\circ = \frac{6}{2} \sin\left(\frac{360}{6}\right)^\circ = 3 \sin 60° = 2.59808$

If $n = 10$, $A = \frac{10}{2} \sin\left(\frac{360}{10}\right)^\circ = 5 \sin 36° = 2.93893$

If $n = 100$, $A = \frac{100}{2} \sin\left(\frac{360}{100}\right)^\circ = 50 \sin 3.6° = 3.13953$

If $n = 1000$, $A = \frac{1000}{2} \sin\left(\frac{360}{1000}\right)^\circ = 500 \sin(0.36)° = 3.14157$

If $n = 10,000$, $A = \frac{10,000}{2} \sin\left(\frac{360}{10,000}\right)^\circ = 5,000 \sin(0.036)° = 3.14159$

n	6	10	100	1000	10,000
A_n	2.59808	2.93893	3.13953	3.14157	3.14159

(B) The area of the circle is $A = \pi r^2 = \pi(1)^2 = \pi$, and A_n seem to approach π, the area of the circle as n increases.

(C) No. An n sided polygon is always a polygon, no matter the size of n, but the inscribed polygon can be made as close to the circle as desired by taking n sufficiently large.

89. From the diagram, we can see that $x = a + \ell$. To determine a, we note that $\cos\theta = \frac{a}{r}$, $r = 1$,

and $\theta = 20\pi t$. Thus,

$$\cos 20\pi t = \frac{a}{1} \text{ and } a = \cos 20\pi t$$

To determine ℓ, we note that triangle PFL is a right triangle. Thus, from the Pythagorean theorem,

$$b^2 + \ell^2 = 5^2$$
$$\ell^2 = 25 - b^2$$

Since $\sin \theta = \dfrac{b}{r}$, $r = 1$, and $\theta = 20\pi t$, we have

$$\sin 20\pi t = \frac{b}{1} \text{ and } b = \sin 20\pi t.$$

Thus,

$$\ell^2 = 25 - (\sin 20\pi t)^2,$$
$$\ell = \sqrt{25 - (\sin 20\pi t)^2}$$
$$x = a + \ell = \cos 20\pi t + \sqrt{25 - (\sin 20\pi t)^2}$$

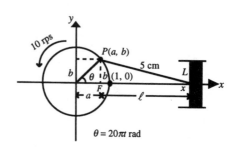

91. Since $I = 35 \sin(48\pi t - 12\pi)$ and $t = 0.13$,

$$= 35 \sin(48\pi(0.13) - 12\pi) \text{ (calculator in radian mode)}$$
$$\approx 35 \sin(-18.095574\ldots) \approx 24 \text{ amperes}$$

93. (A) If the angle of inclination $\theta = 63.5°$, then $m = \tan \theta = \tan 63.5° = 2.01$

If the angle of inclination $\theta = 172°$, then $m = \tan \theta = \tan 172° = -0.14$

(B) If the angle of inclination $\theta = 143°$, then $m = \tan \theta = \tan 143°$. Then the equation of the line is given by

$$y - 6 = \tan 143°(x - (-3))$$
$$y = \tan 143°(x + 3) + 6 = x \tan 143° + 3 \tan 143° + 6$$
$$y = -0.75x + 3.74$$

EXERCISE 2.4   Additional Applications

1. Use $\dfrac{n_2}{n_1} = \dfrac{\sin \alpha}{\sin \beta}$,

   where $n_2 = 1.33$, $n_1 = 1.00$, and $\alpha = 40.6°$

   Solve for β: $\dfrac{1.33}{1.00} = \dfrac{\sin 40.6°}{\sin \beta}$

   $$\sin \beta = \frac{\sin 40.6°}{1.33}$$

   $$\beta = \sin^{-1}\left(\frac{\sin 40.6°}{1.33}\right)$$

   $$= 29.3°$$

3. Use $\dfrac{n_2}{n_1} = \dfrac{\sin \alpha}{\sin \beta}$,

   where $n_2 = 1.66$, $n_1 = 1.33$, and $\alpha = 32.0°$

   Solve for β: $\dfrac{1.66}{1.33} = \dfrac{\sin 32.0°}{\sin \beta}$

   $$\sin \beta = \frac{1.33 \sin 32.0°}{1.66}$$

   $$\beta = \sin^{-1}\left(\frac{1.33 \sin 32.0°}{1.66}\right)$$

   $$= 25.1°$$

5. The index of refraction for diamond is $n_1 = 2.42$ and that for air is 1.00. Find the angle of incidence α such that the angle of refraction β is 90°.

$$\frac{\sin \alpha}{\sin \beta} = \frac{n_2}{n_1}; \quad \sin \alpha = \frac{1.00}{2.42} \sin 90°; \quad \alpha = \sin^{-1}\left[\frac{1.00}{2.42}(1)\right] = 24.4°$$

7. See figure (modified from the figure in the text).

   The eye tends to assume that light travels straight. Thus, the ball at B will be interpreted as a ball at B'. The ball appears to be above the real ball.

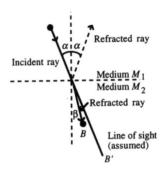

9. Use $\dfrac{n_2}{n_1} = \dfrac{\sin \alpha}{\sin \beta}$,

   where $n_1 = 1.00$.

   $$\alpha = 90° - 38° = 52°$$

   $$\sin \beta = \frac{7.2}{r} = \frac{7.2}{\sqrt{7.2^2 + 12^2}}$$

   Solve for n_2:

   $$\frac{n_2}{1.00} = \sin 52° + \frac{7.2}{\sqrt{7.2^2 + 12^2}}$$

   $$\approx 1.5$$

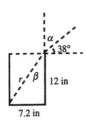

11. We use $\sin \dfrac{\theta}{2} = \dfrac{S_w}{S_b}$, where $\theta = 60°$ and $S_w = 20$ km/hr. Then we solve for S_b:

    $$\sin \frac{60°}{2} = \frac{20}{S_b}; \quad S_b = \frac{20}{\sin 30°} = 40 \text{ km/hr}$$

13. We use $\sin \dfrac{\theta}{2} = \dfrac{S_s}{S_a}$, where $S_a = \text{Mach } 1.5 = 1.5 S_s$ Then we solve for θ:

    $$\sin \frac{\theta}{2} = \frac{S_s}{1.5 S_s}; \quad \sin \frac{\theta}{2} = \frac{1}{1.5}. \text{ Thus } \frac{\theta}{2} = 42°, \ \theta = 84°.$$

15. We use $\sin \dfrac{\theta}{2} = \dfrac{S_1}{S_p}$, where $\theta = 90°$ and $S_1 = 2 \times 10^{10}$ cm/sec. Then we solve for S_p:

    $$\sin \frac{90°}{2} = \frac{2 \times 10^{10}}{S_p}; \quad S_p = \frac{2 \times 10^{10}}{\sin 45°} = 3 \times 10^{10} \text{ cm/sec}$$

17. $d = a + b \sin 4\theta = -2.2 + (-4.5) \sin (4 \cdot 30°) = -2.2 - 4.5 \sin 120° = -6°.$

EXERCISE 2.5   Exact Value for Special Angles and Real Numbers

Note for Problems 1—11: The reference angle α is the angle (always taken positive) between the terminal side of θ and the horizontal axis.

1. $\alpha = \theta = 60°$

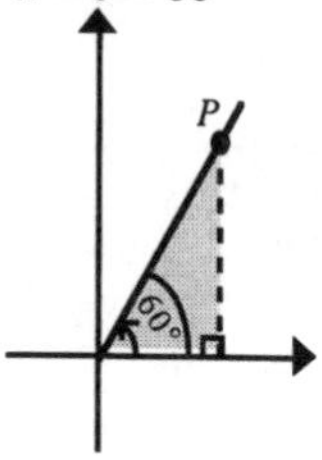

3. $\alpha = |-60°| = 60°$

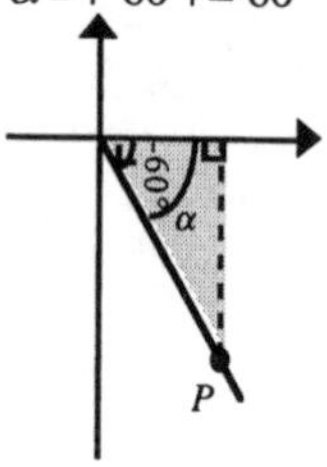

5. $\alpha = |-\pi/3| = \pi/3$

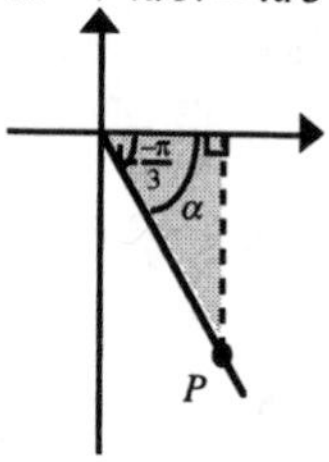

7. $\alpha = \pi - \dfrac{3\pi}{4} = \dfrac{\pi}{4}$

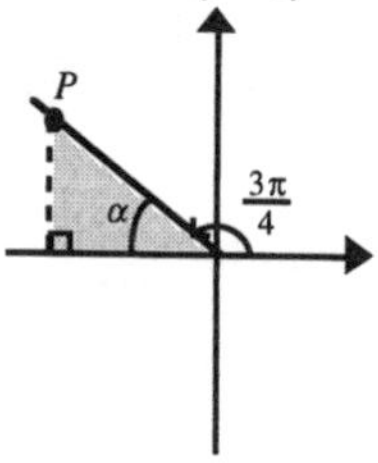

9. $\alpha = 210° - 180° = 30°$

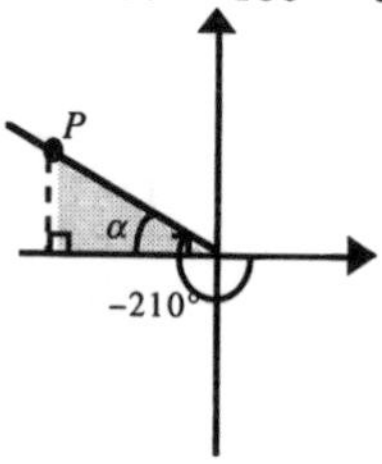

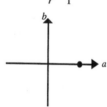

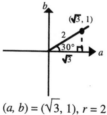

11. $\alpha = \dfrac{5\pi}{4} - \pi = \dfrac{\pi}{4}$

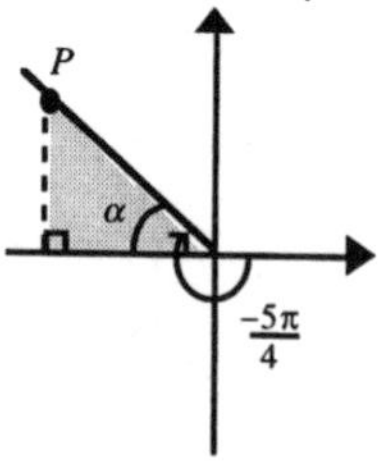

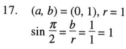

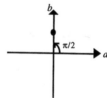

13. $(a, b) = (1, 0), r = 1$

$\cos 0° = \dfrac{a}{r} = \dfrac{1}{1} = 1$

15. Use the special 30°–60° triangle as the reference triangle. Use the sides of the reference triangle to determine $P(a, b)$ and r. Then use Definition 1.

($\sqrt{3}$, 1)

2

30°

$\sqrt{3}$

$(a, b) = (\sqrt{3}, 1), r = 2$

$\sin 30° = \dfrac{b}{r} = \dfrac{1}{2}$

17. $(a, b) = (0, 1), r = 1$

$\sin \dfrac{\pi}{2} = \dfrac{b}{r} = \dfrac{1}{1} = 1$

$\pi/2$

19. Use the special 45° triangle as the reference triangle. Use the sides of the reference triangle to determine $P(a, b)$ and r. Then use Definition 1.

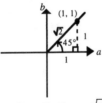

$(a, b) = (1, 1)$, $r = \sqrt{2}$

$\tan 45° = \dfrac{b}{a} = \dfrac{1}{1} = 1$

21. Use the special 30°–60° triangle as the reference triangle. Use the sides of the reference triangle to determine $P(a, b)$. Then use Definition 1.

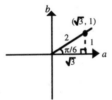

$(a, b) = (\sqrt{3}, 1)$, $r = 2$

$\tan \dfrac{\pi}{6} = \dfrac{b}{a} = \dfrac{1}{\sqrt{3}}$ or $\dfrac{\sqrt{3}}{3}$

23. Locate the 30°–60° reference triangle, determine (a, b) and r, then evaluate.

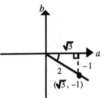

$\sin(-30°) = \dfrac{-1}{2} = -\dfrac{1}{2}$

25.

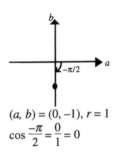

$(a, b) = (0, -1)$, $r = 1$

$\cos \dfrac{-\pi}{2} = \dfrac{0}{1} = 0$

27. Locate the 30°–60° reference triangle, determine (a, b) and r, then evaluate.

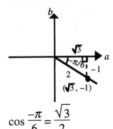

$\cos \dfrac{-\pi}{6} = \dfrac{\sqrt{3}}{2}$

29. Locate the 30°–60° reference triangle, determine (a, b) and r, then evaluate.

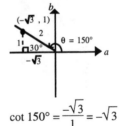

$\cot 150° = \dfrac{-\sqrt{3}}{1} = -\sqrt{3}$

31. Locate the 30°–60° reference triangle, determine (a, b) and r, then evaluate.

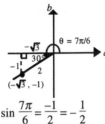

$\sin \dfrac{7\pi}{6} = \dfrac{-1}{2} = -\dfrac{1}{2}$

33.

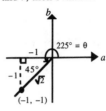

$(a, b) = (0, -1)$, $r = 1$

$\sin \dfrac{3\pi}{2} = \dfrac{-1}{1} = -1$

35. Locate the 45° reference triangle, determine (a, b) and r, then evaluate.

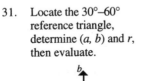

$\sin 225° = \dfrac{-1}{\sqrt{2}} = -\dfrac{1}{\sqrt{2}}$

or $-\dfrac{\sqrt{2}}{2}$

37. Locate the 45° reference triangle, determine (a, b) and r, then evaluate.

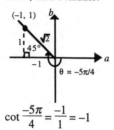

$$\cot \frac{-5\pi}{4} = \frac{-1}{1} = -1$$

39. Locate the 30°–60° reference triangle, determine (a, b) and r, then evaluate.

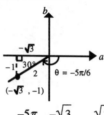

$$\cos \frac{-5\pi}{6} = \frac{-\sqrt{3}}{2} = -\frac{\sqrt{3}}{2}$$

41. Locate the 30°–60° reference triangle, determine (a, b) and r, then evaluate.

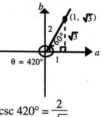

$$\csc 420° = \frac{2}{\sqrt{3}}$$

43. The tangent function is not defined at $\theta = \frac{\pi}{2}$ and $\frac{3\pi}{2}$, because $\tan \theta = \frac{b}{a}$ and $a = 0$ for any point on the vertical axis.

45. The cosecant function is not defined at $\theta = 0$, π, and 2π, because $\csc \theta = \frac{r}{b}$ and $b = 0$ for any point on the horizontal axis.

47. It is known that $\sin\left(\frac{\pi}{6}\right) = \frac{1}{2} = 0.5000...$ and $\cos(0) = 1 = 1.0000...$, so that the calculator displays exact values of these. Thus $\sin(-45°)$ is not given exactly. To find $\sin(-45°)$, locate the 45° reference triangle, determine (a, b) and r, then evaluate.

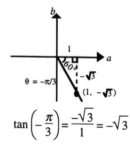

$$\sin(-45°) = \frac{-1}{\sqrt{2}} = -\frac{1}{\sqrt{2}} \text{ or } -\frac{\sqrt{2}}{2}$$

49. It is known that $\tan(45°) = 1 = 1.0000...$ and $\tan(180°) = 0 = 0.0000...$, so that the calculator displays exact values for these.

Thus $\tan\left(-\frac{\pi}{3}\right)$ is not given exactly. To find $\tan\left(-\frac{\pi}{3}\right)$, locate the 30°–60° reference triangle, determine (a, b) and r, then evaluate.

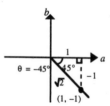

$$\tan\left(-\frac{\pi}{3}\right) = \frac{-\sqrt{3}}{1} = -\sqrt{3}$$

51. Draw a reference triangle in the first quadrant with side opposite reference angle 1 and hypotenuse 2. Observe that this is a special 30°–60° triangle.

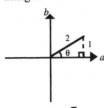

(A) $\theta = 30°$ (B) $\theta = \dfrac{\pi}{6}$

53. Draw a reference triangle in the second quadrant with side adjacent reference angle –1 and hypotenuse 2. Observe that this is a special 30°–60° triangle.

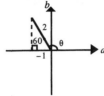

(A) $\theta = 120°$ (B) $\theta = \dfrac{2\pi}{3}$

55. Draw a reference triangle in the second quadrant with side opposite reference angle $\sqrt{3}$ and side adjacent –1. Observe that this is a special 30°–60° triangle.

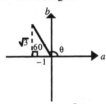

(A) $\theta = 120°$ (B) $\theta = \dfrac{2\pi}{3}$

57. We can draw reference triangles in both quadrants III and IV with side opposite reference angle $-\sqrt{3}$ and hypotenuse 2. Each triangle is a special 30°–60° triangle.

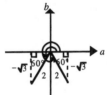

$\theta = 240°$ or $\theta = 300°$

59. We can draw a reference triangle in the second quadrant with side opposite reference angle 1 and side adjacent $-\sqrt{3}$. We can also draw a reference triangle in the fourth quadrant with side opposite reference angle –1 and side adjacent $\sqrt{3}$. Each triangle is a special 30°–60° triangle.

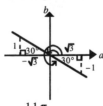

$\theta = \dfrac{5\pi}{6}$ or $\theta = \dfrac{11\pi}{6}$

61. $\cos x = -\dfrac{\sqrt{2}}{2}$

We can draw a reference triangle in the second quadrant with side adjacent reference angle $-\sqrt{2}$ and hypotenuse 2. Observe that this is a special 45° triangle.

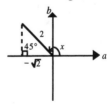

$\pi - x = \dfrac{\pi}{4}$

$x = \dfrac{3\pi}{4}$

63. $\tan x = \dfrac{\sqrt{3}}{3}$

We can draw a reference triangle in the first quadrant
with side opposite reference angle $\sqrt{3}$ and side adjacent
3. Then

$$a^2 + b^2 = r^2$$
$$(\sqrt{3})^2 + 3^2 = r^2$$
$$12 = r^2$$
$$r = \sqrt{12} = 2\sqrt{3}$$

Thus the triangle is a special 30°–60° triangle.

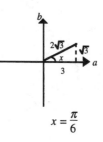

$$x = \dfrac{\pi}{6}$$

65.   (A) Since $\dfrac{7}{x} = \sin 30°$ and $\sin 30° = \dfrac{1}{2}$; $\dfrac{7}{x} = \dfrac{1}{2}$; $x = 14$

Since $\dfrac{7}{y} = \tan 30°$ and $\tan 30° = \dfrac{1}{\sqrt{3}}$; $\dfrac{7}{y} = \dfrac{1}{\sqrt{3}}$; $y = 7\sqrt{3}$

(B) Since $\dfrac{x}{4} = \sin 45°$ and $\sin 45° = \dfrac{1}{\sqrt{2}}$; $\dfrac{x}{4} = \dfrac{1}{\sqrt{2}}$; $x = \dfrac{4}{\sqrt{2}}$

Since $\dfrac{y}{4} = \cos 45°$ and $\cos 45° = \dfrac{1}{\sqrt{2}}$; $\dfrac{y}{4} = \dfrac{1}{\sqrt{2}}$; $y = \dfrac{4}{\sqrt{2}}$

(C) Since $\dfrac{5}{x} = \sin 60°$ and $\sin 60° = \dfrac{\sqrt{3}}{2}$; $\dfrac{5}{x} = \dfrac{\sqrt{3}}{2}$; $x = \dfrac{10}{\sqrt{3}}$

Since $\dfrac{5}{y} = \tan 60°$ and $\tan 60° = \sqrt{3}$; $\dfrac{5}{y} = \sqrt{3}$; $y = \dfrac{5}{\sqrt{3}}$

67. $A = \dfrac{\pi r^2}{2} \sin \dfrac{2\pi}{n}$.

In this problem, $n = 3$, $r = 2$ cm, thus

$$A = \dfrac{3(2)^2}{2} \sin \dfrac{2\pi}{3} = 6 \sin \dfrac{2\pi}{3}$$

To find $\sin \dfrac{2\pi}{3}$, draw a 30°–60° reference triangle,

determine (a, b) and r, then evaluate.

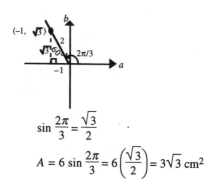

$$\sin \dfrac{2\pi}{3} = \dfrac{\sqrt{3}}{2}$$

$$A = 6 \sin \dfrac{2\pi}{3} = 6\left(\dfrac{\sqrt{3}}{2}\right) = 3\sqrt{3} \text{ cm}^2$$

69. $A = \dfrac{\pi r^2}{2} \sin \dfrac{2\pi}{n}$.

In this problem, $n = 6$, $r = 10$ in., thus

$A = \dfrac{6(10)^2}{2} \sin \dfrac{2\pi}{6} = 300 \sin \dfrac{\pi}{3}$

To find $\sin \dfrac{\pi}{3}$, draw a 30°–60° reference triangle,

determine (a, b) and r, then evaluate.

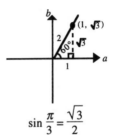

$\sin \dfrac{\pi}{3} = \dfrac{\sqrt{3}}{2}$

$A = 300 \sin \dfrac{\pi}{3} = 300 \left(\dfrac{\sqrt{3}}{2} \right) = 150\sqrt{3}$ in^2

EXERCISE 2.6   Circular Functions

1. (A) Since the circumference of a unit circle is 2π, one-half the circumference is $\dfrac{1}{2} \cdot 2\pi$ or π.

(B) Since the circumference of a unit circle is 2π, three-quarters the circumference is $\dfrac{3}{4} \cdot 2\pi$ or $\dfrac{3\pi}{2}$.

3. (A)   (B)   (C)

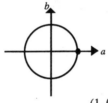   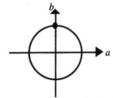   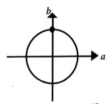

$(1, 0)$ $(0, 1)$ $(0, 1)$

(D)   (E)   (F)

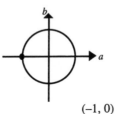   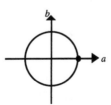   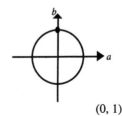

$(-1, 0)$ $(1, 0)$ $(0, 1)$

5. (A) As x varies from 0 to $\dfrac{\pi}{2}$, $y = \sin x$ varies from 0 to 1.

(B) As x varies from $\dfrac{\pi}{2}$ to π, $y = \sin x$ varies from 1 to 0.

(C) As x varies from π to $\dfrac{3\pi}{2}$, $y = \sin x$ varies from 0 to –1.

(D) As x varies from $\dfrac{3\pi}{2}$ to 2π, $y = \sin x$ varies from –1 to 0.

(E) As x varies from 2π to $\dfrac{5\pi}{2}$, $y = \sin x$ varies from 0 to 1.

7. (A) As x varies from 0 to $-\frac{\pi}{2}$, $y = \cos x$ varies from 1 to 0.

   (B) As x varies from $-\frac{\pi}{2}$ to $-\pi$, $y = \cos x$ varies from 0 to -1.

   (C) As x varies from $-\pi$ to $-\frac{3\pi}{2}$, $y = \cos x$ varies from -1 to 0.

   (D) As x varies from $-\frac{3\pi}{2}$ to -2π, $y = \cos x$ varies from 0 to 1.

   (E) As x varies from -2π to $-\frac{5\pi}{2}$, $y = \cos x$ varies from 1 to 0.

9. $\sin x = 1$ requires $b = 1$, thus $P(a, b) = (0, 1)$. This occurs when $\theta = \frac{\pi}{2}$ and when $\theta = \frac{\pi}{2} + 2\pi$ or $\frac{5\pi}{2}$.

11. $\sin x = 0$ requires $b = 0$, thus $P(a, b) = (1, 0)$ or $(-1, 0)$
   $P(a, b) = (1, 0)$ when $\theta = 0$, 2π, or 4π $P(a, b) = (-1, 0)$ when $\theta = \pi$ or 3π

13. $\tan x = 0$ requires $\frac{b}{a} = 0$, thus $b = 0$, thus $P(a, b) = (1, 0)$ or $(-1, 0)$
   $P(a, b) = (1, 0)$ when $\theta = 0$, 2π, or 4π $P(a, b) = (-1, 0)$ when $\theta = \pi$ or 3π

15. $\sin x = -1$ requires $b = -1$, thus $P(a, b) = (0, -1)$. This occurs when $\theta = \frac{3\pi}{2}$ and when $\theta = \frac{3\pi}{2} + 2\pi$ or $\frac{7\pi}{2}$.

17. $\cos x = 1$ requires $a = 1$, thus $P(a, b) = (1, 0)$. This occurs when $\pi = -2\pi$, 0, or 2π.

19. $\cos x = 0$ requires $a = 0$, thus $P(a, b) = (0, 1)$ or $(0, -1)$
   $P(a, b) = (0, 1)$ when $\theta = \frac{\pi}{2}$ and also when $\theta = \frac{\pi}{2} - 2\pi = -\frac{3\pi}{2}$
   $P(a, b) = (0, -1)$ when $\theta = \frac{3\pi}{2}$ and also when $\theta = \frac{3\pi}{2} - 2\pi = -\frac{\pi}{2}$

21. $\tan x$ is not defined when $\frac{b}{a}$ is not defined. This occurs when $a = 0$, thus, $P(a, b) = (0, 1)$ or $(0, -1)$.
   $P(a, b) = (0, 1)$ when $\theta = \frac{\pi}{2}$ and also when $\theta = \frac{\pi}{2} + 2\pi$ or $\frac{5\pi}{2}$
   $P(a, b) = (0, -1)$ when $\theta = \frac{3\pi}{2}$ and also when $\theta = \frac{3\pi}{2} + 2\pi$ or $\frac{7\pi}{2}$

23. $\csc x$ is not defined when $\frac{1}{b}$ is not defined. This occurs when $b = 0$, thus $P(a, b) = (1, 0)$ or $(-1, 0)$.
   $P(a, b) = (1, 0)$ when $\theta = 0$, 2π, or 4π $P(a, b) = (-1, 0)$ when $\theta = \pi$ or 3π

25. Start at $(1, 0)$ and proceed counterclockwise (0.8 is positive) until an arc length of 0.8 has been covered. The point at the terminal end of the arc has coordinates $(0.7, 0.7) = (a, b)$. Thus, $\sin 0.8 = b = 0.7$.

27. Start at $(1, 0)$ and proceed counterclockwise (2.3 is positive) until an arc length of 2.3 has been covered. The point at the terminal end of the arc has coordinates $(-0.7, 0.7) = (a, b)$. Thus, $\cos 2.3 = a = -0.7$.

29. Start at $(1, 0)$ and proceed clockwise (-0.9 is negative) until an arc length of 0.9 has been covered. The point at the terminal end of the arc has coordinates $(0.6, -0.8) = (a, b)$. Thus, $\sin(-0.9) = -0.8$.

31. Start at $(1, 0)$ and proceed counterclockwise (2.2 is positive) until an arc length of 2.2 has been covered. The point at the terminal end of the arc has coordinates $(-0.6, 0.8) = (a, b)$.

    Thus, $\sec 2.2 = \dfrac{1}{a} = \dfrac{1}{-0.6} = -2$ (to one significant digit).

33. Start at $(1, 0)$ and proceed counterclockwise (0.8 is positive) until an arc length of 0.8 has been covered. The point at the terminal end of the arc has coordinates $(0.7, 0.7) = (a, b)$.

    Thus, $\tan 0.8 = \dfrac{b}{a} = \dfrac{0.7}{0.7} = 1$

35. Start at $(1, 0)$ and proceed clockwise (-0.4 is negative) until an arc length of 0.4 has been covered. The point at the terminal end of the arc has coordinates $(0.9, -0.4) = (a, b)$.

    Thus, $\cot(-0.4) = \dfrac{a}{b} = \dfrac{0.9}{-0.4} = -2$ (to one significant digit).

37. $\sin(-0.2103) = \sin(-0.2103 \text{ rad}) = -0.2088$ 39. $\sec 1.432 = \dfrac{1}{\cos 1.432} = \dfrac{1}{\cos(1.432 \text{ rad})} = 7.228$

41. $\tan 4.704 = \tan(4.704 \text{ rad}) = 119.2$ 43. $\cos 105.2 = \cos(105.2 \text{ rad}) = -0.04334$

45. $\cot(-0.03333) = \dfrac{1}{\tan(-0.03333)} = \dfrac{1}{\tan(-0.03333 \text{ rad})} = -29.99$

47. $\csc 6.2 = \dfrac{1}{\sin 6.2} = \dfrac{1}{\sin(6.2 \text{ rad})} = -12.04$

49. $P(a, b) = P(\cos x, \sin x)$ where $x = -0.898$. x is negative since P is moving clockwise. The quadrant in which P lies is determined by the signs of the coordinates of P.
    $P(\cos(-0.898), \sin(-0.898)) = (0.6232, -0.7821)$ and P lies in quadrant IV.

51. $P(a, b) = P(\cos x, \sin x)$ where $x = 26.77$. x is positive since P is moving counterclockwise. The quadrant in which P lies is determined by the signs of the coordinates of P.
    $P(\cos(26.77), \sin(26.77)) = (-0.0664, 0.9978)$ and P lies in quadrant II.

53. $\cos \dfrac{3\pi}{4} = \cos\left(\dfrac{3\pi}{4} \text{ rad}\right)$

    Locate the 45° reference triangle, determine (a, b) and r, then evaluate.

    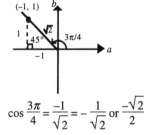

    $\cos \dfrac{3\pi}{4} = \dfrac{-1}{\sqrt{2}} = -\dfrac{1}{\sqrt{2}}$ or $\dfrac{-\sqrt{2}}{2}$

55. $\csc\left(-\dfrac{\pi}{4}\right) = \csc\left(-\dfrac{\pi}{4} \text{ rad}\right)$

    Locate the 45° reference triangle, determine (a, b) and r, then evaluate.

    $\csc\left(-\dfrac{\pi}{4}\right) = \dfrac{\sqrt{2}}{-1} = -\sqrt{2}$

57. $\tan\left(-\dfrac{5\pi}{2}\right) = \tan\left(-\dfrac{5\pi}{2}\,\text{rad}\right) = \dfrac{-1}{0}$

Not defined.

59. (A) – (D) All should equal 0.9525, because the sine function is periodic with period 2π.

61. Calculator in radian mode:

(A) $\tan 1 = 1.6$

$\dfrac{\sin 1}{\cos 1} = 1.6$

(B) $\tan 5.3 = -1.5$

$\dfrac{\sin 5.3}{\cos 5.3} = -1.5$

(C) $\tan(-2.376) = 0.96$

$\dfrac{\sin(-2.376)}{\cos(-2.376)} = 0.96$

63. Calculator in radian mode:

(A) $\sin(-3) = -0.14$

$-\sin 3 = -0.14$

(B) $\sin[-(-12.8)] = 0.23$

$-\sin(-12.8) = 0.23$

(C) $\sin(-407) = 0.99$

$-\sin(407) = 0.99$

Note: Some (very old model) calculators cannot evaluate $\sin(-407)$ and, instead, signal an error. If this occurs, use the periodicity of the sine fucntion and evaluate $\sin(-407 + 2\pi \cdot k)$, where k is an appropriate integer.

65. Calculator in radian mode:

(A) $\sin^2 1 + \cos^2 1 = (0.841\ldots)^2 + (0.540\ldots)^2 = 1.0$

(B) $\sin^2(-8.6) + \cos^2(-8.6) = (-0.734\ldots)^2 + (-0.678\ldots)^2 = 1.0$

(C) $\sin^2(263) + \cos^2(263) = (-0.779\ldots)^2 + (0.626\ldots)^2 = 1.0$

67. $\sin x \csc x = \sin x \dfrac{1}{\sin x}$ Use Identity (1)

$= 1$

69. $\cot x \sec x = \dfrac{\cos x}{\sin x} \cdot \dfrac{1}{\cos x}$ Use Identities (5) and (2)

$= \dfrac{1}{\sin x}$ Use Identity (1)

$= \csc x$

71. $\dfrac{\sin x}{1 - \cos^2 x} = \dfrac{\sin x}{\sin^2 x + \cos^2 x - \cos^2 x}$ Use Identity (9)

$= \dfrac{\sin x}{\sin^2 x}$

$= \dfrac{1}{\sin x} = \csc x$ Use Identity (1)

73. $\cot(-x)\sin(-x) = \dfrac{\cos(-x)}{\sin(-x)}\sin(-x)$ Use Identity (5)

$= \cos(-x)$

$= \cos x$ Use Identity (7)

75. Since (a, b) is on a unit circle with $(a, b) = (0.58064516, 0.81415674) = (\cos S, \sin S)$, we can solve $\cos S = 0.58064516$ or $\sin S = 0.81415674$. Then $s = \cos^{-1}(0.58064516) = 0.951$ or $s = \sin^{-1}(0.81415674) = 0.951$.

77. (A) Identity (4)   (B) Identity (9)   (C) Identity (2)

79. 2π

81. $S_1 = 1$

$S_2 = S_1 + \cos S_1 = 1 + \cos 1 = 1.540302$

$S_3 = S_2 + \cos S_2 = 1.540302 + \cos 1.540302 = 1.570792$

$S_4 = S_3 + \cos S_3 = 1.570792 + \cos 1.570792 = 1.570796$

$S_5 = S_4 + \cos S_4 = 1.570796 + \cos 1.570796 = 1.570796$

$\dfrac{\pi}{2} = 1.570796$

CHAPTER 2   REVIEW EXERCISE

1. (A) $\theta_r = \dfrac{\pi \text{ rad}}{180°}\theta_d$ (B) $\theta_r = \dfrac{\pi \text{ rad}}{180°}\theta_d$ (C) $\theta_r = \dfrac{\pi \text{ rad}}{180°}\theta_d$

$= \dfrac{\pi}{180}60 = \dfrac{\pi}{3}$ $= \dfrac{\pi}{180}45 = \dfrac{\pi}{4}$ $= \dfrac{\pi}{180}90 = \dfrac{\pi}{2}$

2. (A) $\theta_d = \dfrac{180°}{\pi \text{ rad}}\theta_r$ (B) $\theta_d = \dfrac{180°}{\pi \text{ rad}}\theta_r$ (C) $\theta_d = \dfrac{180°}{\pi \text{ rad}}\theta_r$

$= \dfrac{180}{\pi}\cdot\dfrac{\pi}{6}$ $= \dfrac{180}{\pi}\cdot\dfrac{\pi}{2}$ $= \dfrac{180}{\pi}\cdot\dfrac{\pi}{4}$

$= 30°$ $= 90°$ $= 45°$

3. A central angle of radian measure 2 is an angle subtended by an arc with length twice the length of the radius of the circle.

4. An angle of radian measure 1.5 is larger, since the corresponding degree measure of the angle would be $1.5\left(\dfrac{180°}{\pi}\right)$ or, approximately, 85.94°.

5. (A) $\theta_d = \dfrac{180°}{\pi \text{ rad}}\theta_r$ (B) $\theta_r = \dfrac{\pi \text{ rad}}{180°}\theta_d$

$= \dfrac{180}{\pi}\cdot 15.26$ $= \dfrac{\pi}{180}(-389.2)$

$= 874.3°$ $= -6.793 \text{ rad}$

6. $V = r\omega = 25(7.4) = 185 \text{ ft/min}$ 7. $\omega = \dfrac{V}{r} = \dfrac{415}{5.2} = 80 \text{ rad/hr}$

8. $P(a, b) = (-4, 3)$
    $$r = \sqrt{a^2 + b^2} = \sqrt{(-4)^2 + 3^2} = \sqrt{25} = 5$$

    $$\sin \theta = \frac{b}{r} = \frac{3}{5}$$

    $$\tan \theta = \frac{b}{a} = \frac{3}{-4} = -\frac{3}{4}$$

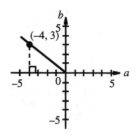

9. No, since csc $x = \dfrac{1}{\sin x}$, then one is positive so is the other.

10. (A) Use the reciprocal relationship cot $\theta = \dfrac{1}{\tan \theta}$.

    Degree mode: cot 53°40' = cot(53.666...°) Convert to decimal degrees, if necessary.

    $$= \frac{1}{\tan(53.666...°)} = 0.7355$$

    (B) Use the reciprocal relationship csc $\theta = \dfrac{1}{\sin \theta}$.

    Degree mode: csc 67°10' = csc(67.166...°) Convert to decimal degrees, if necessary.

    $$= \frac{1}{\sin(67.1666...°)} = 1.085$$

11. (A) Degree mode: cos 23.5° = 0.9171   (B) Degree mode: tan 42.3° = 0.9099

12. (A) Radian mode: cos 0.35 = cos(0.35 rad)   (B) Radian mode: tan 1.38 = tan (1.38 rad)
                                = 0.9394   = 5.177

13. The reference angle α is the angle (always taken positive) between the terminal side of θ and the horizontal axis.

    (A)

    $\alpha = 180° - 120° = 60°$

    (B)

    $$\alpha = 2\pi - \left| -\frac{7\pi}{4} \right| = \frac{\pi}{4}$$

14. **(A)** Use the special 30°–60° triangle as the reference triangle. Use the sides of the reference triangle to determine $P(a, b)$ and r. Then use Definition 1.

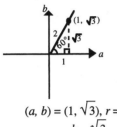

$(a, b) = (1, \sqrt{3}), r = 2$

$\sin 60° = \dfrac{b}{r} = \dfrac{\sqrt{3}}{2}$

(B) Use the special 45° triangle as the reference triangle. Use the sides of the reference triangle to determine $P(a, b)$ and r. Then use Definition 1.

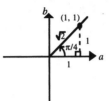

$(a, b) = (1, 1), r = \sqrt{2}$

$\cos \dfrac{\pi}{4} = \dfrac{a}{r} = \dfrac{1}{\sqrt{2}}$ or $\dfrac{\sqrt{2}}{2}$

(C)

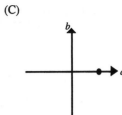

$(a, b) = (1, 0), r = 1$

$\tan 0° = \dfrac{0}{1} = 0$

15. **(A)**

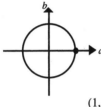

(1, 0)

(B)

(−1, 0)

(C)

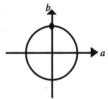

(0, 1)

(D)

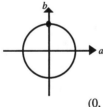

(0, 1)

(E)

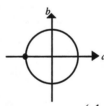

(−1, 0)

(F)

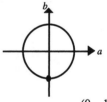

(0, −1)

16. **(A)** As x varies from 0 to $\dfrac{\pi}{2}$, $y = \sin x$ varies from 0 to 1.

(B) As x varies from $\dfrac{\pi}{2}$ to π, $y = \sin x$ varies from 1 to 0.

(C) As x varies from π to $\dfrac{3\pi}{2}$, $y = \sin x$ varies from 0 to −1.

(D) As x varies from $\dfrac{3\pi}{2}$ to 2π, $y = \sin x$ varies from −1 to 0.

(E) As x varies from 2π to $\dfrac{5\pi}{2}$, $y = \sin x$ varies from 0 to 1.

(F) As x varies from $\dfrac{5\pi}{2}$ to 3π, $y = \sin x$ varies from 1 to 0.

17. When the terminal side of the angle is rotated any multiple of a complete revolution (2π rad), in either direction the resulting angle will be coterminal with the original. In this case, for the restricted interval, this happens for $\frac{\pi}{6} \pm 2\pi$. $\frac{\pi}{6} + 2\pi = \frac{13\pi}{6}$; $\frac{\pi}{6} - 2\pi = \frac{-11\pi}{6}$

18. Since the central angle subtended by a circumference has degree measure 360°, the central angle subtended by an arc $\frac{7}{60}$ of a circumference has degree measure $\frac{7}{60}(360°) = 42°$.

19. $s = r\theta = 4(1.5) = 6$ cm.

20. $\theta_r = \frac{\pi \text{ rad}}{180°}$ $\theta_d = \frac{\pi}{180} 212 = \frac{53\pi}{45}$

21. $\theta_d = \frac{180°}{\pi \text{ rad}}$ $\theta_r = \frac{180}{\pi} \cdot \frac{\pi}{12} = 15°$

22. (A) –3.72 (B) 264.71°

23. Yes, since $\theta_d = (180°/\pi \text{ rad})\theta_r$, and if θ_r is tripled, θ_d will also be tripled.

24. No. For example, if $\alpha = \frac{\pi}{6}$ and $\beta = \frac{5\pi}{6}$, α and β are not coterminal, but $\sin \frac{\pi}{6} = \sin \frac{5\pi}{6}$.

25. Use the reciprocal identity: $\csc x = \frac{1}{\sin x} = \frac{1}{0.8594} = 1.1636$.

26. (A)

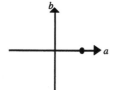

$(a, b) = (1, 0), r = 1$
$\tan 0 = \frac{b}{a} = \frac{0}{1} = 0$

(B)

$(a, b) = (0, 1), r = 1$
$\tan \frac{\pi}{2} = \frac{b}{a} = \frac{1}{0}$
Not defined.

(C)

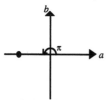

$(a, b) = (-1, 0), r = 1$
$\tan \pi = \frac{b}{a} = \frac{0}{-1} = 0$

(D)

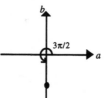

$(a, b) = (0, -1), r = 1$
$\tan \frac{3\pi}{2} = \frac{b}{a} = \frac{-1}{0}$
Not defined.

27. (A) 732° is coterminal with (732 – 360°) = 372°, and thus with (372 – 360)° = 12°. Since 12° is between 0° and 90°, its terminal side lies in quadrant I.

    (B) –7 rad is coterminal with (–7 + 2π)rad ≈ –0.72 rad. Since –0.72 rad is between 0 and $-\dfrac{\pi}{2}$, its terminal side lies in quadrant IV.

28. Degree mode: cos 187.4° = –0.992

29. Use the reciprocal relationship $\sec\theta = \dfrac{1}{\cos\theta}$.

    Degree mode: sec 103°20′ = sec(103.333...°) Convert to decimal degrees, if necessary.

    $$= \frac{1}{\cos(103.333...°)} = -4.34$$

30. Use the reciprocal relationship $\cot\theta = \dfrac{1}{\tan\theta}$.

    Degree mode: cot(–37°40′) = cot(–37.666...°) Convert to decimal degrees, if necessary.

    $$= \frac{1}{\tan(-37.666...°)} = -1.30$$

31. Radian mode: sin 2.39 = 0.683

32. Radian mode: cos 5 = 0.284

33. Use the reciprocal relationship $\cot\theta = \dfrac{1}{\tan\theta}$

    Radian mode: $\cot(-4) = \dfrac{1}{\tan(-4)} = -0.864$

34. Locate the 30°–60° reference triangle, determine (a, b) and r, then evaluate.

    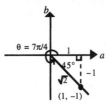

    $\cos\dfrac{5\pi}{6} = \dfrac{-\sqrt{3}}{2} = -\dfrac{\sqrt{3}}{2}$

35. Locate the 45° reference triangle, determine (a, b) and r, then evaluate.

    $\cot\dfrac{7\pi}{4} = \dfrac{1}{-1} = -1$

36.

    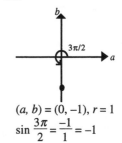

    (a, b) = (0, –1), r = 1

    $\sin\dfrac{3\pi}{2} = \dfrac{-1}{1} = -1$

37. See Problem 36.

$$\cos\frac{3\pi}{2} = \frac{0}{1} = 0$$

38. Locate the 30°–60° reference triangle, determine (a, b) and r, then evaluate.

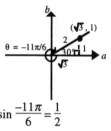

$$\sin\frac{-4\pi}{3} = \frac{\sqrt{3}}{2}$$

39. See Problem 38.

$$\sec\frac{-4\pi}{3} = \frac{2}{-1} = -2$$

40. $(a, b) = (-1, 0), r = 1$

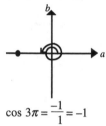

$$\cos 3\pi = \frac{-1}{1} = -1$$

41. See Problem 40.

$$\cot 3\pi = \frac{-1}{0}$$

Not defined.

42. Locate the 30°–60° reference triangle, determine (a, b) and r, then evaluate.

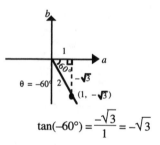

$$\sin\frac{-11\pi}{6} = \frac{1}{2}$$

43. It is known that $\cos\left(\dfrac{\pi}{3}\right) = \dfrac{1}{2} = 0.5000\ldots$ and $\sin(180°) = 0 =$

0.0000..., so that the calculator displays exact values for these. Thus $\tan(-60°)$ is not given exactly. To find $\tan(-60°)$, locate the 30°–60° reference triangle, determine (a, b) and r, then evaluate.

$$\tan(-60°) = \frac{-\sqrt{3}}{1} = -\sqrt{3}$$

44. Degree mode: $\sin 384.0314° = 0.40724$

45. Degree mode: $\tan(-198°43'6'') = \tan(-198.71833\ldots°)$ Convert to decimal degrees, if necessary.
$$= -0.33884$$

46. Radian mode: $\cos 26 = 0.64692$

47. Use the reciprocal relationship $\cot\theta = \dfrac{1}{\tan\theta}$.

Radian mode: $\cot(-68.005) = \dfrac{1}{\tan(-68.005)}$
$$= 0.49639$$

48. If $\sin \theta = -\dfrac{4}{5}$ and the terminal side of θ does not lie in the third quadrant, then it must lie in the fourth quadrant.

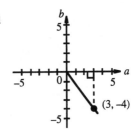

$P(a, b) = (3, -4)$, $r = \sqrt{a^2 + b^2} = \sqrt{3^2 + (-4)^2} = \sqrt{25} = 5$

$\cos \theta = \dfrac{a}{r} = \dfrac{3}{5}$

$\tan \theta = \dfrac{b}{a} = \dfrac{-4}{3} = -\dfrac{4}{3}$

49. Draw a reference triangle in the third quadrant with side opposite reference angle −1 and hypotenuse 2. Observe that this is a special 30°–60° triangle.

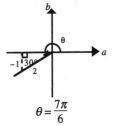

$\theta = \dfrac{7\pi}{6}$

50. Since $\sin \theta < 0$ and $\tan \theta < 0$, the terminal side of θ lies in quadrant IV. We sketch a reference triangle and label what we know.

Since $\sin \theta = \dfrac{b}{r} = -\dfrac{2}{5} = \dfrac{-2}{5}$, we know that $b = -2$ and $r = 5$ (r is never negative). Use the Pythagorean theorem to find a:

$$a^2 + (-2)^2 = 5^2$$
$$a^2 = 25 - 4 = 21$$
$$a = \sqrt{21}$$

a is positive since the terminal side of θ lies in quadrant IV. We can now find the other five functions using Definition 1:

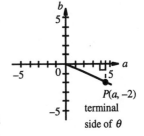

$\csc \theta = \dfrac{r}{b} = \dfrac{5}{-2} = -\dfrac{5}{2}$ $\qquad$ $\sec \theta = \dfrac{r}{a} = \dfrac{5}{\sqrt{21}}$

$\cos \theta = \dfrac{a}{r} = \dfrac{\sqrt{21}}{5}$ $\qquad$ $\cot \theta = \dfrac{a}{b} = \dfrac{\sqrt{21}}{-2} = -\dfrac{\sqrt{21}}{2}$

$\tan \theta = \dfrac{b}{a} = \dfrac{-2}{\sqrt{21}} = -\dfrac{2}{\sqrt{21}}$

51. $\tan \theta = \dfrac{b}{a} = -1 = \dfrac{-1}{1}$ or $\dfrac{1}{-1}$

We can draw a reference triangle in the second quadrant with side opposite reference angle 1 and side adjacent −1. We can also draw a reference triangle in the fourth quadrant with side opposite reference angle −1 and side adjacent 1. Each triangle is a special 45° triangle.

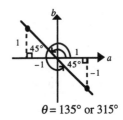

$\theta = 135°$ or $315°$

52. We can draw reference triangles in both quadrants II and III with sides adjacent reference angle $-\sqrt{3}$ and hypotenuse 2. Each triangle is a special 30°–60° triangle.

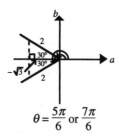

$$\theta = \frac{5\pi}{6} \text{ or } \frac{7\pi}{6}$$

53. (A) $s = r\theta = (12.0)(1.69) \approx 20.3$ cm

(B) $s = \dfrac{\pi}{180} r\theta = \dfrac{\pi}{180} (12.0)(22.5) \approx 4.71$ cm

54. (A) $A = \dfrac{1}{2} r^2\theta$

$$\left[r = \frac{1}{2} D = \frac{1}{2}(80) = 40 \text{ ft} \right]$$

$$A = \frac{1}{2}(40)^2 (0.773) \approx 618 \text{ ft}^2$$

(B) $A = \dfrac{\pi}{360} r^2\theta = \dfrac{\pi}{360} (40)^2 (135) \approx 1{,}880$ ft^2

55. Since the cities have the same longitude, θ is given by their difference in latitude.

$$\theta = 41°28' - 38°21' = 3°7' = \left(3 + \frac{7}{60}\right)^{\circ}$$

$$s = \frac{\pi}{180} r\theta = \frac{\pi}{180} (3{,}964 \text{ miles})\left(3 + \frac{7}{60}\right)$$

$$= 215.6 \text{ miles}$$

56. $\omega = \dfrac{\theta}{t} = \dfrac{6.43}{15.24} = 0.422$ rad/sec

57. A radial line from the axis of rotation sweeps out at an angle at the rate of 12π rad per sec.

58. (A) Calculator in radian mode: $\cos 7 = 0.754$

(B), (C) Both should equal 0.754, because the cosine function is periodic with period 2π.

59. Calculator in radian mode:

(A) $\tan(-7) = -0.871$
    $-\tan 7 = -0.871$

(B) $\tan[-(-17.9)] = -1.40$
    $-\tan(-17.9) = -1.40$

(C) $\tan[-(-2{,}135)] = -3.38$
    $-\tan(-2{,}135) = -3.38$

Note: Some (very old model) calculators cannot evaluate $\tan(-2{,}135)$ and, instead, signal an error. If this occurs, use the periodicity of the tangent function and evaluate $\tan(2{,}135 - 2\pi \cdot k)$, where k is an appropriate integer.

60. (D) is not an identity, since it is not true for all values of the variable x.

61. $(\csc x)(\cot x)(1 - \cos^2 x)$ $= \dfrac{1}{\sin x} \dfrac{\cos x}{\sin x}(1 - \cos^2 x)$ Use Identities (1) and (5)

$= \dfrac{\cos x}{\sin^2 x}(\sin^2 x + \cos^2 x - \cos^2 x)$ Use Identity (9)

$= \dfrac{\cos x}{\sin^2 x} \cdot \sin^2 x = \cos x$

62. $\cot(-x)\sin(-x)$ $= \dfrac{\cos(-x)}{\sin(-x)}\sin(-x)$ Use Identity (5)

$= \cos(-x) = \cos x$ Use Identity (7)

63.   Since $P(a, b)$ is moving clockwise, $x = -29.37$. By the definition of the circular functions, the point has coordinates $P(\cos(-29.37), \sin(-29.37)) = P(-0.4575, 0.8892)$. P lies in quadrant II, since a is negative and b is positive.

64.   The radian measure of a central angle θ subtended by an arc of length s is $\theta = \dfrac{s}{r}$, where r is the radius of the circle. In this case $\theta = \dfrac{1.3}{1} = 1.3$ rad.

65.   Since $\cot x = \dfrac{1}{\tan x}$ and $\csc x = \dfrac{1}{\sin x}$, and $\sin(k\pi) = \tan(k\pi) = 0$ for all integers k, $\cot x$ and $\sec x$ are not defined for these values.

66. $\sin x = -\dfrac{\sqrt{3}}{2}$

We can draw a reference triangle in the third quadrant with side opposite reference angle $-\sqrt{3}$ and hypotenuse 2. Observe that this is a special 30°–60° triangle.

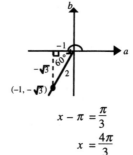

$x - \pi = \dfrac{\pi}{3}$

$x = \dfrac{4\pi}{3}$

67.   Since (a, b) is on a unit circle with $(a, b) = (0.62903226, 0.7773792) = (\cos s, \sin s)$, we can solve $\cos s = 0.62903226$ or $\sin s = 0.7773792$. Then $s = \cos^{-1}(0.62903226) = 0.8905$ or $s = \sin^{-1}(0.7773792) = 0.8905$.

68. Since $A = \dfrac{1}{2} r^2 \theta$ and $s = r\theta$, we can eliminate θ between these two equations as follows:

$$\theta = \dfrac{s}{r}$$

$$A = \dfrac{1}{2} r^2 \left(\dfrac{s}{r}\right) = \dfrac{1}{2} rs.$$

Then, $s = \dfrac{2A}{r} = \dfrac{2(342.5)}{12} \approx 57$ m.

69. From the figure, we note $s = r\theta$.

$$r = \sqrt{a^2 + b^2} = \sqrt{4^2 + 5^2} = \sqrt{41}$$

$$\tan \theta = \dfrac{b}{a} = \dfrac{5}{4}$$

$$\theta = \tan^{-1} \dfrac{5}{4}$$

Thus, $s = \sqrt{41} \, \tan^{-1} \dfrac{5}{4} \approx 5.74$ units.

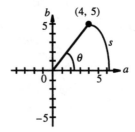

(Calculator in radian mode.)

70. We use $s = r\theta$ with $r = \dfrac{1}{2} d = 5$ cm and $s = 10$ m $= 1000$ cm.

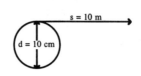

Then, $\theta = \dfrac{s}{r} = \dfrac{1000}{5} = 200$ rad. Since 1 revolution corresponds to 1 circumference $= 2\pi$ radians, 200 radians corresponds to $\dfrac{200}{2\pi} = \dfrac{100}{\pi} \approx 31.8$ revolutions.

71. Since the three gear wheels are coupled together, each must turn through the same distance (arc length). Thus,

$$s = r_1 \theta_1 \qquad s = r_2 \theta_2 \qquad s = r_3 \theta_3 \qquad r_1 = 30 \text{ cm} \qquad r_2 = 20 \text{ cm} \qquad r_3 = 10 \text{ cm}$$

$$r_1 \theta_1 = r_2 \theta_2$$

$$\theta_2 = \dfrac{r_1}{r_2} \theta_1 = \dfrac{30}{20} \text{ (5 revolutions)} = 7.5 \text{ revolutions}$$

$$r_3 \theta_3 = r_1 \theta_1$$

$$\theta_3 = \dfrac{r_1}{r_3} \theta_1 = \dfrac{30}{10} \text{ (5 revolutions)} = 15 \text{ revolutions}$$

72. We use $\omega = \dfrac{V}{r}$ with $V = 70$ ft/sec and $r = \dfrac{27 \text{ in.}}{2} = \dfrac{27 \text{ in.}}{2} \cdot \dfrac{1 \text{ in.}}{12 \text{ ft}} = \dfrac{27}{24}$ ft. Then $\omega = \dfrac{V}{r} = \dfrac{70}{27/24} = 62$ rad/sec.

73. We use $V = r\omega$.

$$r = \text{radius of orbit} = 3{,}964 + 1{,}000 = 4{,}964 \text{ mi}$$

$$\omega = \dfrac{\theta}{t} = \dfrac{2\pi}{114/60 \text{ hr}} = \dfrac{120\pi}{114} \text{ rad/hr}$$

$$V = (4{,}964 \text{ mi})\left(\dfrac{120\pi}{114} \text{ rad/hr}\right) = 16{,}400 \text{ mi/hr}$$

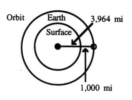

74. $I = 30 \sin(120\pi t - 60\pi) = 30 \sin[120\pi(0.015) - 60\pi] = -17.6$ amp

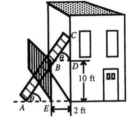

75. (A) We note:
The length of the ladder $AC = AB + BC$

In triangle ABE, $\csc \theta = \dfrac{AB}{BE} = \dfrac{AB}{10 \text{ ft}}$

In triangle BCD, $\sec \theta = \dfrac{BC}{BD} = \dfrac{BC}{2 \text{ ft}}$

Then $AB = 10 \csc \theta$, $BC = 2 \sec \theta$, hence length of ladder
$AC = 10 \csc \theta + 2 \sec \theta$

(B) As θ decreases to 0 rad, L will increase without bound; as θ increases to $\dfrac{\pi}{2}$, L increases without bound. Between these extremes, there appears to be a value of θ that produces a minimum L.

(C)

θ rad	0.70	0.80	0.90	1.00	1.10	1.20	1.30
L ft	18.14	16.81	15.98	15.59	15.63	16.25	17.85

(D) $L = 15.59$ ft for $\theta = 1.00$ rad

76. Use $\dfrac{n_2}{n_1} = \dfrac{\sin \alpha}{\sin \beta}$, where $n_2 = 1.33$, $n_1 = 1.00$, and $\alpha = 31.7°$

Solve for β: $\dfrac{1.33}{1.00} = \dfrac{\sin 31.7°}{\sin \beta}$

$\sin \beta = \dfrac{\sin 31.7°}{1.33}$

$\beta = \sin^{-1}\left(\dfrac{\sin 31.7°}{1.33}\right)$

$= 23.3°$

77. Find the angle of incidence α such that the angle of refraction is $90°$.

$\dfrac{\sin \alpha}{\sin \beta} = \dfrac{n_2}{n_1}$ $\sin \alpha = \dfrac{n_2}{n_1} \sin \beta = \dfrac{1.00}{1.52} \sin 90°$ $\alpha = \sin^{-1}\left[\dfrac{1.00}{1.52}(1)\right] = 41.1°$

78. We use $\sin \dfrac{\theta}{2} = \dfrac{S_w}{S_b}$, where $\theta = 51°$ and $S_b = 25$ mph. Then we solve for S_w:

$\sin \dfrac{51°}{2} = \dfrac{S_w}{25}$ $S_w = 25 \sin 25.5° = 11$ mph

Chapter 3   Graphing Trigonometric Functions

EXERCISE 3.1   Basic Graphs

1. $2\pi,\ 2\pi,\ \pi$ 3. (A) 1 unit   (B) Indefinitely far   (C) Indefinitely far

5. $-\dfrac{3\pi}{2},\ -\dfrac{\pi}{2},\ \dfrac{\pi}{2},\ \dfrac{3\pi}{2}$ 7. $-2\pi,\ -\pi,\ 0,\ \pi,\ 2\pi$ 9. The graph has no x intercepts; sec x is never 0.

11. (A) None; sin x is always defined. (B) $-2\pi,\ -\pi,\ 0,\ \pi,\ 2\pi$ (C) $-\dfrac{3\pi}{2},\ -\dfrac{\pi}{2},\ \dfrac{\pi}{2},\ \dfrac{3\pi}{2}$

13.

x	0	0.1	0.2	0.3	0.4	0.5	0.6	0.7	0.8
cos x	1	1.0	0.98	0.96	0.92	0.88	0.83	0.76	0.70

x	0.9	1.0	1.1	1.2	1.3	1.4	1.5	1.6
cos x	0.62	0.54	0.45	0.36	0.27	0.17	0.07	−0.03

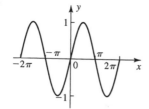

15. 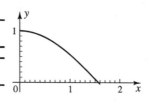   17.

19.   The dashed line shows $y = \sin x$ in this interval. The solid line is $y = \csc x$.

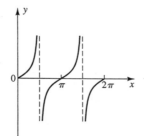

21. (A) $y = 3 \sin x$ $y = \sin x$
        $y = -2 \sin x$

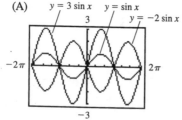

(B) No, all three graphs have the same x intercepts, at -2π, $-\pi$, 0, π, 2π.

(C) The highest point on the graph of $y = \sin x$ has y coordinate 1. The lowest point has y coordinate -1. The highest point on the graph of $y = -2 \sin x$ has y coordinate 2. The lowest point has y coordinate -2. The highest point on the graph of $y = 3 \sin x$ has y coordinate 3. The lowest point has y coordinate -3.

(D) The deviation of the graph from the x axis is changed by changing A. The deviation appears to be $|A|$.

23. (A)

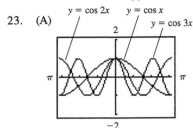

(B) One period of $y = \cos x$ appears.
($-\pi$ to π)
Two periods of $y = \cos 2x$ appear.
($-\pi$ to 0, 0 to π)
Three periods of $y = \cos 3x$ appear.
$$\left(-\pi \text{ to } -\frac{\pi}{3}, \ -\frac{\pi}{3} \text{ to } \frac{\pi}{3}, \ \frac{\pi}{3} \text{ to } \pi\right)$$

(C) n periods of $y = \cos nx$ would appear.

25. (A)

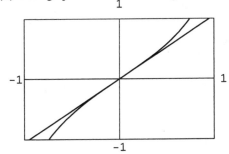

(B) The graph of $y = \sin x$ is shifted $|C|$ units to the right if $C < 0$ and $|C|$ units to the left if $C > 0$.

27. Depending on the particular calculator used, either an error message will occur or some very large number will occur because of round-off error, because:

(A) 0 is not in the domain of the cotangent function.

(B) $\frac{\pi}{2}$ is not in the domain of the tangent function.

(C) π is not in the domain of the cosecant function.

29. (A) Both graphs are almost indistinguishable the closer x is to the origin.

x	−0.3	−0.2	−0.1	0.0	0.1	0.2	0.3
$\tan x$	−0.309	−0.203	−0.100	0.000	0.100	0.203	0.309

(B)

(C) It is not valid to replace tan x with x for small x if x is in degrees, as is clear from the graph.

31.   For a given value of T, the y value on the unit circle and the corresponding y value on the sine curve are the same. This is a graphing utility illustration of how the sine function is defined as a circular function. See Figure 3 in this section of the text.

EXERCISE 3.2   Graphing $y = k + A \sin Bx$ and $y = k + A \cos Bx$

1.

3. $y = -2 \sin x$. Amplitude $= |-2| = 2$. Period $= \dfrac{2\pi}{B} = \dfrac{2\pi}{1} = 2\pi$.

Since $A = -2$ is negative, the basic curve for $y = \sin x$ is turned upside down. One full cycle of the graph is completed as
x goes from 0 to 2π. Block out this interval, divide it into four equal parts, locate high and low points, and locate x intercepts. Then complete the graph.

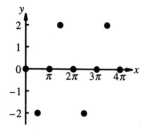 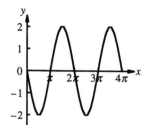

5. $y = \dfrac{1}{2} \sin x$. Amplitude $= \left| \dfrac{1}{2} \right| = \dfrac{1}{2}$. Period $= \dfrac{2\pi}{B} = \dfrac{2\pi}{1} = 2\pi$.

One full cycle of the graph is completed as x goes from 0 to 2π. Block out this interval, divide it into four equal parts, locate high and low points, and locate x intercepts. Then complete the graph.

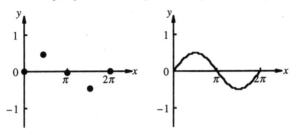

7. $y = \sin 2\pi x$. Amplitude $= |A| = |1| = 1$. Period $= \dfrac{2\pi}{2\pi} = 1$.

One full cycle of this graph is completed as x goes from 0 to 1. Block out this interval, divide it into four equal parts, locate high and low points, and locate x intercepts. Then complete the graph.

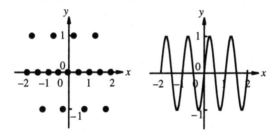

9. $y = \cos \dfrac{x}{4}$. Amplitude $= |A| = |1| = 1$. Period $= \dfrac{2\pi}{1/4} = 8\pi$.

One full cycle of this graph is completed as x goes from 0 to 8π. Block out this interval, divide it into four equal parts, locate high and low points, and locate x intercepts. Then complete the graph.

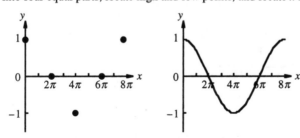

11. $y = 2 \sin 4x$. Amplitude $= |2| = 2$. Period $= \dfrac{2\pi}{4} = \dfrac{\pi}{2}$.

One full cycle of this graph is completed as x goes from 0 to $\dfrac{\pi}{2}$. Block out this interval, divide it into four equal parts, locate high and low points, and locate x intercepts. Then complete the graph.

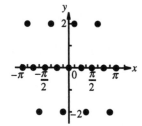 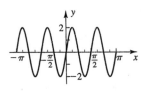

13. $y = \dfrac{1}{3} \cos 2\pi x$. Amplitude $= \left| \dfrac{1}{3} \right| = \dfrac{1}{3}$. Period $= \dfrac{2\pi}{2\pi} = 1$.

One full cycle of this graph is completed as x goes from 0 to 1. Block out this interval, divide it into four equal parts, locate high and low points, and locate x intercepts. Then complete the graph.

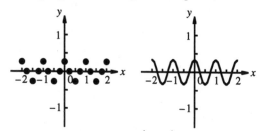

15. $y = -\dfrac{1}{4} \sin \dfrac{x}{2}$. Amplitude $= \left| -\dfrac{1}{4} \right| = \dfrac{1}{4}$. Period $= \dfrac{2\pi}{1/2} = 4\pi$.

Since $A = -\dfrac{1}{4}$ is negative, the basic curve for $y = \sin x$ is turned upside down. One full cycle of this graph is completed as x goes from 0 to 4π. Block out this interval, divide it into four equal parts, locate high and low points, and locate x intercepts. Then complete the graph.

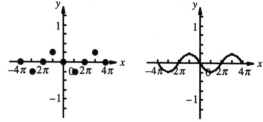

17.   Since the displacement is 0 ft when t is 0, the equation should be of the form $y = A \sin Bt$. Since the amplitude is 2 ft, $|A| = 2$, hence $A = 2$ or $A = -2$. Since the period is 2 sec, write

$$\dfrac{2\pi}{B} = 2$$
$$2\pi = 2B$$
$$B = \pi$$

Thus, the required equation is $y = 2 \sin \pi t$ or $y = -2 \sin \pi t$.

19.   Since $P = \dfrac{2\pi}{B}$, P tends to zero as B increases without bound.

21. $y = -1 + \dfrac{1}{3}\cos 2\pi x$. Amplitude $= \left| \dfrac{1}{3} \right| = \dfrac{1}{3}$. Period $= \dfrac{2\pi}{2\pi} = 1$.

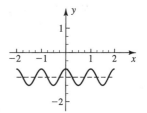

$y = \dfrac{1}{3}\cos 2\pi x$ was graphed in Problem 13. This graph is the graph

of $y = \dfrac{1}{3}\cos 2\pi x$ moved down $|k| = |-1| = 1$ unit. We start by

drawing a horizontal broken line 1 unit below the x axis, then

graph $y = \dfrac{1}{3}\cos 2\pi x$ relative to the broken line and the original

y axis.

23. $y = 2 - \dfrac{1}{4}\sin \dfrac{x}{2}$. Amplitude $= \left| -\dfrac{1}{4} \right| = \dfrac{1}{4}$. Period $= \dfrac{2\pi}{1/2} = 4\pi$.

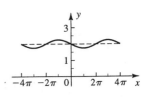

$y = -\dfrac{1}{4}\sin \dfrac{x}{2}$ was graphed in Problem 15. This graph is the graph

of $y = -\dfrac{1}{4}\sin \dfrac{x}{2}$ moved up $k = 2$ units. We start by drawing a

horizontal broken line 2 units above the x axis, then graph

$y = -\dfrac{1}{4}\sin \dfrac{x}{2}$ relative to the broken line and the original y axis.

25.   Amplitude $= 5 = |A|$. Period $= \dfrac{2\pi}{B} = \pi$. Thus, $B = 2$. The form of the graph is that of the basic sine
curve. Thus, $y = |A| \sin Bx = 5 \sin 2x$.

27.   Amplitude $= 4 = |A|$. Period $= \dfrac{2\pi}{B} = 4$. Thus, $B = \dfrac{2\pi}{4} = \dfrac{\pi}{2}$. The form of the graph is that of the basic

sine curve turned upside down. Thus, $y = -|A| \sin Bx = -4 \sin\left(\dfrac{\pi x}{2}\right)$.

29.   Amplitude $= 8 = |A|$. Period $= \dfrac{2\pi}{B} = 8\pi$. Thus, $B = \dfrac{2\pi}{8\pi} = \dfrac{1}{4}$. The form of the graph is that of the

basic sine curve. Thus, $y = |A| \cos Bx = 8 \cos\left(\dfrac{1}{4} x\right)$.

31.   Amplitude $= 1 = |A|$. Period $= \dfrac{2\pi}{B} = 6$. Thus, $B = \dfrac{2\pi}{6} = \dfrac{\pi}{3}$. The form of the graph is that of the basic

cosine curve turned upside down. Thus, $y = -|A| \cos Bx = -\cos\left(\dfrac{\pi x}{3}\right)$.

33. The graph of $y = \sin x \cos x$ is shown in the figure.
    Amplitude $= 0.5 = |A|$. Period $= \dfrac{2\pi}{B} = \pi$. Thus, $B = \dfrac{2\pi}{\pi} = 2$.
    The form of the graph is that of the basic sine curve.
    Thus, $y = |A| \sin Bx = 0.5 \sin 2x$.

    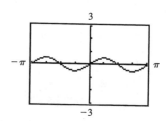

35. The graph of $y = 2 \cos^2 x$ is shown in the figure.
    Amplitude $= \dfrac{1}{2}$ (y coordinate of highest point $- y$ coordinate of
    lowest point) $= \dfrac{1}{2}(2 - 0) = 1 = |A|$; Period $= \dfrac{2\pi}{B} = \pi$.
    Thus, $B = \dfrac{2\pi}{\pi} = 2$. The form of the graph is that of the basic
    cosine curve shifted up 1 unit.
    Thus, $y = 1 + |A| \cos Bx = 1 + \cos 2x$.

    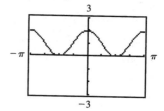

37. The graph of $y = 2 - 4 \sin^2 2x$ is shown in the figure.
    Amplitude $= 2 = |A|$. Period $= \dfrac{2\pi}{B} = \dfrac{\pi}{2}$.
    Thus, $B = 2\pi \div \dfrac{\pi}{2} = 4$. The form of the graph is that of the
    basic cosine curve.
    Thus, $y = |A| \cos Bx = 2 \cos 4x$.

    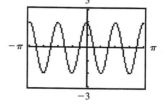

39. (A) $C = 0$ and $-\dfrac{\pi}{2}$ $C = 0$ and $\dfrac{\pi}{2}$

    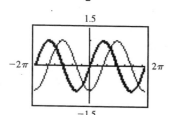   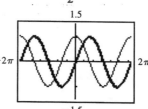

    (B) If $C < 0$, then the graph of $y = \sin x$ is shifted $|C|$ units to the right. If $C > 0$, then the graph of
    $y = \sin x$ is shifted C units to the left.

41. The graph of $f(x) = \cos^2 x$, $-2\pi \le x \le 2\pi$, is shown in the figure.

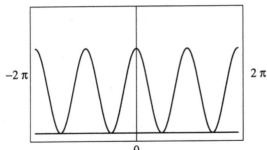

It appears that four full periods of the graph occur in the interval from -2π to 2π, thus, the period is the length of this interval divided by 4.

$$P = \frac{2\pi - (-2\pi)}{4} = \frac{4\pi}{4} = \pi.$$

43. Since the maximum value occurs at the end points of the interval, it would appear that a function of the form $y = A \cos Bx$ would be required, with A positive.

Since the maximum value of the function seems to be 2, and the minimum value seems to be -2,

$A = \frac{2 - (-2)}{2} = \frac{4}{2} = 2$. Since the maximum value is achieved at $x = 0$ and $x = 3$, the period of the

function is 3. Hence $\frac{2\pi}{B} = 3$ and $B = \frac{2\pi}{3}$. Thus the required function is $y = 2 \cos \frac{2\pi x}{3}$.

45. $E = 110 \sin 120\pi t$. Amplitude $= |110| = 110$. Period $= \frac{2\pi}{120\pi} = \frac{1}{60}$ sec.

Frequency $f = \frac{1}{p} = \frac{1}{1/60} = 60$ Hz

One full cycle of this graph is completed as t goes from 0 to $\frac{1}{60}$. Block out this interval, divide it

into four equal parts, locate high and low points, and locate t intercepts. Then complete the graph.

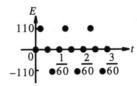
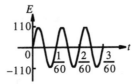

47. *Find A:* The amplitude $|A|$ is given to be 12. Since $E = 12$ when $t = 0$, $A = 12$ (and not -12).
Find B: We are given that the frequency, f, is 40 Hz. Hence, the period is found using the reciprocal formula:

$P = \frac{1}{f} = \frac{1}{40}$ sec. But, $P = \frac{2\pi}{B}$. Thus, $B = \frac{2\pi}{P} = \frac{2\pi}{1/40} = 80\pi$.

Write the equation: $E = 12 \cos 80\pi t$

49. (A) We use the formula: Period $= \dfrac{2\pi}{\sqrt{1{,}000\ gA/M}}$ with $g = 9.75$ m/sec^2,

$A = 3$ m $\times$ 3 m $= 9$ m^2, and Period $= 1$ sec, and solve for M. Thus, $1 = \dfrac{2\pi}{\sqrt{(1{,}000)(9.75)(9)/M}}$;

$M = \dfrac{(1{,}000)(9.75)(9)}{4\pi^2} \approx 2{,}220$ kg

(B) $D = 0.2$, $B = \dfrac{2\pi}{\text{Period}}$, $y = 0.2 \sin 2\pi t$

(C) One full cycle of this graph is completed as t goes from 0 to 1. Block out this interval, divide it
into four equal parts, locate high and low points, and locate t intercepts. Then complete the
graph.

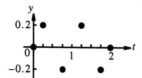 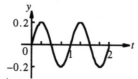

51. (A) max vol $= 85\ \ell$, min vol $= 0.05\ \ell$; $0.40 \cos \dfrac{\pi t}{2}$ is maximum when $\cos \dfrac{\pi t}{2}$ is 1 and is minimum

when $\cos \dfrac{\pi t}{2}$ is -1.

Therefore, max vol $= 0.45 + 0.40 = 0.85\ \ell$ and min vol $= 0.45 - 0.40 = 0.05\ \ell$.

(B) The period $= \dfrac{2\pi}{B} = 2\pi \div \left(\dfrac{\pi}{2}\right) = 2\pi \cdot \dfrac{2}{\pi} = 4$ seconds

(C) A breath is taken every 4 sec. Since there are 60 seconds per minute, there are $\dfrac{60}{4} = 15$ breaths
per minute.

(D)

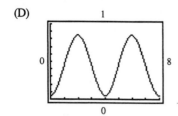

From the graph, the maximum volume
occurs at $t = 2$ and $t = 6$ and appears to be
0.85ℓ. The minimum volume occurs at
$t = 0, 4, 8$, and appears to be 0.05ℓ.

53. Since the rotation is at 4 revolutions per minute, in 1 minute it covers 4 revolutions, or 8π radians.
In t minutes, it covers $8\pi t$ radians, thus $\theta = 8\pi t$. To see that $x = 20 \sin 8\pi t$, it might help to look
sideways at the wheel. Below, we have indicated a new coordinate system in which θ is in standard
position. Then, $\sin \theta = \dfrac{b}{20}$, so $b = 20 \sin \theta = 20 \sin 8\pi t$. Thus, the coordinate of the shadow $= b = x$
in the author's coordinate system. That is, $x = 20 \sin 8\pi t$.

To graph this, note:

Amplitude $= |A| = |20| = 20$

Period $= \dfrac{2\pi}{8\pi} = \dfrac{1}{4}$.

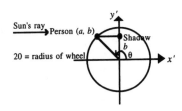

20 = radius of wheel

One full cycle of the graph is completed as t goes from 0 to $\dfrac{1}{4}$.

Block out this interval, divide it into four equal parts, locate high and low points, and locate t intercepts. Then complete the graph.

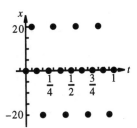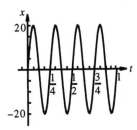

55. (A) The data for θ repeats every 2 seconds, so the period is $P = 2$ sec. The angle θ deviates from 0 by $25°$ in each direction, so the amplitude is $|A| = 25°$.

   (B) $\theta = A \sin Bt$ is not suitable, because, for example, for $t = 0$, $A \sin Bt = 0$ no matter what the choice of A and B. $\theta = A \cos Bt$ appears suitable, because, for example if $t = 0$ and $A = -25$, then we can get the first value in the table, a good start. Choose $A = -25$ and $B = \dfrac{2\pi}{P} = \pi$, which yields $\theta = -25 \cos \pi t$. A calculator can be used to check that this equation produces (or comes close to producing) all the values in the table.

   (C) One full cycle of this graph is completed as t goes from 0 to 2. Block out this interval, divide it into four equal parts, locate high and low points, and locate t intercepts. Then complete the graph.

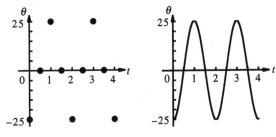

57. (A)

t sec	0	3	6	9	12	15	18	21	24
h ft	28	13	-2	13	28	13	-2	13	28

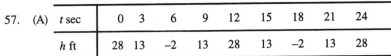

(B)

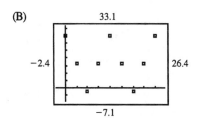

(C) Since the maximum value of h is repeated at intervals of 12 sec, the period = 12 sec. Because the maximum value of $h = k + A \cos Bt$ occurs when $t = 0$ (assuming A is positive), and this corresponds to a maximum value in the table 1 when $t = 0$, a function of this form appears to be a better model for the data.

(D) $|A| = \dfrac{\max h - \min h}{2} = \dfrac{28 - (-2)}{2} = 15$

(E)

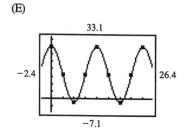

$B = \dfrac{2\pi}{\text{Period}} = \dfrac{2\pi}{12} = \dfrac{\pi}{6}$

$k = |A| + \min h = 15 + (-2) = 13$

Thus, $h = 13 + 15 \cos \dfrac{\pi}{6} t$ or $h = 13 - 15 \cos \dfrac{\pi}{6} t$.

Since $h = 28$ when $t = 0$, $h = 13 + 15 \cos \dfrac{\pi}{6} \cdot 0 = 28$ indicates that

$\qquad h = 13 + 15 \cos \dfrac{\pi}{6} t$ is the correct equation.

EXERCISE 3.3   Graphing $y = k + A \sin(Bx + C)$ and $y = k + A \cos(Bx + C)$

1. If the graph of a simple harmonic having no phase shift is moved two units to the left, the result is a simple harmonic having a phase shift of –2.

3. Amplitude $= |A| = |1| = 1$
    Phase Shift and Period: Solve
    $\qquad Bx + C = 0$ and $Bx + C = 2\pi$
    $\qquad x + \dfrac{\pi}{2} = 0 \qquad x + \dfrac{\pi}{2} = 2\pi$
    $\qquad\quad x = -\dfrac{\pi}{2} \qquad\quad x = -\dfrac{\pi}{2} + 2\pi$
    $\qquad\qquad \uparrow \qquad\qquad\qquad \uparrow \quad \uparrow$
    $\qquad\quad$ Phase Shift $\qquad$ Period $= 2\pi$

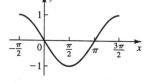

Phase Shift $= -\dfrac{\pi}{2}$. Graph one cycle over the interval from $-\dfrac{\pi}{2}$ to $\left(-\dfrac{\pi}{2} + 2\pi\right) = \dfrac{3\pi}{2}$.

5. Amplitude $= |A| = |1| = 1$
    Phase Shift and Period: Solve
    $\qquad Bx + C = 0$ and $Bx + C = 2\pi$
    $\qquad x - \dfrac{\pi}{4} = 0 \qquad x - \dfrac{\pi}{4} = 2\pi$
    $\qquad\quad x = \dfrac{\pi}{4} \qquad\quad x = \dfrac{\pi}{4} + 2\pi$
    $\qquad\qquad \uparrow \qquad\qquad\qquad \uparrow \quad \uparrow$
    $\qquad\quad$ Phase Shift $\qquad$ Period $= 2\pi$

Phase Shift $= \dfrac{\pi}{4}$. Graph one cycle over the interval from $\dfrac{\pi}{4}$ to $\left(\dfrac{\pi}{4} + 2\pi\right) = \dfrac{9\pi}{4}$.

Extend the graph from $-\pi$ to $\dfrac{\pi}{4}$ and delete

the portion of the graph from 2π to $\dfrac{9\pi}{4}$,

since this was not required.

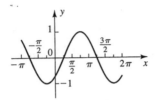

7. Amplitude $= |A| = |4| = 4$
    Phase Shift and Period: Solve
    $$Bx + C = 0 \quad \text{and} \quad Bx + C = 2\pi$$
    $$\pi x + \frac{\pi}{4} = 0 \qquad \pi x + \frac{\pi}{4} = 2\pi$$
    $$x = -\frac{1}{4} \qquad x = -\frac{1}{4} + 2$$
    $\qquad\quad\uparrow \qquad\qquad\qquad \uparrow \quad \uparrow$
    $\qquad$ Phase Shift $\qquad\quad$ Period $= 2$

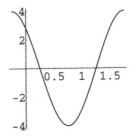

Phase Shift $= -\dfrac{1}{4}$. Graph one cycle over the

interval from $-\dfrac{1}{4}$ to $\left(-\dfrac{1}{4} + 2\right) = \dfrac{7}{4}$.

Extend the graph from -1 to 3.

9. Amplitude $= |A| = |-2| = 2$
    Phase Shift and Period: Solve
    $$Bx + C = 0 \quad \text{and} \quad Bx + C = 2\pi$$
    $$2x + \pi = 0 \qquad 2x + \pi = 2\pi$$
    $$x = -\frac{\pi}{2} \qquad x = -\frac{\pi}{2} + \pi$$
    $\qquad\quad\uparrow \qquad\qquad\qquad \uparrow \quad \uparrow$
    $\qquad$ Phase Shift $\qquad\quad$ Period $= \pi$

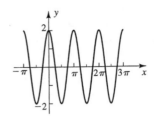

Phase Shift $= -\dfrac{\pi}{2}$. Graph one cycle (the

basic cosine curve turned upside down) over

the interval from $-\dfrac{\pi}{2}$ to $\left(-\dfrac{\pi}{2} + \pi\right) = \dfrac{\pi}{2}$.

Extend the graph from $-\pi$ to 3π.

Chapter 3 Graphing Trigonometric Functions

11. We sketch one period of the graph of $y = \sin x$ below. It has amplitude 1, period 2π, and

phase shift 0. To graph $y = \cos\left(x - \dfrac{\pi}{2}\right)$ we compute its amplitude, period, and phase shift:

Amplitude $= |A| = |1| = 1$
Phase Shift and Period: Solve

$$Bx + C = 0 \quad \text{and} \quad Bx + C = 2\pi$$

$$x - \frac{\pi}{2} = 0 \qquad x - \frac{\pi}{2} = 2\pi$$

$$x = \frac{\pi}{2} \qquad x = \frac{\pi}{2} + 2\pi$$

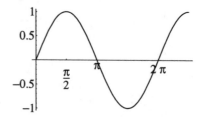

$$\underset{\text{Phase Shift}}{\uparrow} \qquad \underset{\text{Period} = 2\pi}{\uparrow \quad \uparrow}$$

If we sketch one period of the graph starting at $x = \dfrac{\pi}{2}$ (the phase shift) and ending at $x = \dfrac{\pi}{2} + 2\pi = \dfrac{5\pi}{2}$

(the phase shift plus one period), we see that it is the same as that of $y = \sin x$ over the interval from

$x = \dfrac{\pi}{2}$ to $x = 2\pi$. Since both curves can be extended indefinitely in both directions, we conclude that

the graphs are the same, thus $\cos\left(x - \dfrac{\pi}{2}\right) = \sin x$ for all x.

13. $y = -2 + 4 \cos\left(\pi x + \dfrac{\pi}{4}\right)$.

$y = 4 \cos\left(\pi x + \dfrac{\pi}{4}\right)$ was graphed in Problem 7. This graph is the graph of $y = 4 \cos\left(\pi x + \dfrac{\pi}{4}\right)$

moved down $|k| = |-2| = 2$ units. We start by drawing a horizontal

broken line 2 units below the x axis, then graph $y = 4 \cos\left(\pi x + \dfrac{\pi}{4}\right)$

relative to the broken line and the original y axis.

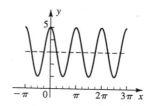

15. $y = 3 - 2 \cos(2x + \pi)$.
$y = -2 \cos(2x + \pi)$ was graphed in Problem 9.
This graph is the graph of $y = -2 \cos(2x + \pi)$
moved up $k = 3$ units. We start by drawing a
horizontal broken line 3 units above the x axis,
then graph $y = -2 \cos(2x + \pi)$ relative to the
broken line and the original y axis.

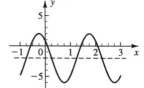

17. Compute the period and phase shift for $y = 2 \sin\left(\pi x - \dfrac{\pi}{2}\right)$:

Solve $Bx + C = 0$ and $Bx + C = 2\pi$

$$\pi x - \frac{\pi}{2} = 0 \qquad \pi x - \frac{\pi}{2} = 2\pi$$

$$\pi x = \frac{\pi}{2} \qquad \pi x = \frac{\pi}{2} + 2\pi$$

$$x = \frac{1}{2} \qquad x = \frac{1}{2} + 2$$

$$\uparrow \qquad\qquad \uparrow \ \uparrow$$

$$\text{Phase Shift} \qquad \text{Period}$$

Thus, the graph of the equation is a sine curve with a period of 2 and a phase shift of $\dfrac{1}{2}$, which means the sine curve is shifted $\dfrac{1}{2}$ unit to the right. This matches (b).

19. Compute the period and phase shift for $y = 2 \cos\left(2x + \dfrac{\pi}{2}\right)$:

Solve $Bx + C = 0$ and $Bx + C = 2\pi$

$$2x + \frac{\pi}{2} = 0 \qquad 2x + \frac{\pi}{2} = 2\pi$$

$$2x = -\frac{\pi}{2} \qquad 2x = -\frac{\pi}{2} + 2\pi$$

$$x = -\frac{\pi}{4} \qquad x = -\frac{\pi}{4} + \pi$$

$$\uparrow \qquad\qquad \uparrow \ \uparrow$$

$$\text{Phase Shift} \qquad \text{Period}$$

Thus, the graph of the equation is a cosine curve with a period of π and a phase shift of $-\dfrac{\pi}{4}$, which means the cosine curve is shifted $-\dfrac{\pi}{4}$ unit to the left. This matches (a).

21. Since the maximum deviation from the x axis is 5, we can write:
Amplitude $= |A| = 5$. Thus, $A = 5$ or -5. Since the period is $3 - (-1) = 4$, we can write:
Period $= \dfrac{2\pi}{B} = 4$. Thus, $B = \dfrac{2\pi}{4} = \dfrac{\pi}{2}$. Since we are instructed to choose the phase shift between 0 and 2, we can regard this graph as containing the basic sine curve with a phase shift of 1. This requires us to choose A positive, since the graph shows that as x increases from 1 to 2, y *increases* like the basic sine curve (not the upside down sine curve). So $A = 5$. Then, $-\dfrac{C}{B} = 1$. Thus,

$$C = -B = -\frac{\pi}{2} \text{ and } y = A \sin(Bx + C) = 5 \sin\left(\frac{\pi}{2}x - \frac{\pi}{2}\right).$$

Check: When $x = 0$, $y = 5 \sin\left(\dfrac{\pi}{2} \cdot 0 - \dfrac{\pi}{2}\right) = 5 \sin\left(-\dfrac{\pi}{2}\right) = -5$

When $x = 1$, $y = 5 \sin\left(\dfrac{\pi}{2} \cdot 1 - \dfrac{\pi}{2}\right) = 5 \sin 0 = 0$

23. Since the maximum deviation from the x axis is 2, we can write:

Amplitude $= |A| = 2$. Thus, $A = 2$ or -2. Since the period is $\dfrac{5\pi}{2} - \left(-\dfrac{3\pi}{2}\right) = 4\pi$, we can write:

Period $= \dfrac{2\pi}{B} = 4\pi$. Thus, $B = \dfrac{2\pi}{4\pi} = \dfrac{1}{2}$. Since we are instructed to choose $-2\pi < -\dfrac{C}{B} < 0$, that is, the

phase shift to the left, we regard this graph as containing the upside down cosine curve, with a phase

shift of $-\dfrac{\pi}{2}$. This requires us to choose A negative, since the graph shows that as x increases from

$-\dfrac{\pi}{2}$ to 0, y *increases* like the upside down cosine curve (not the basic cosine curve). So, $A = -2$.

Then, $-\dfrac{C}{B} = -\dfrac{\pi}{2}$.

Thus, $C = \dfrac{\pi}{2} B = \dfrac{\pi}{2} \cdot \dfrac{1}{2} = \dfrac{\pi}{4}$ and $y = A \cos(Bx + C) = -2 \cos\left(\dfrac{1}{2}x + \dfrac{\pi}{4}\right)$.

Check: When $x = 0$, $y = -2 \cos\left(\dfrac{1}{2} \cdot 0 + \dfrac{\pi}{4}\right) = -2 \cos\dfrac{\pi}{4} = -\sqrt{2}$

When $x = \dfrac{\pi}{2}$, $y = -2 \cos\left(\dfrac{1}{2} \cdot \dfrac{\pi}{2} + \dfrac{\pi}{4}\right) = -2 \cos\dfrac{\pi}{2} = 0$

25. Amplitude $= |A| = |2| = 2$
Phase Shift and Period: Solve
$Bx + C = 0$ and $Bx + C = 2\pi$

$3x - \dfrac{\pi}{2} = 0 \qquad 3x - \dfrac{\pi}{2} = 2\pi$

$x = \dfrac{\pi}{6} \qquad x = \dfrac{\pi}{6} + \dfrac{2\pi}{3}$
↑   ↑   ↑

Phase Shift   Period $= \dfrac{2\pi}{3}$

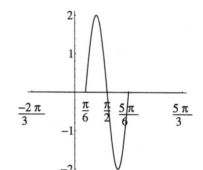

Phase Shift $= \dfrac{\pi}{6}$. Graph one cycle over the interval

from $\dfrac{\pi}{6}$ to $\left(\dfrac{\pi}{6} + \dfrac{2\pi}{3}\right) = \dfrac{5\pi}{6}$.

Extend the graph from $-\dfrac{2\pi}{3}$ to $\dfrac{5\pi}{3}$.

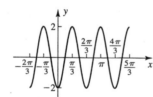

27. $y = 4 + 2 \sin\left(3x - \dfrac{\pi}{2}\right)$.

$y = 2 \sin\left(3x - \dfrac{\pi}{2}\right)$ was graphed in Problem 25. This graph is the graph

of $y = 2 \sin\left(3x - \dfrac{\pi}{2}\right)$ moved up $k = 4$ units. We start by drawing a

horizontal broken line 4 units above the x axis, then graph

$y = 2 \sin\left(3x - \dfrac{\pi}{2}\right)$ relative to the broken line and the original y axis.

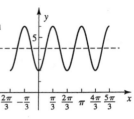

29. Amplitude $= |A| = |2.3| = 2.3$
     Phase Shift and Period: Solve

$$\dfrac{\pi}{1.5}(x - 2) = 0 \qquad \dfrac{\pi}{1.5}(x - 2) = 2\pi$$

$$x - 2 = 0 \qquad\qquad x - 2 = 3$$

$$x = 2 \qquad\qquad\quad x = 2 + 3$$
     ↑   ↑ ↑
     Phase Shift   Period

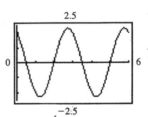

31. Amplitude $= |A| = |18| = 18$
     Phase Shift and Period: Solve

$$4\pi(x + 0.137) = 0 \qquad 4\pi(x + 0.137) = 2\pi$$

$$x + 0.137 = 0 \qquad\qquad x + 0.137 = \dfrac{1}{2}$$

$$x = -0.137 \qquad\qquad x = -0.137 + \dfrac{1}{2}$$
     ↑   ↑   ↑
     Phase Shift   Period

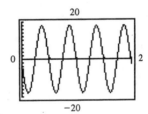

33.

The graph of $y = \sin x + \sqrt{3} \cos x$ is shown in the figure. This graph appears to be a sine wave with amplitude 2 and period 2π that has been shifted to the left. Thus, we conclude that $A = 2$ and $B = \dfrac{2\pi}{2\pi} = 1$. To determine C, we use the zoom feature or the built-in approximation routine to locate the x intercept closest to the origin at $x = -1.047$. This is the phase-shift for the graph.

Substitute $B = 1$ and $x = -1.047$ into the phase-shift equation

$$x = -\dfrac{C}{B}$$

$$-1.047 = -\dfrac{C}{1}$$

$$C = 1.047$$

Thus, the equation required is $y = 2 \sin(x + 1.047)$.

35.

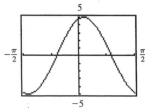

The graph of $y = \sqrt{2}\sin x - \sqrt{2}\cos x$ is shown in the figure. This graph appears to be a sine wave with amplitude 2 and period 2π that has been shifted to the right. Thus, we conclude that $A = 2$ and $B = \dfrac{2\pi}{2\pi} = 1$. To determine C, we use the zoom feature or the built-in approximation routine to locate the x intercept closest to the origin at $x = 0.785$. This is the phase-shift for the graph.

Substitute $B = 1$ and $x = 0.785$ into the phase-shift equation

$$x = -\frac{C}{B}$$

$$0.785 = -\frac{C}{1}$$

$$C = -0.785$$

Thus, the equation required is $y = 2\sin(x - 0.785)$.

37.

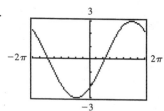

The graph of $y = 1.4\sin 2x + 4.8\cos 2x$ is shown in the figure. This graph appears to be a sine wave with amplitude 5 and period π that has been shifted to the left. Thus, we conclude that $A = 5$ and $B = \dfrac{2\pi}{\pi} = 2$. To determine C, we use the zoom feature or the built-in approximation routine to locate the x intercept closest to the origin at $x = -0.644$. This is the phase-shift for the graph.

Substitute $B = 2$ and $x = -0.644$ into the phase-shift equation

$$x = -\frac{C}{B}$$

$$-0.644 = -\frac{C}{2}$$

$$C = 1.288$$

Thus, the equation required is $y = 5\sin(x + 1.288)$.

39.

The graph of $y = 2\sin\dfrac{x}{2} - \sqrt{5}\cos\dfrac{x}{2}$ is shown in the figure. This graph appears to be a sine wave with amplitude 3 and period 4π that has been shifted to the right. Thus, we conclude that $A = 3$ and $B = \dfrac{2\pi}{4\pi} = \dfrac{1}{2}$. To determine C, we use the zoom feature or the built-in approximation routine to locate the x intercept closest to the origin at $x = 1.682$. This is the phase-shift for the graph.

Substitute $B = \dfrac{1}{2}$ and $x = 1.682$ into the phase-shift equation $x = -\dfrac{C}{B}$

$$1.682 = -\frac{C}{1/2}$$

$$C = -\frac{1}{2}(1.682) = -0.841$$

Thus, the equation required is $y = 3\sin\left(\dfrac{x}{2} - 0.841\right)$.

41. Amplitude $= |A| = |5| = 5$
     Phase Shift and Period: Solve
         $Bx + C = 0$ and $Bx + C = 2\pi$

     $\dfrac{\pi}{6}(t + 3) = 0$ $\dfrac{\pi}{6}(t + 3) = 2\pi$

     $\quad t = -3$ $\quad t = -3 + 12$
     $\quad\quad \uparrow$ $\quad\quad\quad \uparrow \quad\quad \uparrow$
     $\quad\quad$ Phase Shift $\quad$ Period

     Phase Shift $= -3$. Period $= 12$ sec
     Graph one cycle over the interval from
     -3 to $(-3 + 12) = 9$.

     Extend the graph from 9 to 39, and delete the
     portion of the graph from -3 to 0, since this was
     not required.

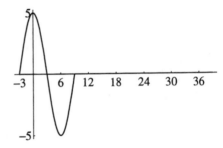

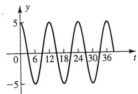

43. Amplitude $= |A| = |30| = 30$
     Phase Shift and Period: Solve
         $Bx + C = 0$ and $\quad Bx + C = 2\pi$
         $120\pi t - \pi = 0$ $\quad 120\pi t - \pi = 2\pi$

     $\quad t = \dfrac{1}{120}$ $\quad t = \dfrac{1}{120} + \dfrac{1}{60}$
     $\quad\quad \uparrow$ $\quad\quad\quad \uparrow \quad\quad \uparrow$
     $\quad\quad$ Phase Shift $\quad$ Period

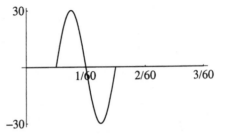

Phase Shift $= \dfrac{1}{120}$. Period $= \dfrac{1}{60}$.

Frequency $= \dfrac{1}{\text{Period}} = \dfrac{1}{1/60} = 60$ Hz

Graph one cycle over the interval

from $\dfrac{1}{120}$ to $\left(\dfrac{1}{120} + \dfrac{1}{60}\right) = \dfrac{3}{120}$.

Extend the graph from 0 to $\dfrac{3}{60}$.

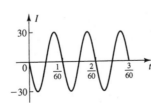

45. (A)

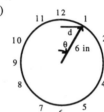

Since the second hand sweeps out 2π radians in 60 seconds, it sweeps out an angle of $\dfrac{2\pi}{12}$ or $\dfrac{\pi}{6}$ radians every 5 seconds. Since the required distance d satisfies $\dfrac{d}{6} = \sin\theta$ or $d = 6\sin\theta$, we can complete the table by adding $\dfrac{\pi}{6}$ radians every 5 seconds to θ starting at $\theta = 0$ when $t = 0$ and the hand points to 12. Thus:

$t = 5 \quad d = 6\sin\dfrac{\pi}{6} = 3.0$ $\qquad t = 35 \quad d = 6\sin\dfrac{7\pi}{6} = -3$

$t = 10 \quad d = 6\sin\dfrac{2\pi}{6} = 5.2$ $\qquad t = 40 \quad d = 6\sin\dfrac{8\pi}{6} = -5.2$

$t = 15 \quad d = 6\sin\dfrac{3\pi}{6} = 6$ $\qquad t = 45 \quad d = 6\sin\dfrac{9\pi}{6} = -6$

$t = 20 \quad d = 6\sin\dfrac{4\pi}{6} = 5.2$ $\qquad t = 50 \quad d = 6\sin\dfrac{10\pi}{6} = -5.2$

$t = 25 \quad d = 6\sin\dfrac{5\pi}{6} = 3$ $\qquad t = 55 \quad d = 6\sin\dfrac{11\pi}{6} = -3.0$

$t = 30 \quad d = 6\sin\dfrac{6\pi}{6} = 0$ $\qquad t = 60 \quad d = 6\sin\dfrac{12\pi}{6} = 0$

Thus, we can complete the table 1 as follows:

TABLE 1 (Distance d, t sec after the second hand points to 12.)

t sec	0	5	10	15	20	25	30	35	40	45	50	55	60
d in	0.0	3.0	5.2	6.0	5.2	3.0	0.0	–3.0	–5.2	–6.0	–5.2	–3.0	0.0

TABLE 2 (has the same values for d, but starting with $d = 6\sin\dfrac{9\pi}{6} = -6$)

t sec	0	5	10	15	20	25	30	35	40	45	50	55	60
d in	–6.0	–5.2	–3.0	0.0	3.0	5.2	6.0	5.2	3.0	0.0	–3.0	–5.2	–6.0

(B) From the table values and the position of the hand on the clock, since the same values are found in Table 2 15 seconds later than in Table 1, we see that relation (2) is 15 seconds out of phase with relation (1).

(C) Clearly the relations repeat the values every 60 seconds, hence this is the period. Since the largest value is 6.0 and the smallest value is –6.0, the amplitude is 6.0.

(D) For relation (1), $|A| = 6.0$

$$\text{Period} = 60, \text{ thus } \dfrac{2\pi}{B} = 60, \text{ hence, } B = \dfrac{2\pi}{60} = \dfrac{\pi}{30}.$$

Since $d = 0$ when $t = 0$, there is no phase shift and $C = 0$. Thus, $y = 6.0\sin\dfrac{\pi}{30}t$ or $y = -6.0\sin\dfrac{\pi}{30}t$.

Since $d = 15$ when $t = 6.0$, the equation must be

$$y = 6.0\sin\dfrac{\pi}{30}t$$

For relation (2), $A = 6.0$ and $B = \dfrac{\pi}{30}$ again. Since the phase shift is 15, $-\dfrac{C}{B} = 15$ and

$C = -15B = -15\left(\dfrac{\pi}{30}\right) = -\dfrac{\pi}{2}$. Hence the equation must be

$$y = 6.0 \sin\left(\dfrac{\pi}{30}t - \dfrac{\pi}{2}\right)$$

The student should check that these equations give the values in the tables above.

(E) Here the equation will be again of the form $y = 6.0 \sin\left(\dfrac{\pi}{30}t + C\right)$. The second hand will point

to 3 after 15 sec, hence the phase shift will be -15. $-\dfrac{C}{B} = -15$, hence $C = +15B = 15\left(\dfrac{\pi}{30}\right) = \dfrac{\pi}{2}$.

Hence the equation must be

$$y = 6.0 \sin\left(\dfrac{\pi}{30}t + \dfrac{\pi}{2}\right)$$

(F)

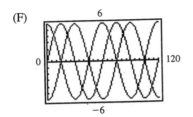

47. (A)

(B) $|A| = \dfrac{\max y - \min y}{2} = \dfrac{85 - 50}{2} = 17.5$; Period = 12 months, therefore, $B = \dfrac{2\pi}{P} = \dfrac{\pi}{6}$;

$k = |A| + \min y = 17.5 + 50 = 67.5$. From the scatter plot it appears that if we use $A = 17.5$,
then the phase shift will be about 3.0--which can be adjusted for a better visual fit later, if

necessary. Thus, using $PS = -\dfrac{C}{B}$, $C = -\dfrac{\pi}{6}(3.0) = -1.6$, and $y = 67.5 + 17.5 \sin\left(\dfrac{\pi t}{6} - 1.6\right)$.

Graphing this equation in the same viewing window
as the scatter plot, we see that adjusting C to -2.1
produces a little better visual fit. Thus, with this
adjustment, the equation and graph are:

$$y = 67.5 + 17.5 \sin\left(\dfrac{\pi t}{6} - 2.1\right)$$

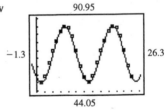

(C) $y = 68.7 + 17.1 \sin(0.5t - 2.1)$

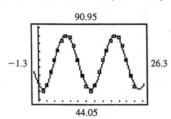

(D) The regression equation differs slightly in k, A, and B, but not in C. Both equations appear to fit the data very well.

49.   No answer is provided here. The steps are as in the previous problem, but the data should be collected by the student for the relevant city.

EXERCISE 3.4   Additional Applications

1.   (A) Amplitude $|A| = |10| = 10$. Phase Shift: Solve $Bx + C = 0$. $120\pi t - \dfrac{\pi}{2} = 0$ $t = \dfrac{1}{240}$

Phase Shift $= \dfrac{1}{240}$. Frequency $= \dfrac{B}{2\pi} = \dfrac{120\pi}{2\pi} = 60$ Hz

(B) The maximum current is the amplitude, which is 10 amperes.

(C)

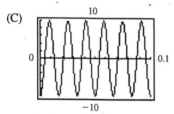

There are 60 cycles per second, hence there are $60(0.1) = 6$ cycles in 0.1 second, as shown.

3.   *Find A:* The amplitude $|A|$ is given to be 20. Since $A > 0$, $A = 20$.

Find B: We are given that the frequency $f = 30$ Hz. But $f = \dfrac{B}{2\pi}$. Thus, $\dfrac{B}{2\pi} = 30$, $B = 60\pi$

Write the equation: $I = 20 \cos 60\pi t$

5.   The height of the wave from trough to crest is the difference in height between the crest (height A) and the trough (height $-A$). In this case, $A = 15$ ft. $A - (-A) = 2A = 2(15$ ft$) = 30$ ft. To find the wavelength γ, we note: $\gamma = 5.12T^2$, $T = \dfrac{2\pi}{B}$, $B = \dfrac{\pi}{8}$. Thus, $T = \dfrac{2\pi}{\pi/8} = 16$ sec,
$\gamma = 5.12(16)^2 \approx 1311$ ft. To find the speed S, we use

$$S = \sqrt{\dfrac{g\gamma}{2\pi}} \qquad g = 32 \text{ ft/sec}^2$$

$$= \sqrt{\dfrac{32(1311)}{2\pi}} \approx 82 \text{ ft/sec}$$

7. To graph $y = 15 \sin \frac{\pi}{8} t$, we note: Amplitude $= |A| = 15$ ft. Period $= \frac{2\pi}{B} = 16$ sec. One full cycle of
    the graph is completed as t goes from 0 to 16. Block out this interval, divide it into four equal parts,
    locate high and low points, and locate t intercepts. Then complete the graph.

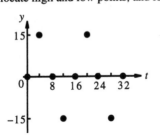 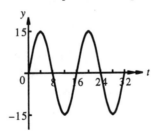

9. (A) *Find A:* The amplitude $|A|$ is given to be 2. Although, on the basis of the given information, A
        could be either 2 or –2, it is natural to choose $A = 2$.
        Find B: Since the variable in this problem is r, the distance from the source, the length of one
        cycle = wavelength = $\lambda = \frac{2\pi}{B}$. Thus, $B = \frac{2\pi}{\lambda} = \frac{2\pi}{150} = \frac{\pi}{75}$
        Write the equation: $y = 2 \sin \frac{\pi}{75} r$

    (B) To find the period T, we use: $\lambda = 5.12 T^2$ (λ in feet), $T = \sqrt{\dfrac{\lambda}{5.12}}$

        Substituting $\lambda = 150$ miles $= (150)(5280)$ feet, we have $T = \sqrt{\dfrac{(150)(5280)}{5.12}} \approx 393$ sec

11. (A) The equation $y = 25 \sin 2\pi \left(\dfrac{t}{10} + \dfrac{1024}{512} \right) = 25 \sin 2\pi \left(\dfrac{t}{10} + 2 \right)$ models the vertical motion of
        the wave at the fixed point $r = 1{,}024$ ft from the source relative to time in seconds.

    (B) The appropriate choice is period, since period is defined in terms of time and wavelength is
        defined in terms of distance. The period of $25 \sin 2\pi \left(\dfrac{t}{10} + 2 \right) = 25 \sin \left(\dfrac{2\pi}{10} t + 4\pi \right)$ is $\dfrac{2\pi}{B}$,
        where $B = \dfrac{2\pi}{10}$. Thus, period $= 2\pi \div \dfrac{2\pi}{10} = 10$ seconds.

    (C)

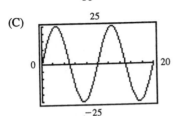

13. Period $= \dfrac{1}{v} = \dfrac{1}{10^8} = 10^{-8}$ sec. To find the wavelength λ, we use the formula $\lambda v = c$ with $c \approx 3 \times 10^8$
    m/sec, $\lambda = \dfrac{c}{v} = \dfrac{3 \times 10^8 \text{ m/sec}}{10^8 \text{ Hz}} = 3$ m

15. We first use $\lambda v = c$ to find the frequency and then use $v = \dfrac{B}{2\pi}$ to find B:

$$v = \frac{c}{\lambda} = \frac{3 \times 10^8 \text{ m/sec}}{3 \times 10^{-10}} = 10^{18} \text{ Hz}; \quad B = 2\pi v = 2\pi \times 10^{18}$$

17. If $y = A \sin 2\pi \times 10^6 \, t$, then $B = 2\pi \times 10^6$. Since $v = \dfrac{B}{2\pi}$, we have $v = \dfrac{2\pi \times 10^6}{2\pi} = 10^6$ Hz

Since Period $= \dfrac{2\pi}{B}$, we have Period $= \dfrac{2\pi}{2\pi \times 10^6} = 10^{-6}$ sec

Figure 2 (text) shows atmospheric adsorption, for waves of frequency 10^6 Hz, as total. No, such waves cannot pass through the atmosphere.

19. (A)

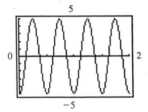

(B)

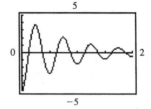

Since the amplitude remains constant over time, this represents simple harmonic motion.

Since the amplitude decreases to 0 as time increases, this represents damped harmonic motion.

EXERCISE 3.5   Graphing Combined Forms

1. We form $y_1 = x$ and $y_2 = \cos x$. We sketch the graph of each equation in the same coordinate system (dashed lines), then add the ordinates $y_1 + y_2$ (solid curve).

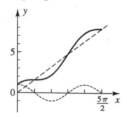

3. We form $y_1 = \dfrac{x}{2}$ and $y_2 = \cos \pi x$. We sketch the graph of each equation in the same coordinate system (dashed lines), then add the ordinates $y_1 + y_2$ (solid curve).

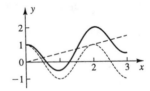

5. We form $y_1 = 3 \cos x$ and $y_2 = \sin 2x$. We sketch the graph of each equation in the same coordinate system (dashed curves), then add the ordinates $y_1 + y_2$ (solid curve).

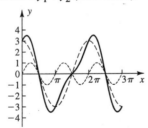

7. We form $y_1 = \sin x$ and $y_2 = 2 \cos 2x$. We sketch the graph of each equation in the same coordinate system (dashed curves), then add the ordinates $y_1 + y_2$ (solid curve).

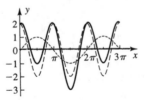

9.

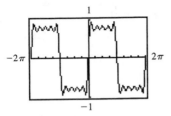

11. (A)

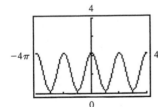 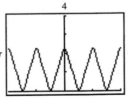 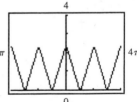

(B) The graphs approach a saw-tooth wave form.

(C)

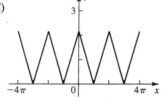

Chapter 3   Graphing Trigonometric Functions

13.

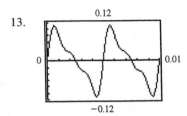

15. *Determine k:* The dashed line indicates that high and low points of the curve are equal distances from the line $V = 0.45$. Hence, $k = 0.45$.

Determine A: The maximum deviation from the line $V = 0.45$ is seen at points such as $t = 2$, $V = 0.8$ or $t = 4$, $V = 0.1$. Thus, $|A| = 0.8 - 0.45$ or $|A| = 0.45 - 0.1$. From either statement, we see, $|A| = 0.35$. So, $A = 0.35$ or $A = -0.35$. Since the portion of the curve as t increases from 0 to 1 has the form of an upside down basic cosine curve (V increasing), A must be negative. $A = -0.35$.

Determine B: One full cycle of the curve is completed as t varies from 0 to 4 seconds. Hence, the period $P = 4$ sec. Since

$$P = \frac{2\pi}{B}, \text{ we have } 4 = \frac{2\pi}{B}, \text{ or } B = \frac{2\pi}{4} = \frac{\pi}{2}.$$

$$V = k + A \cos Bt = 0.45 - 0.35 \cos \frac{\pi}{2} t.$$

17. (A) We form $S_1 = 5 + \frac{t}{52}$ and $S_2 = -4 \cos \frac{\pi t}{26}$. We sketch the graph of each equation in the same coordinate system (dashed lines), then add the ordinates $S_1 + S_2$ (solid curve).

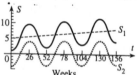

(B) In the 26th week of the 3rd year,
$t = 2 \cdot 52 + 26 = 130$. The sales are given by
$$S = 5 + \frac{130}{52} - 4 \cos \frac{\pi \cdot 130}{26}$$
$$= 5 + 2.5 - 4 \cos 5\pi = 7.5 - 4(-1) = \$11.5 \text{ million}$$

(C) In the 52nd week of the 3rd year, $t = 3 \cdot 52 = 156$. The sales are given by
$$S = 5 + \frac{156}{52} - 4 \cos \frac{\pi \cdot 156}{26} = 5 + 3 - 4 \cos 6\pi = 8 - 4(1) = \$4 \text{ million}$$

19. (A)

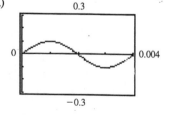

(B)

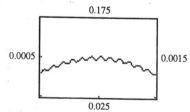

21. (A)

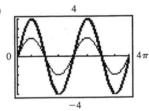

y1 = y2 = 2 sin t
y3 = 4 sin t
Since the amplitude of y3 (darker line) is greater than that of y1 and y2, this represents constructive interference.

(B)

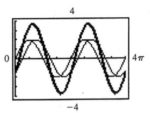

$y1 = 2 \sin t$ $y2 = 2 \sin\left(t - \dfrac{\pi}{4}\right)$

$y1 + y2 = 2 \sin t + 2 \sin\left(t - \dfrac{\pi}{4}\right)$

Since the amplitude of y3 (darker line) is greater than that of y1 and y2, this represents constructive interference.

(C)

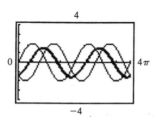

$y1 = 2 \sin t$ $y2 = 2 \sin\left(t - \dfrac{3\pi}{4}\right)$

$y3 = 2 \sin t + 2 \sin\left(t - \dfrac{3\pi}{4}\right)$

Since the amplitude of y3 (darker line) is less than that of y1 and y2, this represents destructive interference.

(D)

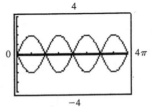

y1 = 2 sin t   y2 = 2 sin(t – π)
y3 = 2 sin t + 2 sin(t – π)
Since the amplitude of y3 (dark line along x axis) is less than that of y1 and y2 (in fact, 0), this represents destructive interference.

23. (A) y2 must have the same amplitude and period as y1, but must be completely out of phase with y2. Hence C = π or –π. Since C is required to be negative, y2 = 65 sin(400πt – π).

(B) y2 added to y1 produces a sound wave of zero amplitude--no noise.

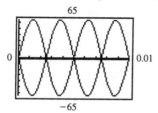

Chapter 3   Graphing Trigonometric Functions

EXERCISE 3.6   Tangent, Cotangent, Secant, and Cosecant Functions Revisited

1.

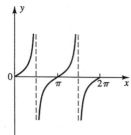

3.   The dashed line shows $y = \sin x$ in this interval. The solid line is $y = \csc x$.

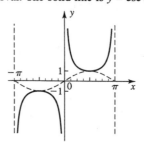

5.   Period $= \dfrac{\pi}{B} = \dfrac{\pi}{2}$

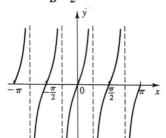

7.   Period $= \dfrac{\pi}{1/2} = 2\pi$

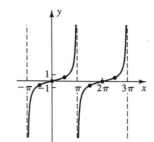

9.   Period $= \dfrac{2\pi}{1/2} = 4\pi$

The dashed line shows $y = \dfrac{1}{2}\sin\left(\dfrac{x}{2}\right)$ in this

interval. The solid line is $y = 2\csc\left(\dfrac{x}{2}\right)$.

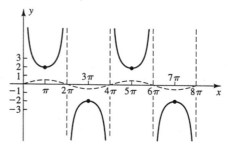

11.   We find the period and phase shift by solving $2x - \pi = 0$ and $2x - \pi = \pi$

$$2x = \pi \qquad\qquad 2x = \pi + \pi$$

$$x = \frac{\pi}{2} \qquad\qquad x = \frac{\pi}{2} + \frac{\pi}{2}$$

$$\text{Period} = \frac{\pi}{2}, \quad \text{Phase Shift} = \frac{\pi}{2}$$

We then sketch one period of the graph starting at $x = \dfrac{\pi}{2}$ (the phase shift) and ending at

$x = \dfrac{\pi}{2} + \dfrac{\pi}{2} = \pi$ (the phase shift plus one period). Note that vertical asymptotes are at $x = \dfrac{\pi}{2}$ and $x = \pi$.

We then extend the graph from $-\frac{\pi}{2}$ to $\frac{\pi}{2}$ and delete the portion of the graph from $\frac{\pi}{2}$ to π, since that was not required.

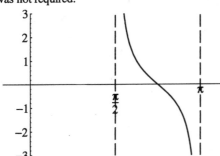 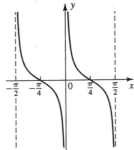

13. We find the period and phase shift by solving

$$\pi x - \frac{\pi}{2} = 0 \quad \text{and} \quad \pi x - \frac{\pi}{2} = 2\pi$$

$$x = \frac{1}{2} \qquad\qquad x = \frac{1}{2} + 2$$

$$\text{Period} = 2, \ \text{Phase Shift} = \frac{1}{2}$$

Now, since $\csc\left(\pi x - \frac{\pi}{2}\right) = \dfrac{1}{\sin\left(\pi x - \dfrac{\pi}{2}\right)}$, we graph $y = \sin\left(\pi x - \frac{\pi}{2}\right)$ for one cycle from

$\frac{1}{2}$ to $\frac{1}{2} + 2 = \frac{5}{2}$ with a broken line graph, then take reciprocals. We also place vertical asymptotes through the x intercepts of the sine graph to guide us when we sketch the cosecant function. We then extend the one cycle over the required interval from $-\frac{1}{2}$ to $\frac{5}{2}$.

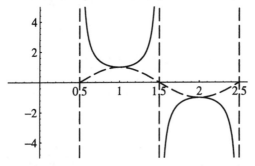 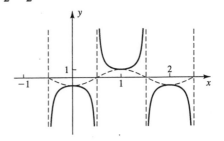

15. The graph of $y = \csc x - \cot x$ is shown in the figure. This graph appears to have vertical asymptotes at $x = -\pi$ and $x = \pi$, and period 2π. It appears, therefore, to be the same as the graph of $y = \tan Bx$, with $\frac{\pi}{B} = 2\pi$, that is

$$B = \frac{1}{2}.$$

The required equation is $y = \tan \frac{1}{2} x$.

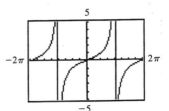

17. The graph of $y = \cot x + \tan x$ is shown in the figure.
     This graph appears to have vertical asymptotes at $x = -\pi$
     and $x = -\dfrac{\pi}{2}$, $x = 0$, $x = \dfrac{\pi}{2}$, and $x = \pi$, and period π. Its
     high and low points appear to have y coordinates of -2 and
     2, respectively. It appears, therefore, to be the same as the
     graph of $y = 2 \csc Bx$, with $\dfrac{2\pi}{B} = \pi$, that is $B = 2$.

     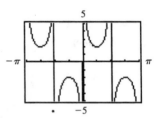

     The required equation is $y = 2 \csc 2x$.

19. We find the period and phase shift by solving $\quad \pi x - \pi = 0$ and $\quad \pi x - \pi = \pi$
     $$x = 1 \qquad\qquad x = 1 + 1$$
     $$\text{Period} = 1, \ \text{Phase Shift} = 1$$
     We then sketch one period of the graph starting at $x = 1$ (the phase shift) and ending at $x = 1 + 1 = 2$
     (the phase shift plus one period). Note that the graph of the basic cotangent curve is reflected with
     respect to the x axis. We then extend the graph from -2 to 2.

     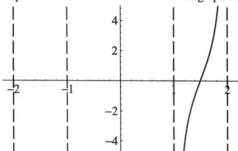
     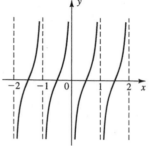

21. We find the period and phase shift by solving $\quad \pi x - \dfrac{\pi}{2} = 0$ and $\quad \pi x - \dfrac{\pi}{2} = 2\pi$
     $$x = \frac{1}{2} \qquad\qquad x = \frac{1}{2} + 2$$
     $$\text{Period} = 2, \ \text{Phase Shift} = \frac{1}{2}$$

     Now, since $2 \sec\left(\pi x - \dfrac{\pi}{2}\right) = \dfrac{1}{\dfrac{1}{2}\cos\left(\pi x - \dfrac{\pi}{2}\right)}$, we graph $y = \dfrac{1}{2}\cos\left(\pi x - \dfrac{\pi}{2}\right)$ for one cycle from $\dfrac{1}{2}$

     to $\dfrac{1}{2} + 2 = \dfrac{5}{2}$ with a broken line graph, then take reciprocals. We also place vertical asymptotes
     through the x intercepts of the cosine graph to guide us when we sketch the secant function.

We then extend the one cycle over the required interval from –1 to 3.

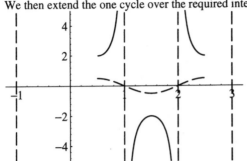 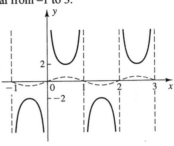

23. The graph of $y = \cos 2x + \sin 2x \tan 2x$ is shown in the figure. This graph appears to have vertical asymptotes at $x = -\dfrac{3\pi}{4}$, $x = -\dfrac{\pi}{4}$, $x = \dfrac{\pi}{4}$, and $x = \dfrac{3\pi}{4}$, and period π. Its high and low points appear to have y coordinates of –1 and 1, respectively. It appears, therefore, to be the same as the graph of $y = \sec Bx$, $\dfrac{2\pi}{B} = \pi$, that is $B = 2$. The required equation is $y = \sec 2x$.

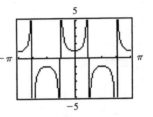

25. The graph of $y = \dfrac{\sin 6x}{1 - \cos 6x}$ is shown in the figure. This graph appears to have vertical asymptotes at $x = -\pi$, $-\dfrac{2\pi}{3}$, $-\dfrac{\pi}{3}$, 0, $\dfrac{\pi}{3}$, and π, and period $\dfrac{\pi}{3}$. It appears, therefore, to be the same as the graph of $y = \cot Bx$, $\dfrac{\pi}{B} = \dfrac{\pi}{3}$, that is $B = 3$. The required equation is $y = \cot 3x$.

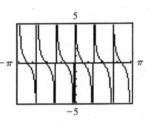

27. (A) In triangle ABC, we can write
$$\tan \theta = \frac{a}{b} = \frac{a}{15}$$
Thus, $a = 15 \tan \theta$, or $a = 15 \tan 2\pi t$.

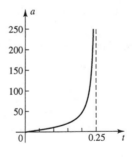

(B) Period $= \dfrac{\pi}{2\pi} = \dfrac{1}{2}$

One period of the graph would therefore extend from 0 to $\dfrac{1}{2}$, with a vertical asymptote at $t = \dfrac{1}{4}$, or 0.25. We sketch half of one period, since the required interval is from 0 to 0.25 only. Ordinates can be determined from a calculator, thus:

t	0	0.05	0.10	0.15	0.20	0.24
$15 \tan 2\pi t$	0	4.9	11	21	46	240

(C) a increases without bound as t approaches 0.25 (the graph has a vertical asymptote at $t = 0.25$.)

CHAPTER 3   **REVIEW EXERCISE**

1.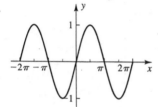

2.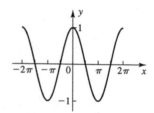

3.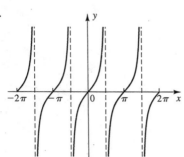

4.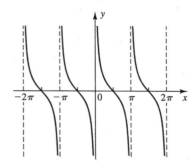

5.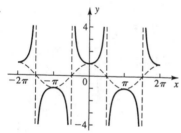

6.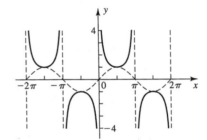

7.   Amplitude $= |3| = 3$. Period $= \dfrac{2\pi}{1/2} = 4\pi$.

One full cycle of this graph is completed as x goes from 0 to 4π. Block out this interval, divide it into four equal parts, locate high and low points, and locate x intercepts. Then complete the graph.

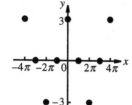

8. Amplitude $= \left| \frac{1}{2} \right| = \frac{1}{2}$. Period $= \frac{2\pi}{2} = \pi$.

    One full cycle of this graph is completed as x goes from 0 to π. Block out this interval, divide it into four equal parts, locate high and low points, and locate x intercepts, then complete the graph.

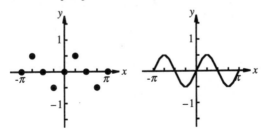

9. We form $y_1 = 4$ and $y_2 = \cos x$. We sketch the graph of each equation in the same coordinate system (dashed lines), then add the ordinates $y_1 + y_2$ (solid curve).

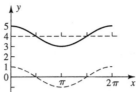

10. All six trigonometric functions are periodic. This is the key property shared by all.

11. Since the period of $y = A \cos Bx$ is given by $\frac{2\pi}{B} =$ period, if B is increased, the period decreases; if B is decreased, the period increases.

12. If a positive number is added to C, the graph is moved to the left, since the graph of $y = f(x + h)$ is shifted left h units from the graph of $y = f(x)$.
    If a negative number is added to C, the graph is moved to the right, since the graph of $y = f(x - h)$ is shifted right h units from the graph of $y = f(x)$.

    Thus, for example, the graph of $y = A \sin(3x + \pi)$ is shifted $\frac{\pi}{3}$ units to the left of the graph of $y = A \sin 3x$, while the graph of $y = A \sin(3x - \pi)$ is shifted $\frac{\pi}{3}$ units to the right of the graph of $y = A \sin 3x$.

13. (A) The basic functions with period 2π are sine, cosine, secant and cosecant. Of these, sine and cosine are always defined, and secant is undefined at $x = \frac{\pi}{2} + n\pi$, n an integer. The correct match is with cosecant.

    (B) The basic functions with period π are tangent and cotangent. Of these, tangent is undefined at $x = \frac{\pi}{2} + n\pi$, n an integer. The correct match is with cotangent.

    (C) The basic functions with amplitude 1 are sine and cosine. Since $\sin 0 = 0$ and $\cos 0 = 1$, only the graph of sine passes through $(0, 0)$.

14. Amplitude $= \left| -\dfrac{1}{3} \right| = \dfrac{1}{3}$. Period $= \dfrac{2\pi}{2\pi} = 1$.

Since $A = -\dfrac{1}{3}$ is negative, the basic curve for $y = \cos x$ is turned upside down. One full cycle of the graph is completed as x goes from 0 to 1. Block out this interval, divide it into four equal parts, locate high and low points, and locate x intercepts, then complete the graph.

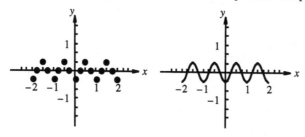

15. $y = -1 + \dfrac{1}{2} \sin 2x$. Amplitude $= \left| \dfrac{1}{2} \right| = \dfrac{1}{2}$. Period $= \dfrac{2\pi}{2} = \pi$,

$y = \dfrac{1}{2} \sin 2x$ was graphed in Problem 8. This graph is the

graph of $y = \dfrac{1}{2} \sin 2x$ moved down $|k| = |-1| = 1$ unit.

We start by drawing a horizontal broken line 1 unit below the

x axis, then graph $y = \dfrac{1}{2} \sin 2x$ relative to the broken line and

the original y axis.

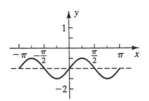

16. This graph is the graph of $y = -2 \sin(\pi x - \pi)$ moved up 4 units. To graph $y = -2 \sin(\pi x - \pi)$, we work as follows:

Amplitude $= |A| = |-2| = 2$. Phase Shift and Period: Solve $\quad Bx + C = 0 \quad$ and $\quad Bx + C = 2\pi$

$$\pi x - \pi = 0 \qquad\qquad \pi x - \pi = 2\pi$$
$$x = 1 \qquad\qquad x = 1 + 2$$
$$\text{Phase Shift} = 1 \qquad\qquad \text{Period} = 2$$

Graph one cycle (the basic sine curve turned upside down) over the interval from 1 to $(1 + 2) = 3$. Then extend the graph from 0 to 1 and delete the portion of the graph from 2 to 3 since this was not required.

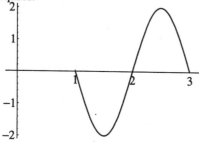
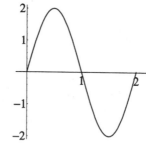

We then shift this graph up 4 units by drawing a horizontal broken line 4 units above the x axis, then graphing $y = -2 \sin(\pi x - \pi)$ relative to the broken line and the original y axis, $y = 4 - 2 \sin(\pi x - \pi)$

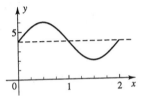

17. Period $= \dfrac{\pi}{2}$

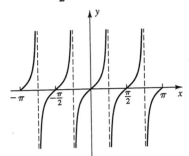

18. Period $= \dfrac{\pi}{\pi} = 1$

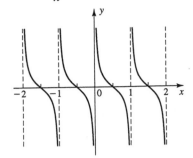

19. Period $= \dfrac{2\pi}{\pi} = 2$. We first sketch a graph of $y = \dfrac{1}{3} \sin \pi x$ from -1 to 2, which has amplitude $\dfrac{1}{3}$ and period 2. This curve (dashed curve) can serve as a guide for $y = 3 \csc \pi x = \dfrac{1}{(1/3) \sin \pi x}$ by taking reciprocals of ordinates.

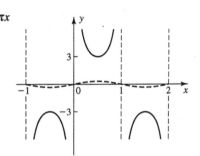

20. Period $= \dfrac{2\pi}{1/2} = 4\pi$. We first sketch a graph of $y = \dfrac{1}{2} \cos \dfrac{x}{2}$ from $-\pi$ to 3π, which has amplitude $\dfrac{1}{2}$ and period 4π. This curve (dashed curve) can serve as a guide for $y = 2 \sec \dfrac{x}{2} = \dfrac{1}{(1/2) \cos (x/2)}$ by taking reciprocals of ordinates.

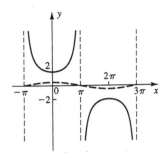

21. We find the period and phase shift by solving $x + \dfrac{\pi}{2} = 0$ and $x + \dfrac{\pi}{2} = \pi$

$$x = -\dfrac{\pi}{2} \qquad\qquad x = -\dfrac{\pi}{2} + \pi$$

$$\text{Period} = \pi \qquad \text{Phase Shift} = -\dfrac{\pi}{2}$$

We then sketch one period of the graph starting at $x = -\frac{\pi}{2}$ (the phase shift) and ending at $x = -\frac{\pi}{2} + \pi$ $= \frac{\pi}{2}$ (the phase shift plus one period). Note that a vertical asymptote is at $x = 0$. We then extend the graph from $-\pi$ to π.

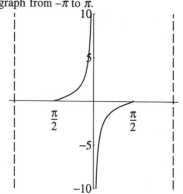
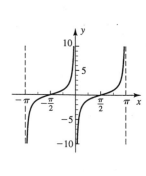

22. See Problem 14

23. See Problems 18 and 19.

24. Amplitude $= |A| = |-3| = 3$

   To find the period and phase shift, we solve

   $\pi x + \pi = 0$ and $\pi x + \pi = 2\pi$
   $x = -1$ $x = -1 + 2$
   Period $= 2$ Phase Shift $= -1$

25. To find the period and phase shift, we solve

   $\frac{\pi}{2}x + \frac{\pi}{2} = 0$ and $\frac{\pi}{2}x + \frac{\pi}{2} = \pi$
   $x = -1$ $x = -1 + 2$
   Period $= 2$ Phase Shift $= -1$

26. (A) y2 is y1 with half the period.

   (B) y2 is y1 reflected across the x axis with twice the amplitude.

   (C) The amplitudes of y1 and y2 are both 1. Hence $|A| = 1$. Since both y1 and y2 have the form of a basic sine curve, and not an upside down sine curve, $A = 1$. The period of y1 is 2, hence $\frac{2\pi}{B} = 2$ and $B = \frac{2\pi}{2} = \pi$. Hence y1 $= \sin \pi x$. The period of y2 is 1, hence $\frac{2\pi}{B} = 1$ and $B = 2\pi$. Hence y2 $= \sin 2\pi x$.

   (D) The amplitude of y1 is 1. Hence $|A| = 1$. Since y1 has the form of a basic cosine curve and not an upside down cosine curve, $A = 1$. The period of y1 is 2, hence $\frac{2\pi}{B} = 2$ and $B = \frac{2\pi}{2} = \pi$. Hence y1 $= \cos \pi x$. Since y2 is y1 reflected across the x axis with twice the amplitude, but the same period, $A = -2$ and y2 $= -2 \cos \pi x$.

27. Since the period is 4, we can write $\frac{2\pi}{B} = 4$, hence $B = \frac{2\pi}{4} = \frac{\pi}{2}$. Since the $y_{max} = 7$ and $y_{min} = -1$, we can write amplitude $= |A| = \frac{y_{max} - y_{min}}{2} = \frac{7 - (-1)}{2} = 4$ and $k = \frac{y_{max} - y_{min}}{2} = \frac{7 + (-1)}{2} = 3$.

   Thus $y = 3 + 4 \sin \frac{\pi}{2}x$ or $y = 3 - 4 \sin \frac{\pi}{2}x$. Since the portion of the graph for $0 \le x \le 4$ has the form of the basic sine curve, and not the upside down sine curve, we have $y = 3 + 4 \sin \frac{\pi}{2}x$.

Check: When $x = 0$, $y = 3 + 4 \sin\left(\dfrac{\pi}{2} \cdot 0\right) = 3 + 4 \cdot 0 = 3$

When $x = 1$, $y = 3 + 4 \sin\left(\dfrac{\pi}{2} \cdot 1\right) = 3 + 4 \cdot 1 = 7$

28. Since the displacement is 0 when $t = 0$, the equation has the form $y = A \sin Bt$, and not
$y = A \cos Bt$. Since the amplitude is 65, we can choose $A = 65$ or $A = -65$. Since the period is 0.01,
we can write $\dfrac{2\pi}{B} = 0.01$, hence $B = \dfrac{2\pi}{0.01} = 200\pi$. We choose A positive for the sake of simplicity,
thus $y = 65 \sin 200\pi t$.

29. Since, for each value of x, y2 is the sum of the ordinate values for y1 and y3, y2 = y1 + y3.

30. 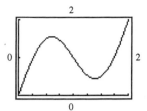   31.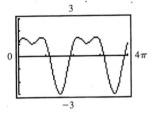

32. The graph of $y = \dfrac{1}{1 + \tan^2 x}$ is shown in the figure.

Amplitude $= \dfrac{1}{2}$ (y coordinate of highest point $-$ y coordinate of lowest point) $= \dfrac{1}{2}(1 - 0) = \dfrac{1}{2} = |A|$

Period $= \dfrac{2\pi}{2} = \pi$.

Thus, $B = \dfrac{2\pi}{\pi} = 2$. The form of the graph is that of the

basic cosine curve shifted up $\dfrac{1}{2}$ unit.

Thus, $y = \dfrac{1}{2} + |A| \cos Bx = \dfrac{1}{2} + \dfrac{1}{2} \cos 2x$.

33.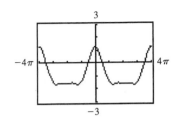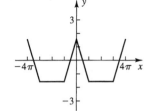

34. (A) The graph of $y = \dfrac{2 \sin x}{\sin 2x}$ is shown in the figure. This graph appears to have vertical

asymptotes at $x = -\dfrac{3\pi}{4}$, $x = -\dfrac{\pi}{4}$, $x = \dfrac{\pi}{4}$, and $x = \dfrac{3\pi}{4}$,

and period 2π. Its high and low points appear to have y coordinates of -1 and 1, respectively. It appears, therefore, to be the same as the graph of $y = \sec Bx$,

$\dfrac{2\pi}{B} = 2\pi$, that is, $B = 1$.

The required equation is $y = \sec x$.

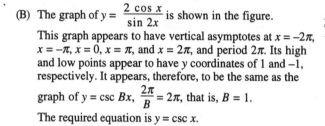

(B) The graph of $y = \dfrac{2 \cos x}{\sin 2x}$ is shown in the figure.

This graph appears to have vertical asymptotes at $x = -2\pi$, $x = -\pi$, $x = 0$, $x = \pi$, and $x = 2\pi$, and period 2π. Its high and low points appear to have y coordinates of 1 and -1, respectively. It appears, therefore, to be the same as the

graph of $y = \csc Bx$, $\dfrac{2\pi}{B} = 2\pi$, that is, $B = 1$.

The required equation is $y = \csc x$.

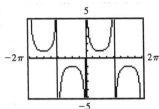

(C) The graph of $y = \dfrac{2 \cos^2 x}{\sin 2x}$ is shown in the figure. This

graph appears to have vertical asymptotes at $x = -2\pi$, $x = -\pi$, $x = 0$, $x = \pi$, and $x = 2\pi$, and period π. It appears, therefore, to be the same as the graph of $y = \cot Bx$,

$\dfrac{\pi}{B} = \pi$, that is, $B = 1$. The required equation is $y = \cot x$.

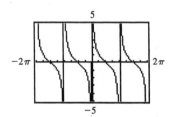

(D) The graph of $y = \dfrac{2 \sin^2 x}{\sin 2x}$ is shown in the figure. This

graph appears to have vertical asymptotes at $x = -\dfrac{3\pi}{2}$,

$x = -\dfrac{\pi}{2}$, $x = \dfrac{\pi}{2}$, and $x = \dfrac{3\pi}{2}$, and period π. It appears,

therefore, to be the same as the graph of $y = \tan Bx$, with

$\dfrac{\pi}{B} = \pi$, that is, $B = 1$.

The required equation is $y = \tan x$.

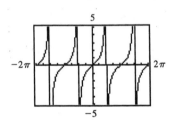

35. A horizontal shift of $\dfrac{\pi}{2}$ to the left (or right) combined with a reflection in the x axis transforms the graph of $y = \cot x$ into the graph of $y = \tan x$.

36. We find the period and phase shift by solving $\quad \pi x + \dfrac{\pi}{2} = 0 \quad$ and $\quad \pi x + \dfrac{\pi}{2} = \pi$

$$x = -\dfrac{1}{2} \qquad\qquad x = -\dfrac{1}{2} + 1$$

$$\text{Period} = 1 \qquad \text{Phase Shift} = -\dfrac{1}{2}$$

We then sketch one period of the graph starting at $x = -\frac{1}{2}$ (the phase shift) and ending at

$x = -\frac{1}{2} + 1 = \frac{1}{2}$ (the phase shift plus one period). Note that the y axis ($x = 0$) is a vertical asymptote. We then extend the graph from -1 to 1.

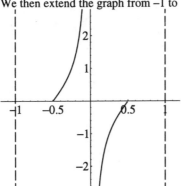
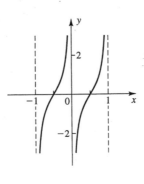

37. We find the period and phase shift by solving $2x - \pi = 0$ and $2x - \pi = 2\pi$

$$x = \frac{\pi}{2} \qquad x = \frac{\pi}{2} + \pi$$

$$\text{Period} = \pi \qquad \text{Phase Shift} = \frac{\pi}{2}$$

Now, since $2\sec(2x - \pi) = \dfrac{1}{(1/2)\cos(2x - \pi)}$, we graph $y = \frac{1}{2}\cos(2x - \pi)$ for one cycle from $\frac{\pi}{2}$ to

$\frac{\pi}{2} + \pi = \frac{3\pi}{2}$ with a broken line graph, then take reciprocals. We also place vertical asymptotes through the x intercepts of the cosine graph to guide us when we sketch the secant function.

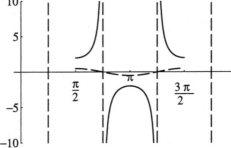

We then extend the one cycle over the required interval from 0 to $\frac{5\pi}{4}$, and delete the portion of the graph from $\frac{5\pi}{4}$ to $\frac{3\pi}{2}$, since this was not required.

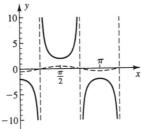

38. Since the function values follow the pattern 0, maximum, 0, minimum, 0, it would appear that a
     function of the form $y = A \sin Bx$ would be required, with A positive.
     Since the maximum value of the function seems to be 2, and the minimum value seems to be –2,
     $A = \dfrac{2 - (-2)}{2} = \dfrac{4}{2} = 2.$

     Since the maximum value is achieved at $x = 0.25$ and $x = 1.25$, the period of the function is 1.
     Hence $\dfrac{2\pi}{B} = 1$ and $B = 2\pi$. Thus, the required function is $y = 2 \sin 2\pi x$.

39. Since the maximum deviation from the x axis is 1, we can write: Amplitude = $|A| = 1$.
     Thus, $A = 1$ or -1.

     Since the period is $\dfrac{5}{4} - \left(-\dfrac{3}{4}\right) = 2$, we can write: Period $= \dfrac{2\pi}{B} = 2$. Thus, $B = \dfrac{2\pi}{2} = \pi$.

     Since we are instructed to choose the phase shift between 0 and 1, we can regard this graph as

     containing the upside down sine curve, with a phase shift of $\dfrac{1}{4}$. This requires us to choose A

     negative, since the graph shows that as x increases from $\dfrac{1}{4}$ to $\dfrac{3}{4}$, y *decreases* like the upside down

     sine curve. So, $A = -1$. Then, $-\dfrac{C}{B} = \dfrac{1}{4}$. Thus, $C = -\dfrac{1}{4}B = -\dfrac{\pi}{4}$.

     $y = A \sin(Bx + C) = -\sin\left(\pi x - \dfrac{\pi}{4}\right).$

     Check: When $x = 0$, $y = -\sin\left(\pi \cdot 0 - \dfrac{\pi}{4}\right) = -\sin\left(-\dfrac{\pi}{4}\right) = \dfrac{\sqrt{2}}{2}$

     When $x = \dfrac{1}{4}$, $y = -\sin\left(\pi \cdot \dfrac{1}{4} - \dfrac{\pi}{4}\right) = -\sin 0 = 0.$

40. The graph of $y = 1.2 \sin 2x + 1.6 \cos 2x$ is shown in the figure. This graph appears to be a sine
     wave with amplitude 2 and period π that has been shifted to the left. Thus, we conclude that $A = 2$
     and $B = \dfrac{2\pi}{\pi} = 2$. To determine C, we use the zoom feature
     or the built-in approximation routine to locate the x
     intercept closest to the origin at $x = -0.464$. This is the
     phase-shift for the graph. Substitute $B = 2$ and $x = -0.464$
     into the phase-shift equation $x = -\dfrac{C}{B}$; $-0.464 = -\dfrac{C}{2}$;
     $C = 0.928.$
     Thus, the equation required is $y = 2 \sin(2x + 0.928)$.

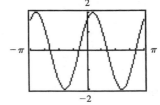

41. (A)

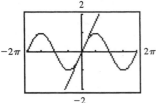

     (B)

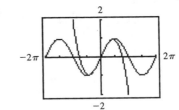

(C)

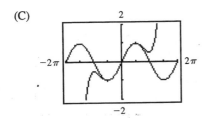

(D) As more terms of the series are used, the resulting approximation of sin x improves over a wider interval.

42. Here is a computer-generated graph of $f(x) = |\sin x|$, $-2\pi \le x \le 2\pi$

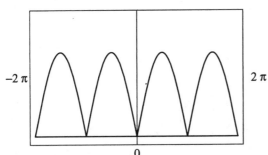

From the graph, it appears that $f(x)$ repeats its value after an interval of π, thus, $|\sin(x + \pi)| = |\sin x|$, and the period of $f(x)$ is π.

43. *Find A:* The amplitude $|A|$ is given to be 4. Since $y = -4$ (4 cm *below* position at rest) when $t = 0$, $A = -4$ (and not 4).
    Find B: We are given that the frequency, f, is 8 Hz. Hence, the period is found using the reciprocal formula: $P = \dfrac{1}{f} = \dfrac{1}{8}$ sec. But $P = \dfrac{2\pi}{B}$. Thus, $B = \dfrac{2\pi}{P} = \dfrac{2\pi}{1/8} = 16\pi$.
    Write the equation: $y = -4 \cos 16\pi t$.
    An equation of the form $y = A \sin Bt$ cannot be used to model the motion, because when $t = 0$, y cannot equal -4 for any values of A and B.

44. *Determine K:* The dashed line indicates that high and low points of the curve are equal distances from the line $P = 1$. Hence, $K = 1$.

    Determine A: The maximum deviation from the line $P = 1$ is seen at points such as $n = 0$, $P = 2$ or $n = 26$, $P = 0$. Thus, $|A| = 2 - 1$ or $|A| = 1 - 0$. From either statement, we see $|A| = 1$. So, $A = 1$ or $A = -1$. Since the portion of the curve as n increases from 0 to 26 has the form of the basic cosine curve (P decreasing)--and not the upside down cosine curve--A must be positive. $A = 1$.

    Determine B: one full cycle of the curve is completed as n varies from 0 to 52 weeks. Hence, the Period = 52 weeks.
    Since Period $= \dfrac{2\pi}{B}$, we have $52 = \dfrac{2\pi}{B}$, or $B = \dfrac{2\pi}{52} = \dfrac{\pi}{26}$.

    $P = K + A \cos Bn = 1 + \cos \dfrac{\pi n}{26}$, $0 \le n \le 104$.

    An equation of the form $P = k + A \sin Bn$ will not work: The curve starts at $(0, 2)$ and oscillates 1 unit above and below the line $P = 1$; a sine curve would have to start at $(0, 1)$.

45. (A) We use the formula:

$$\text{Period} = \frac{2\pi}{\sqrt{1000\ gA/M}} \text{ with } g = 9.75 \text{ m/sec}^2,\ A = \pi\left(\frac{1.2}{2}\text{ m}\right)^2, \text{ and Period} = 0.8 \text{ sec, and solve}$$

for M. Thus,

$$0.8 = \frac{2\pi}{\sqrt{(1000)(9.75)\ \pi\ (0.6)^2/M}} \qquad\qquad M = \frac{(1000)(9.75)\ \pi\ (0.6)^2\ (0.8)^2}{4\pi^2} \approx 179 \text{ kg}$$

(B) $D = \text{amplitude} = 0.6$ $B = \dfrac{2\pi}{\text{Period}} = \dfrac{2\pi}{0.8} = 2.5\pi$ $y = 0.6 \sin(2.5\pi t)$

(C) One full cycle of this graph is completed as t goes from 0 to 0.8. block out this interval, divide it into four equal parts, locate high and low points, and locate t intercepts. Then complete the graph.

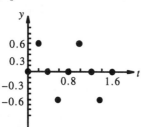 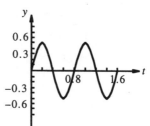

46. (A) Use $\text{Period} = \dfrac{1}{v}$ and $B = \dfrac{2\pi}{\text{Period}}$

         $\text{Period} = \dfrac{1}{280} \text{ sec}$ $B = \dfrac{2\pi}{1/280} = 560\pi$

(B) Use $v = \dfrac{1}{\text{Period}}$ and $B = \dfrac{2\pi}{\text{Period}}$

         $v = \dfrac{1}{.0025 \text{ sec}} = 400 \text{ Hz}$ $B = \dfrac{2\pi}{.0025} = 800\pi$

(C) Use $\text{Period} = \dfrac{2\pi}{B}$ and $v = \dfrac{1}{\text{Period}}$

         $\text{Period} = \dfrac{2\pi}{700\pi} = \dfrac{1}{350} \text{ sec}$ $v = \dfrac{1}{1/350 \text{ sec}} = 350 \text{ Hz}$

47. *Find A:* The amplitude $|A|$ is given to be 18. Since $E = 12$ when $t = 0$, $A = 18$ (and not -18).
Find B: We are given that the frequency, v, is 30 Hz. Hence, the period is found using the reciprocal formula:

$$P = \frac{1}{v} = \frac{1}{30} \text{ sec. But, } P = \frac{2\pi}{B}. \text{ Thus, } B = \frac{2\pi}{P} = \frac{2\pi}{1/30} = 60\pi.$$

Write the equation: $y = 18 \cos 60\pi t$

     $y = 6 \cos \dfrac{\pi}{10}(t - 5)$

48.   We compute amplitude, period, and phase shift as follows:
Amplitude $= |A| = |6| = 6$
Phase Shift and Period: Solve
$$Bx + C = 0 \quad \text{and} \quad Bx + C = 2\pi$$
$$\frac{\pi}{10}(t - 5) = 0 \qquad \frac{\pi}{10}(t - 5) = 2\pi$$
$$t = 5 \qquad\qquad t = 5 + 20$$
$$\uparrow \qquad\qquad\qquad \uparrow \quad \uparrow$$
Phase Shift   Period

Graph one cycle over the interval from 5 to 25.

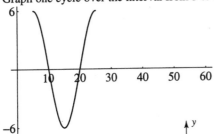

Extend the graph from 0 to 60.

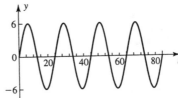

49.   The height of the wave from trough to crest is the difference in height between the crest (height A) and the trough (height $-A$). In this case, $A = 12$ ft. $A - (-A) = 2A = 2(12 \text{ ft}) = 24$ ft. To find the wavelength λ, we note: $\lambda = 5.12T^2$,
$$T = \frac{2\pi}{B}, B = \frac{\pi}{3}. \text{ Thus, } T = \frac{2\pi}{\pi/3} = 6 \text{ sec}, \lambda = 5.12(6)^2 \approx 184 \text{ ft. To find the speed } S, \text{ we use}$$
$$S = \sqrt{\frac{g\lambda}{2\pi}} \qquad g = 32 \text{ ft/sec}^2$$
$$= \sqrt{\frac{32(184)}{2\pi}} \approx 31 \text{ ft/sec}$$

50.   Period $= \frac{1}{v} = \frac{1}{10^{15}} = 10^{-15}$ sec. To find the wavelength λ, we use the formula $\lambda v = c$ with
$$c \approx 3 \times 10^8 \text{ m/sec}, \lambda = \frac{c}{v} = \frac{3 \times 10^8 \text{ m/sec}}{10^{15} \text{ Hz}} = 3 \times 10^{-7} \text{ m}$$

Chapter 3   Graphing Trigonometric Functions

51. (A)

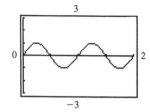

Since the amplitude remains contant over time, this represents simple harmonic motion.

(B)

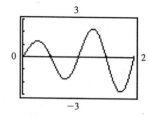

Since the amplitude increases as time increases, this represents resonance.

(C)

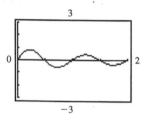

Since the amplitude decreases to 0 as time increases, this represents damped harmonic motion.

52. (A)

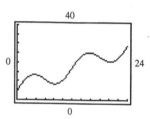

(B) It shows that sales have an overall upward trend with seasonal variations.

53. (A) The triangle shown in the text figure is a right triangle. Hence,

$\tan \theta = \dfrac{a}{b} = \dfrac{h}{1000}$.

Thus, $h = 1000 \tan \theta$.

(B)

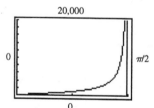

(C) As θ approaches $\dfrac{\pi}{2}$, h increases without bound.

54. (A) The data for θ repeats every second, so the period is $P = 1$ sec. The angle θ deviates from 0 by $36°$ in each direction, so the amplitude is $|A| = 36°$.

(B) $\theta = A \cos Bt$ is not suitable, because, for example, for $t = 0$, $A \cos Bt = A$, which will be $36°$ and not $0°$. $\theta = A \sin Bt$ appears suitable, because, for example, if $t = 0$, then $\theta = 0°$, a good start. Choose $A = 36$ and $B = \dfrac{2\pi}{P} = 2\pi$, which yields $\theta = 36 \sin 2\pi t$. The student should check that this equation produces (or comes close to producing) all the values in the table.

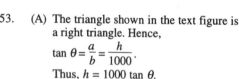

98

(C) Amplitude = 36. Period = 1.
One full cycle of this graph is completed as x goes from 0 to 1. Block out this interval, divide it into four equal parts, locate high and low points, and locate t intercepts. The points that result are precisely the points given in the table. Then complete the graph.

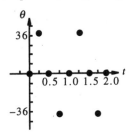
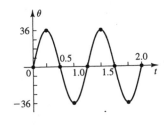

55. (A)

x (months)	1, 13	2, 14	3, 15	4, 16	5, 17	6, 18	7, 19	8, 20	9, 21
y (decimal hours)	17.08	17.63	18.12	18.60	19.07	19.48	19.58	19.25	18.57

x	10, 22	11, 23	12, 24
y	17.78	17.12	16.85

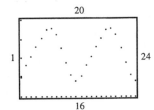

(B) From the table, Max y = 19.58 and Min y = 16.85. Then,

$$A = \frac{(\text{Max } y - \text{Min } y)}{2} = \frac{(19.58 - 16.85)}{2} = 1.37$$

$$B = \frac{2\pi}{\text{Period}} = \frac{2\pi}{12} = \frac{\pi}{6}$$

$$k = \text{Min } y + A = 16.85 + 1.37 = 18.22$$

From the plot in (A) or the table, we estimate the smallest positive value of x for which $y = k = 18.22$ to be approximately 3.2. Then this is the phase-shift for the graph. Substitute $B = \frac{\pi}{6}$ and $x = 3.2$ into the phase-shift equation $x = -\frac{C}{B}$; $3.2 = -\frac{C}{\pi/6}$; $C = -\frac{3.2\pi}{6} \approx -1.7$.

Thus, the equation required is $y = 18.22 + 1.37 \sin\left(\frac{\pi}{6}x - 1.7\right)$.

(C)

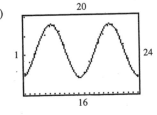

CUMULATIVE REVIEW EXERCISE CHAPTERS 1—3

1. Since one complete revolution has measure $360°$, $\frac{1}{4}$ revolution has measure $\frac{1}{4}(360°) = 90°$.

   Since $90°$ is one-half of $180°$, the corresponding radian measure must be $\frac{1}{2}$ of π, or $\frac{\pi}{2}$ rad.

2. Since $47' = \frac{47°}{60}$, then $21°47' = \left(21 + \frac{47}{60}\right)° \approx 21.78°$

3. $\theta_d = \frac{180°}{\pi \text{ rad}}\, \theta_r$. If $\theta_r = 1.67$, $\theta_d = \frac{180}{\pi}(1.67) = 95.68°$

4. $\theta_r = \frac{\pi \text{ rad}}{180°}\, \theta_d$. If $\theta_d = -715.3°$, $\theta_r = \frac{\pi}{180}(-715.3) = -12.48$ rad

5. *Solve for the complementary angle:*
   $90° - \theta = 90° - 25° = 65°$
   Solve for b: We will use the sine. Thus,
   $\sin \theta = \frac{b}{c}$
   $b = c \sin \theta = (34 \text{ in.})(\sin 25°) = 14$ in.

   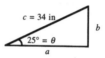

   Solve for a: We will use the cosine. Thus, $\cos \theta = \frac{a}{c}$, $a = c \cos \theta = (34 \text{ in.})(\cos 25°) = 31$ in.

6. $P(a, b) = (8, -15)$
   $r = \sqrt{a^2 + b^2} = \sqrt{8^2 + (-15)^2} = 17$
   $\sin \theta = \frac{b}{r} = \frac{-15}{17} = -\frac{15}{17}$
   $\tan \theta = \frac{b}{a} = \frac{-15}{8} = -\frac{15}{8}$

   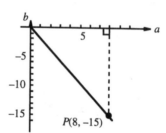

7. (A) Degree mode: $\sin(23°12') = \sin(23.2°) = 0.3939$ Convert to decimal degrees, if necessary.

   (B) Use the reciprocal relationship $\sec \theta = \frac{1}{\cos \theta}$. Degree mode: $\sec 145.6° = \frac{1}{\cos 145.6°} = -1.212$

   (C) Use the reciprocal relationship $\cot \theta = \frac{1}{\tan \theta}$. Radian mode: $\cot 0.88 = \frac{1}{\tan 0.88} = 0.8267$

8. The reference angle α is the angle (always taken positive) between the terminal side of θ and the horizontal axis.

   (A) $\alpha = 2\pi - \frac{11\pi}{6} = \frac{\pi}{6}$

   (B) $\alpha = |-225°| - 180° = 45°$

   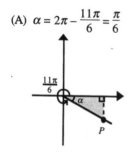

   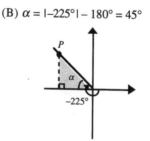

9. (A)

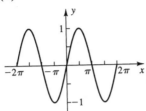

(B)

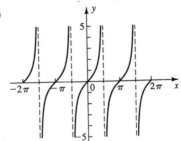

(C)

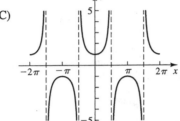

10. The central angle of a circle subtended by an arc that is one and one-half the length of the radius of the circle.

11. Yes. For example, for any x such that $\frac{\pi}{2} < x < \pi$, cos x is negative and csc $x = \frac{1}{\sin x}$ is positive.

12. No. The sum of all three angles in any triangle is 180°. An obtuse angle is one that has a measure between 90° and 180°, and a triangle with more than one obtuse angle would have angles whose measure would add up to more than 180°, which would be a contradiction to the first statement.

13. If tan $\theta = 0.9465$, then
$$\theta = \tan^{-1} 0.9465 = 43°30'$$

14. The tip of the second hand travels 1 revolution, or 2π radian, in 60 seconds. In 40 seconds it travels $\frac{40}{60}$ revolution, or $\frac{40}{60} \cdot 2\pi$ radian, that is, $\frac{4\pi}{3}$ radian. Since the distance traveled $s = R\theta$, we have

$$s = R\theta = (5.00 \text{ cm})\left(\frac{4\pi}{3} \text{ rad}\right) = 20.94 \text{ cm}$$

15. The tip of the second hand travels 1 circumference in 1 minute. Thus, its speed is given by
$$V = \frac{d}{t} = \frac{2\pi r}{t} = \frac{2\pi(5.00 \text{ cm})}{1 \text{ min}} = 31.4 \text{ cm/min}$$

16. Label the sides of the triangle as shown at the right.

Since triangles ABC and ADE are similar, we have

$$\frac{DE}{AE} = \frac{BC}{AC}$$

$$\frac{x}{10} = \frac{12}{10 + 5}$$

$$x = 10 \cdot \frac{12}{15} = 8$$

17. $\theta_r = \dfrac{\pi \text{ rad}}{180°} = \theta_d$

$$= \frac{\pi}{180}(48) = \frac{4\pi}{15} \text{ rad}$$

18. Since $\sin(3.78°) = 0.0659$ and $\tan(-76.25°) = -4.0867$, the first display (a) is the result of the calculator being set in degree mode. Since $\sin(3.78 \text{ rad}) = -0.5959$ and $\tan(-76.25 \text{ rad}) = -1.1424$, the second display (b) is the result of the calculator being set in radian mode.

19. Yes. Each is a different measure of the same angle, and if one is doubled the other must be doubled. this can be seen using the conversion formula $\theta_r = \dfrac{\pi}{180}\theta_d$: double θ_d, then θ_r is doubled.

20. Use the identity: $\cot x = \dfrac{1}{\tan x}$. Thus $\cot x = \dfrac{1}{0.5453} = 1.8339$

21. (A) Locate the 45° reference triangle, determine (a, b) and r, then evaluate.

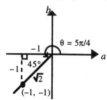

$$\sin \frac{5\pi}{4} = \frac{-1}{\sqrt{2}} = -\frac{1}{\sqrt{2}}$$

(B) Locate the 30°–60° reference triangle, determine (a, b) and r, then evaluate.

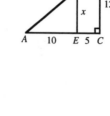

$$\cos \frac{7\pi}{6} = \frac{-\sqrt{3}}{2} = -\frac{\sqrt{3}}{2}$$

(C) Locate the 30°–60° reference triangle, determine (a, b) and r, then evaluate.

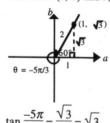

$$\tan \frac{-5\pi}{3} = \frac{\sqrt{3}}{1} = \sqrt{3}$$

(D) $(a, b) = (-1, 0)$, $r = 1$

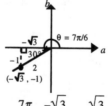

$$\csc 3\pi = \frac{1}{0}$$

Not defined

22. Since $\cos\theta < 0$ and $\tan\theta < 0$, the terminal side of θ lies in quadrant II. We sketch a reference triangle and label what we know.

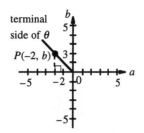

Since $\cos\theta = \dfrac{a}{r} = -\dfrac{2}{3} = \dfrac{-2}{3}$, we know that $a = -2$ and $r = 3$ (r is never negative). Use the Pythagorean theorem to find b:
$$(-2)^2 + b^2 = 3^2$$
$$b^2 = 9 - 4 = 5$$
$$b = \sqrt{5}$$

b is positive since the the terminal side of θ lies in quadrant II. We can now find the other five functions using their definitions.

$$\sin\theta = \frac{b}{r} = \frac{\sqrt{5}}{3} \qquad \sec\theta = \frac{r}{a} = \frac{3}{-2} = -\frac{3}{2}$$

$$\tan\theta = \frac{b}{a} = \frac{\sqrt{5}}{-2} = -\frac{\sqrt{5}}{2} \qquad \csc\theta = \frac{r}{b} = \frac{3}{\sqrt{5}} \qquad \cot\theta = \frac{a}{b} = \frac{-2}{\sqrt{5}} = -\frac{2}{\sqrt{5}}$$

23. It is known that $\cos(0°) = 1 = 1.0000...$ and $\tan(135°) = -1 = -1.0000...$, so that the calculator displays exact values for these.

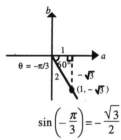

Thus $\sin\left(-\dfrac{\pi}{3}\right)$ is not given exactly. To find $\sin\left(-\dfrac{\pi}{3}\right)$, locate the $30°$–$60°$ reference triangle, determine (a, b) and r, then evaluate.

$$\sin\left(-\frac{\pi}{3}\right) = -\frac{\sqrt{3}}{2}$$

24. $y = 1 - \dfrac{1}{2}\cos 2x$. Amplitude $= \left|-\dfrac{1}{2}\right| = \dfrac{1}{2}$.

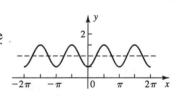

Period $= \dfrac{2\pi}{2} = \pi$. This graph is the graph of $y = -\dfrac{1}{2}\cos 2x$ moved up 1 unit. We start by drawing a horizontal broken line 1 unit above the x axis, then graph $y = -\dfrac{1}{2}\cos 2x$ (an upside down cosine curve with amplitude $\dfrac{1}{2}$ and period π) relative to the broken line and the original y axis.

25. Amplitude $= |A| = |2| = 2$. Phase Shift and Period: Solve
$$Bx + C = 0 \qquad \text{and} \qquad Bx + C = 2\pi$$
$$x - \frac{\pi}{4} = 0 \qquad\qquad x - \frac{\pi}{4} = 2\pi$$
$$x = \frac{\pi}{4} \qquad\qquad x = \frac{\pi}{4} + 2\pi$$
$$\text{Phase Shift} = \frac{\pi}{4} \qquad\qquad \text{Period} = 2\pi$$

Graph one cycle over the interval from $\frac{\pi}{4}$ to $\left(\frac{\pi}{4} + 2\pi\right) = \frac{9\pi}{4}$. Then extend the graph from $-\pi$ to 3π.

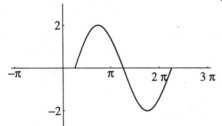 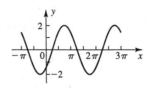

26. Period $= \frac{\pi}{4}$

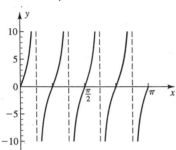

27. Period $= \frac{2\pi}{1/2} = 4\pi$

The dashed line shows $y = \sin\frac{x}{2}$ in this interval. The solid line is $y = \csc\frac{x}{2}$.

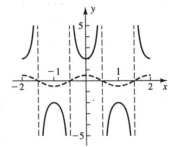

28. Period $= \frac{2\pi}{\pi} = 2$.

We first sketch a graph of $y = \frac{1}{2}\cos \pi x$ from -2 to 2, which has amplitude $\frac{1}{2}$ and period 2. This curve (dashed curve) can serve as guide for $y = 2 \sec \pi x = \dfrac{1}{(1/2)\cos \pi x}$ by taking reciprocals of ordinates.

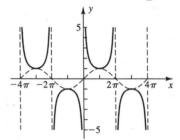

29. We first find the period and phase shift by solving $\pi x + \frac{\pi}{2} = 0$ and $\pi x + \frac{\pi}{2} = \pi$

$$\pi x = -\frac{\pi}{2} \qquad\qquad \pi x = -\frac{\pi}{2} + \pi$$

$$x = -\frac{1}{2} \qquad\qquad x = -\frac{1}{2} + 1$$

$$\text{Period} = 1 \qquad\qquad \text{Phase Shift} = -\frac{1}{2}$$

We then sketch one period of the graph starring at $x = -\frac{1}{2}$ (the phase shift) and ending at

$x = -\frac{1}{2} + 1 = \frac{1}{2}$ (the phase shift plus one period). We then extend the graph from -1 to 3.

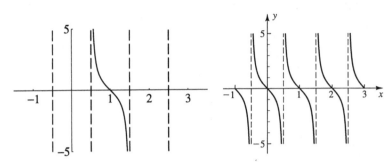

30. Since the maximum deviation from the x axis is 2, we can write: Amplitude $= |A| = 2$. Thus, $A = 2$ or -2. Since the period is 1, we can write: Period $= \frac{2\pi}{B} = 1$. Thus, $B = 2\pi$. As x increases from 0 to 0.25, y *decreases* like the upside down sine curve. So, A is negative. $A = -2$; $y = -2 \sin(2\pi x)$.

31. Since the period is 2, we can write $\frac{2\pi}{B} = 2$, hence $B = \frac{2\pi}{2} = \pi$. Since the $y_{max} = 3$ and $y_{min} = -1$, we

   can write amplitude $|A| = \frac{y_{max} - y_{min}}{2} = \frac{3 - (-1)}{2} = 2$, and $k = \frac{y_{max} + y_{min}}{2} = \frac{3 + (-1)}{2} = 1$. Thus,

   $y = 1 + 2 \cos \pi x$ or $y = 1 - 2 \cos \pi x$. Since the portion of the graph for $0 \le x \le 2$ has the form of the basic cosine curve, and not the upside-down cosine curve, we have $y = 1 + 2 \cos \pi x$.

   Check: When $x = 0$ $y = 1 + 2 \cos \pi \cdot 0 = 1 + 2 \cdot 1 = 3$
   When $x = 1$ $y = 1 + 2 \cos \pi(1) = 1 + 2(-1) = -1$

32. $(\tan x)(\sin x) + \cos x = \frac{\sin x}{\cos x} \sin x + \cos x = \frac{\sin^2 x}{\cos x} + \frac{\cos x}{1} = \frac{\sin^2 x}{\cos x} + \frac{\cos^2 x}{\cos x} = \frac{\sin^2 x + \cos^2 x}{\cos x}$

   $= \frac{1}{\cos x} = \sec x$

33. We can draw reference triangles in both quadrants III and IV with sides opposite reference angle -1 and hypotenuse 2. Each triangle is a special $30°–60°$ triangle.

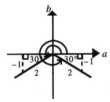

34. *Solve for θ:* We will use the tangent. Thus,

$$\tan \theta = \frac{b}{a} = \frac{23.5 \text{ in.}}{37.3 \text{ in.}} = 0.6300$$

$$\theta = \tan^{-1} 0.6300 = 32.2°$$

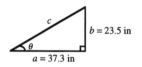

Solve for the complementary angle:

$$90° - \theta = 90° - 32.2° = 57.8°$$

Solve for c: We use the Pythagorean theorem.
Since $c^2 = a^2 + b^2$,

$$c = \sqrt{a^2 + b^2} = \sqrt{(37.3)^2 + (23.5)^2} = 44.1 \text{ in.}$$

35.

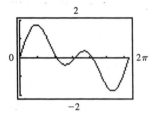

36. The graph of $y = \dfrac{\tan^2 x}{1 + \tan^2 x}$ is shown in the figure.

Amplitude $= \dfrac{1}{2}$ (y coordinate of highest point – y coordinate of lowest point)

$$= \frac{1}{2}(1 - 0) = \frac{1}{2} = |A|$$

Period $= \dfrac{2\pi}{2} = \pi.$

Thus, $B = \dfrac{2\pi}{\pi} = 2.$ The form of the graph is that of the
upside down cosine
curve shifted up $\dfrac{1}{2}$ unit.

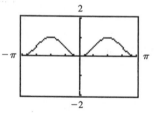

Thus, $y = \dfrac{1}{2} - |A| \cos Bx = \dfrac{1}{2} - \dfrac{1}{2} \cos 2x.$

37.

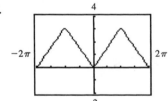

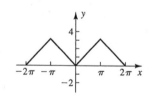

38. Using the fundamental identities, we can write $\tan \theta = a$, hence, $\dfrac{\sin \theta}{\cos \theta} = a$, or $\sin \theta = a \cos \theta$. Also, $\sin^2 \theta + \cos^2 \theta = 1$. Substituting $a \cos \theta$ for $\sin \theta$ and solving for $\cos \theta$, we obtain:

$$(a \cos \theta)^2 + \cos^2 \theta = 1$$
$$(a^2 + 1) \cos^2 \theta = 1$$
$$\cos^2 \theta = \frac{1}{1 + a^2}$$

Since θ is a first quadrant angle, $\cos \theta$ is positive. Hence, $\cos \theta = \dfrac{1}{\sqrt{1 + a^2}}$. It follows that

$\sin \theta = a \cos \theta = \dfrac{a}{\sqrt{1 + a^2}}$. Using the reciprocal identities, we can write

$$\cot \theta = \frac{1}{\tan \theta} = \frac{1}{a}$$

$$\sec \theta = \frac{1}{\cos \theta} = \frac{1}{1/\sqrt{1 + a^2}} = \sqrt{1 + a^2} \qquad \csc \theta = \frac{1}{\sin \theta} = \frac{1}{a/\sqrt{1 + a^2}} = \frac{\sqrt{1 + a^2}}{a}$$

39. In the figure, we note, $s = r\theta$

$$r = \sqrt{a^2 + b^2} = \sqrt{8^2 + 15^2} = 17;$$

$$\tan \theta = \frac{b}{a} = \frac{15}{8}; \ \theta = \tan^{-1} \frac{15}{8} \text{ radians}$$

Therefore,

$$s = r\theta = 17 \tan^{-1} \frac{15}{8} \approx 18.37 \text{ units (calculator in radian mode).}$$

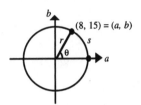

40. Since the point moves clockwise, x is negative, and the coordinates of the point are: $P(\cos(-53.077)$, $\sin(-53.077)) = (-0.9460, -0.3241)$. The quadrant in which P lies is determined by the signs of the coordinates. In this case, P lies in quadrant III, because both coordinates are negative.

41. Since (a, b) is on a unit circle with $(a, b) = (0.5796, 0.8145) = (\cos s, \sin s)$, we can solve $\cos s = 0.5796$ or $\sin s = 0.8149$. Then $s = \cos^{-1} (0.5796) = 0.9526$ or $s = \sin^{-1} (0.8148) = 0.9526$.

42. Since the maximum deviation from the x axis is 4, we can write:

Amplitude $= |A| = 4$. Thus, $A = 4$ or -4. Since the period is $\dfrac{4}{3} - \left(-\dfrac{2}{3}\right) = 2$, we can write:

Period $= \dfrac{2\pi}{B} = 2$. Thus, $B = \dfrac{2\pi}{2} = \pi$. Since we are instructed to choose the phase shift between 0 and 1, we can regard this graph as containing the basic sine curve with a phase shift of $\dfrac{1}{3}$. This requires us to choose A positive, since the graph shows that as x increases from $\dfrac{1}{3}$ to $\dfrac{5}{6}$, y *increases* like the basic sine curve (not the upside down sine curve). So $A = 4$. Then, $-\dfrac{C}{B} = \dfrac{1}{3}$. Thus,

$$C = -\frac{1}{3} B = -\frac{1}{3} \pi \qquad y = A \sin(Bx + C) = 4 \sin \left(\pi x - \frac{1}{3} \pi \right).$$

Check: When $x = 0$, $y = 4 \sin\left(\pi \cdot 0 - \frac{1}{3}\pi\right) = 4 \sin\left(-\frac{1}{3}\pi\right) = -2\sqrt{3}$

When $x = \frac{1}{3}$, $y = 4 \sin\left(\pi \cdot \frac{1}{3} - \frac{1}{3}\pi\right) = 4 \sin 0 = 0$

43. The graph of $y = 2.4 \sin\frac{x}{2} - 1.8 \cos\frac{x}{2}$ is shown in the figure. This graph appears to be a sine wave with amplitude 3 and period 4π that has been shifted to the right. Thus, we conclude that $A = 3$ and $B = \frac{2\pi}{4\pi} = \frac{1}{2}$. To determine C, we use the zoom feature or the built-in approximation routine to locate the x intercept closest to the origin at $x = 1.287$. This is the

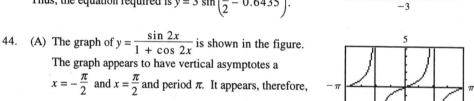

phase-shift for the graph. Substitute $B = \frac{1}{2}$ and $x = 1.287$

into the phase-shift equation $x = -\frac{C}{B}$; $1.287 = -\frac{C}{1/2}$;

$C = -0.6435$.

Thus, the equation required is $y = 3 \sin\left(\frac{x}{2} - 0.6435\right)$.

44. (A) The graph of $y = \dfrac{\sin 2x}{1 + \cos 2x}$ is shown in the figure.

The graph appears to have vertical asymptotes a

$x = -\frac{\pi}{2}$ and $x = \frac{\pi}{2}$ and period π. It appears, therefore,

to be the same as the graph of $y = \tan Bx$, with $\frac{\pi}{B} = \pi$,

that is, $B = 1$. The required equation is $y = \tan x$.

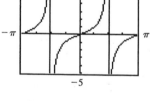

(B) The graph of $y = \dfrac{2 \cos x}{1 + \cos 2x}$ is shown in the figure.

The graph appears to have vertical asymptotes at

$x = -\frac{\pi}{2}$ and $x = \frac{\pi}{2}$ and period 2π. Its high and low

points appear to have y coordinates of -1 and 1,

respectively. It appears, therefore, to be the same as

the graph of $y = \sec Bx$, $\dfrac{2\pi}{B} = 2\pi$, that is, $B = 1$.

The required equation is $y = \sec x$.

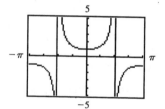

(C) The graph of $y = \dfrac{2 \sin x}{1 - \cos 2x}$ is shown in the figure.

The graph appears to have vertical asymptotes at

$x = -\pi$, $x = 0$, and $x = \pi$, and period 2π. Its high and

low points appear to have y coordinates of -1 and 1,

respectively. It appears, therefore, to be the same as

the graph of $y = \csc Bx$, $\dfrac{2\pi}{B} = 2\pi$, that is, $B = 1$.

The required equation is $y = \csc x$.

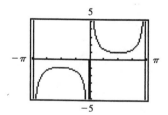

(D) The graph of $y = \dfrac{\sin 2x}{1 - \cos 2x}$ is shown in the figure.

The graph appears to have vertical asymptotes at
$x = -\pi$, $x = 0$, and $x = \pi$, and period π. It appears,
therefore, to be the same as the graph of $y = \cot Bx$,

with $\dfrac{\pi}{B} = \pi$, that is, $B = 1$.

The required equation is $y = \cot x$.

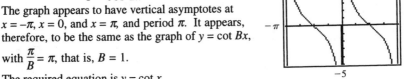

45. We use the diagram and reason as follows: Since the cities
have the same longitude, θ is given by their difference in

latitude $\theta = 41°36' - 30°25' = 11°11' = \left(11 + \dfrac{11}{60}\right)°$.

Since $\dfrac{s}{C} = \dfrac{\theta}{360°}$ and $C = 2\pi r$, then $\dfrac{s}{2\pi r} = \dfrac{\theta}{360°}$.

$s = 2\pi r \cdot \dfrac{\theta}{360°} \approx 2(\pi)(3960 \text{ mi})\dfrac{11 + \frac{11}{60}}{360} \approx 773$ mi.

46. Since the two right triangles shown in the figure are similar, we can write $\dfrac{r}{4} = \dfrac{6}{9}$, thus, $r = \dfrac{8}{3}$ cm.

Then, we have $V = \dfrac{1}{3}\pi r^2 h = \dfrac{1}{3}\pi\left(\dfrac{8}{3}\right)^2 6 = \dfrac{128\pi}{9} \approx 45$ cm^3

47. Labeling the diagram as shown, we can write:

$\cos\theta = \dfrac{AC}{AB}$ θ = angle of elevation

$\theta = \cos^{-1}\dfrac{AC}{AB} = \cos^{-1}\dfrac{30}{40} = 41°$

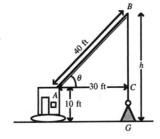

To find h, the altitude of the tip, we note

$h = BC + CG$

$BC^2 = AB^2 - AC^2$ (Pythagorean theorem)

$BC = \sqrt{AB^2 - AC^2} = \sqrt{40^2 - 30^2}$

$= 26$ ft

Then, $h = 26 + 10 = 36$ ft

48. The two right triangles shown in the text figure have corresponding angles equal, hence they are
similar. Thus, we can write $\dfrac{x}{2} = \dfrac{4 - x}{4}$; $2x = 4 - x$; $x = \dfrac{4}{3}$ ft.

49. From the figure it is clear that $\tan\theta = \dfrac{a}{b}$.

Given: percentage of inclination $\dfrac{a}{b} = 3\% = 0.03$, then $\tan\theta = 0.03$; $\theta = 1.7°$

Given: angle of inclination $\theta = 3°$, then $\dfrac{a}{b} = \tan 3°$; $\dfrac{a}{b} = 0.05$ or 5%

50.　Labeling the text figure as shown, we note:
We are asked for $BT = x + 10$, the height of the office building, and d, the width of the street.

In right triangle SCT, $\cot 68° = \dfrac{d}{x}$

In right triangle ABT, $\cot 72° = \dfrac{d}{x + 10}$

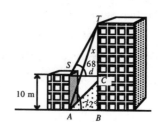

We solve the system of equations

$$\cot 68° = \frac{d}{x} \quad \cot 72° = \frac{d}{x + 10}$$

by clearing of fractions, then eliminating d.

(1)　$d = x \cot 68°$　$d = (x + 10) \cot 72°$

$x \cot 68° = (x + 10) \cot 72°$

$= x \cot 72° + 10 \cot 72°$

$x \cot 68° - x \cot 72° = 10 \cot 72°$

$$x = \frac{10 \cot 72°}{\cot 68° - \cot 72°} = 41 \text{ m}$$

Then the height of the office building $= x + 10 = 51$ m. Substituting in (1),
$$d = x \cot 68° = (41 \text{ m}) \cot 68° = 17 \text{ m}$$

51.　We redraw and label the figure in the text.
(A) We note: $AC = AB - BC = 5 - x$. We are to find x.

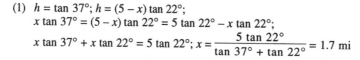

In right triangle BCF, $\tan 37° = \dfrac{h}{x}$

In right triangle ACF, $\tan 22° = \dfrac{h}{5 - x}$

We solve the system of equations
$$\tan 37° = \frac{h}{x} \quad \tan 22° = \frac{h}{5 - x}$$
by clearing of fractions, then eliminating h.

(1)　$h = \tan 37°$;　$h = (5 - x) \tan 22°$;
$x \tan 37° = (5 - x) \tan 22° = 5 \tan 22° - x \tan 22°$;
$$x \tan 37° + x \tan 22° = 5 \tan 22°; \quad x = \frac{5 \tan 22°}{\tan 37° + \tan 22°} = 1.7 \text{ mi}$$

(B) We are to find h. From (1) in part (A), $h = x \tan 37° = 1.7 \tan 37° = 1.3$ mi

52.　(A) For the drive motor, we note: 300 rpm $= 300 \cdot 2\pi$ rad/min $= 600\pi$ rad/min.

$$V = r\omega = (15 \text{ in.})(600\pi \text{ rad/min}) = 9000\pi \text{ in./min}$$

This is the linear velocity of the chain, hence the linear velocity of the smaller wheel of the saw.

Then, for the saw,
$$\omega = \frac{V}{r} = \frac{9000\pi \text{ in./min}}{30 \text{ in.}} = 300\pi \text{ rad/min} \approx 942 \text{ rad/min}$$

(B) For the saw itself, ω is also 300π rad/min

$$V = r\omega = (68 \text{ in.})(300\pi \text{ rad/min}) = 20{,}400\pi \text{ in./min} \approx 64{,}088 \text{ in./min}$$

53. Use $\dfrac{n_2}{n_1} = \dfrac{\sin \alpha}{\sin \beta}$, where $n_2 = 1.33$, $n_1 = 1.00$, and $\alpha = 38.4°$.

Solve for β: $\dfrac{1.33}{1.00} = \dfrac{\sin 38.4°}{\sin \beta}$; $\sin \beta = \dfrac{\sin 38.4°}{1.33}$; $\beta = \sin^{-1}\left(\dfrac{\sin 38.4°}{1.33}\right) = 27.8°$

54. We use $\sin \dfrac{\theta}{2} = \dfrac{S_s}{S_p}$, where S_s is the speed of sound and S_p = speed of the plane = $1.5 S_s$. Thus,

$\sin \dfrac{\theta}{2} = \dfrac{S_s}{1.5 S_s} = \dfrac{1}{1.5}$; $\dfrac{\theta}{2} \approx 42°$; $\theta \approx 84°$

55. (A) $A_1 = \dfrac{1}{2}$ (base)(height) $= \dfrac{1}{2}(1)(\sin x) = \dfrac{1}{2}\sin x$ (the height is the perpendicular distance from P to

the x axis, thus, $\sin x$).

$A_2 = \dfrac{1}{2}$ (radius)^2 (angle) $= \dfrac{1}{2}(1)^2 x = \dfrac{1}{2}x$

$A_3 = \dfrac{1}{2}$ (base)(height) $= \dfrac{1}{2}(1)\,h$. In triangle OAB, $\tan x = \dfrac{h}{1}$, hence, $h = \tan x$. Thus, $A_3 = \dfrac{1}{2}\tan x$.

(B) Since $A_1 < A_2 < A_3$, we can write $\dfrac{1}{2}\sin x < \dfrac{1}{2}x < \dfrac{1}{2}\tan x$

Multiplying by 2, we can write $\sin x < x < \tan x$

Applying a fundamental identity, we can write $\sin x < x < \dfrac{\sin x}{\cos x}$

As long as $x > 0$, these quantities are positive. For positive quantities, $a < b < c$ is equivalent to

$\dfrac{1}{c} < \dfrac{1}{b} < \dfrac{1}{a}$.

Hence, $\dfrac{\cos x}{\sin x} < \dfrac{1}{x} < \dfrac{1}{\sin x}$. If $x > 0$, $\sin x > 0$, and we can multiply all parts of this double

inequality by $\sin x$ without altering the sense of inequalities. Thus,

$\sin x \cdot \dfrac{\cos x}{\sin x} < \sin x \cdot \dfrac{1}{x} < \sin x \cdot \dfrac{1}{\sin x}$

$\cos x < \dfrac{\sin x}{x} < 1$, $\quad x > 0$

(C) $\cos x$ approaches 1 as x approaches 0, and $\dfrac{\sin x}{x}$ is between $\cos x$ and 1, therefore, $\dfrac{\sin x}{x}$ must

approach 1 as x approaches 0.

56. (A)

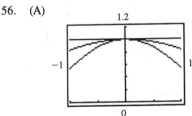

(B) $y1 < y2 < y3$

(C) $\cos x$ approaches 1 as x approaches 0, and $\dfrac{\sin x}{x}$ is between $\cos x$ and 1, therefore $\dfrac{\sin x}{x}$ must

approach 1 as x approaches 0. This can also be observed using $\boxed{\text{TRACE}}$.

57. Since the amplitude is 3.6 cm, $|A| = 3.6$, and since the position when $t = 0$ sec is taken positive, $A = 3.6$. Since the frequency is 6 Hz, write

$$\text{Frequency} = \frac{1}{\text{Period}} = \frac{1}{2\pi/B} = \frac{B}{2\pi}$$

$$6 = \frac{B}{2\pi}$$

$$12\pi = B$$

Thus, the required equation is $y = 3.6 \cos 12\pi t$.
An equation of the form $y = A \sin Bt$ cannot be used to model the motion, because then, for $t = 0$, y will be 0 no matter what values are assigned to A and B.

58. (A) $I = 12 \sin(60\pi t - \pi)$
We compute amplitude, period, frequency, and phase shift as follows:
Amplitude $= |A| = |12| = 12$.
Phase Shift and Period: Solve

$$
\begin{array}{ll}
Bx + C = 0 \quad \text{and} & Bx + C = 2\pi \\
60\pi t - \pi = 0 & 60\pi t - \pi = 2\pi \\
60\pi t = \pi & 60\pi t = \pi + 2\pi \\
t = \dfrac{1}{60} & t = \dfrac{1}{60} + \dfrac{1}{30} \\
\quad\uparrow & \quad\uparrow \quad \uparrow \\
\text{Phase Shift} & \text{Period}
\end{array}
$$

$$\text{Frequency} = \frac{1}{\text{Period}} = \frac{1}{1/30} = 30 \text{ Hz}$$

(B)

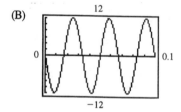

59. To find the period T, we use the formula $\lambda v = c$, with $v = \dfrac{1}{\text{Period}} = \dfrac{1}{T}$ and solve for T.

$\lambda \dfrac{1}{T} = c$; $\lambda = Tc$; $T = \dfrac{\lambda}{c}$. Hence, $T = \dfrac{6 \times 10^{-5} \text{ m}}{3 \times 10^8 \text{ m/sec}} = 2 \times 10^{-13}$ sec.

60. (A) In the figure, note that ABC and CDE are right triangles, and that $AE = AC + CE = a + b$. Then,

$\csc \theta = \dfrac{a}{100}$ from triangle ABC, so $a = 100 \csc \theta$,

$\sec \theta = \dfrac{b}{70}$ from triangle CDE, so $b = 70 \sec \theta$.

Thus, $AE = 100 \csc \theta + 70 \sec \theta$.

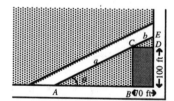

(B) For small θ (near 0°), L is extremely large. As θ increases from 0°, L decreases to some minimum value, then increases again beyond all bounds as θ approaches 90°.

(C)

θ rad	0.50	0.60	0.70	0.80	0.90	1.00	1.10
L ft	288.3	261.9	246.7	239.9	240.3	248.4	266.5

According to the table, the minimum value is 239.9 ft when $\theta = 0.80$ rad.

(D) According to the graph, the minimum $L = 239.16$ ft when $\theta = 0.84$ rad.

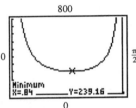

61. (A) In the right triangle in the figure, $\tan \theta = \dfrac{d}{50}$. Hence, $d = 50 \tan \theta$.

(B) 20 rpm $= 20(2\pi)$ rad/min $= 40\pi$ rad/min. Since $\theta = \omega t$, and $\omega = 40\pi$ rad/min, $\theta = 40\pi t$.

(C) Substituting the expression for θ from part (B) into $d = 50 \tan \theta$, we obtain $d = 50 \tan 40\pi t$.

(D)

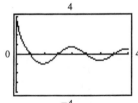

d increases without bound as t approaches $\dfrac{1}{80}$ min.

62. (A)

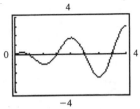

Since the amplitude decreases to 0 as time increases, this represents damped harmonic motion.

(B)

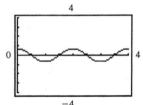

Since the amplitude remains constant over time, this represents simple harmonic motion.

(C)

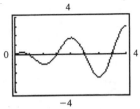

Since the amplitude increases as time increases, this represents resonance.

63. (A)

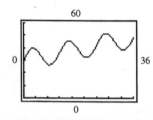

   (B) The sales trend is up but with the expected seasonal variations.

64. (A)

x(months)	1, 13	2, 14	3, 15	4, 16	5, 17	6, 18	7, 19	8, 20	9, 21
y (temperatures)	19	23	32	45	55	65	71	69	62

x	10, 22	11, 23	12, 24
y	51	37	25

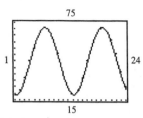

(B) From the table, Max $y = 71$ and Min $y = 19$. Then,

$$A = \frac{(\text{Max } y - \text{Min } y)}{2} = \frac{(71 - 19)}{2} = 26$$

$$B = \frac{2\pi}{\text{Period}} = \frac{2\pi}{12} = \frac{\pi}{6}$$

$$k = \text{Min } y + A = 19 + 26 = 45$$

From the plot in (A) or the table, we estimate the smallest positive value of x for which $y = k = 45$ to be approximately 4.2. Then this is the phase-shift for the graph. Substitute $B = \frac{\pi}{6}$ and $x = 4.2$ into the phase-shift equation $x = -\frac{C}{B}$; $4.2 = \frac{-C}{\pi/6}$; $C = -\frac{4.2\pi}{6} \approx -2.2$.

Thus, the equation required is $y = 45 + 26 \sin\left(\frac{\pi x}{6} x - 2.2\right)$.

(C)

114

Chapter 4   Identities

EXERCISE 4.1   Fundamental Identities and Their Use

1. Reciprocal Identities:

$$\csc x = \frac{1}{\sin x} \qquad \sec x = \frac{1}{\cos x} \qquad \cot x = \frac{1}{\tan x}$$

 Identities for Negatives:

$$\sin(-x) = -\sin x \qquad \cos(-x) = \cos x \qquad \tan(-x) = -\tan x$$

3. (1) is equivalent to $6x - 9 = 9 - 6x$, or $12x = 18$, or $x = 1.5$, by the algebra rules for equivalent equations. These statements are true for only one value of x (that is, 1.5) and false for all other values of x for which both sides are defined.

 (2) is equivalent to $6x - 9 = 6x - 9$. This is true for all values of x for which both sides are defined.

 Hence, (1) is a conditional equation and (2) is an identity.

5. *Find tan x:* $\tan x = \dfrac{\sin x}{\cos x} = \dfrac{-2/3}{\sqrt{5/3}} = -\dfrac{2}{\sqrt 5}$ *Find cot x:* $\cot x = \dfrac{1}{\tan x} = \dfrac{1}{-2/\sqrt 5} = -\dfrac{\sqrt 5}{2}$

 Find csc x: $\csc x = \dfrac{1}{\sin x} = \dfrac{1}{-2/3} = -\dfrac{3}{2}$ *Find sec x:* $\sec x = \dfrac{1}{\cos x} = \dfrac{1}{\sqrt{5/3}} = \dfrac{3}{\sqrt 5}$

7. *Find cos x:* Since $\tan x = \dfrac{\sin x}{\cos x}$, we can write $\cos x = \dfrac{\sin x}{\tan x}$; $\cos x = \dfrac{\sin x}{\tan x} = \dfrac{-2/\sqrt 5}{2} = -\dfrac{1}{\sqrt 5}$

 Find csc x: $\csc x = \dfrac{1}{\sin x} = \dfrac{1}{-2/\sqrt 5} = -\dfrac{\sqrt 5}{2}$

 Find sec x: $\sec x = \dfrac{1}{\cos x} = \dfrac{1}{-1/\sqrt 5} = -\sqrt 5$ *Find cot x:* $\cot x = \dfrac{1}{\tan x} = \dfrac{1}{2}$

9. $\tan u \cot u = \tan u \, \dfrac{1}{\tan u}$ Reciprocal identity

 $\qquad\qquad = 1$ Algebra

11. $\tan x \csc x = \dfrac{\sin x}{\cos x} \dfrac{1}{\sin x}$ Quotient and reciprocal identities

 $\qquad\qquad = \dfrac{1}{\cos x}$ Algebra

 $\qquad\qquad = \sec x$ Reciprocal identity

13. $\dfrac{\sec^2 x - 1}{\tan x} = \dfrac{\tan^2 x + 1 - 1}{\tan x}$ Pythagorean identity

    $= \dfrac{\tan^2 x}{\tan x}$ Algebra

    $= \tan x$ Algebra

15. $\dfrac{\sin^2 \theta}{\cos \theta} + \cos \theta = \dfrac{\sin^2 \theta}{\cos \theta} + \dfrac{\cos \theta}{1}$ Algebra

    $= \dfrac{\sin^2 \theta}{\cos \theta} + \dfrac{\cos^2 \theta}{\cos \theta}$ Algebra

    $= \dfrac{\sin^2 \theta + \cos^2 \theta}{\cos \theta}$ Algebra

    $= \dfrac{1}{\cos \theta}$ Pythagorean idenrtity

    $= \sec \theta$ Reciprocal identity

    Key algebraic steps: $\dfrac{a^2}{b} + b = \dfrac{a^2}{b} + \dfrac{b}{1} = \dfrac{a^2}{b} + \dfrac{b^2}{b} = \dfrac{a^2 + b^2}{b}$

17. $\dfrac{1}{\sin^2 \beta} - 1 = \left(\dfrac{1}{\sin \beta}\right)^2 - 1$ Algebra

    $= \csc^2 \beta - 1$ Reciprocal identity
    $= 1 + \cot^2 \beta - 1$ Pythagorean identity
    $= \cot^2 \beta$ Algebra

19. $\dfrac{(1 - \cos x)^2 + \sin^2 x}{1 - \cos x} = \dfrac{1 - 2 \cos x + \cos^2 x + \sin^2 x}{1 - \cos x}$ Algebra

    $= \dfrac{1 - 2 \cos x + 1}{1 - \cos x}$ Pythagorean identity

    $= \dfrac{2 - 2 \cos x}{1 - \cos x}$ Algebra

    $= \dfrac{2(1 - \cos x)}{1 - \cos x}$ Algebra

    $= 2$ Algebra

    Key algebraic step: $(1 - a)^2 = 1 - 2a + a^2$

21. Not necessarily. For example, $\sin x = 0$ for infinitely many values ($x = k\pi$, k any integer), but the equation is not an identity. The left side is not equal to the right side for any other value of x than $x = k\pi$, k an integer; for example, if $x = \dfrac{\pi}{2}$, $\sin \dfrac{\pi}{2} = 1$, hence $\sin \dfrac{\pi}{2} = 0$ is false.

23. *Find cos x:* We start with the Pythagorean identity $\sin^2 x + \cos^2 x = 1$ and solve for $\cos x$.
$$\cos x = \pm\sqrt{1 - \sin^2 x}$$
Since $\sin x$ is positive and $\tan x$ is negative, x is associated with the second quadrant, where $\cos x$ is

negative; hence, $\cos x = -\sqrt{1 - \sin^2 x} = -\sqrt{1 - \left(\dfrac{1}{4}\right)^2} = -\sqrt{\dfrac{15}{16}} = -\dfrac{\sqrt{15}}{4}$

Find sec x: $\sec x = \dfrac{1}{\cos x} = \dfrac{1}{-\sqrt{15}/4} = -\dfrac{4}{\sqrt{15}}$

Find csc x: $\csc x = \dfrac{1}{\sin x} = \dfrac{1}{1/4} = 4$

Find tan x: $\tan x = \dfrac{\sin x}{\cos x} = \dfrac{1/4}{-\sqrt{15}/4} = -\dfrac{1}{\sqrt{15}}$

Find cot x: $\cot x = \dfrac{1}{\tan x} = \dfrac{1}{-1/\sqrt{15}} = -\sqrt{15}$

25. *Find sec x:* We start with the Pythagorean identity $\tan^2 x + 1 = \sec^2 x$ and solve for $\sec x$.
$$\sec x = \pm\sqrt{\tan^2 x + 1}$$

Since $\sin x$ and $\tan x$ are both negative, x is associated with the fourth quadrant, where $\sec x$ is

positive; hence, $\sec x = \sqrt{\tan^2 x + 1} = \sqrt{(-2)^2 + 1} = \sqrt{5}$

Find cos x: $\cos x = \dfrac{1}{\sec x} = \dfrac{1}{\sqrt{5}}$

Find sin x: Since $\tan x = \dfrac{\sin x}{\cos x}$, we can write $\sin x = \cos x \tan x = \dfrac{1}{\sqrt{5}}(-2) = -\dfrac{2}{\sqrt{5}}$

Find csc x: $\csc x = \dfrac{1}{\sin x} = \dfrac{1}{-2/\sqrt{5}} = -\dfrac{\sqrt{5}}{2}$

Find cot x: $\cot x = \dfrac{1}{\tan x} = \dfrac{1}{-2} = -\dfrac{1}{2}$

27. *Find sin x:* Since $\csc x = \dfrac{1}{\sin x}$, we can write $\sin x = \dfrac{1}{\csc x} = \dfrac{1}{3/2} = \dfrac{2}{3}$

Find cos x: We start with the Pythagorean identity $\sin^2 x + \cos^2 x = 1$ and solve for $\cos x$.
$$\cos x = \pm\sqrt{1 - \sin^2 x}$$
Since $\sin x$ is positive and $\tan x$ is negative, x is associated with the second quadrant, where $\cos x$ is

negative; hence $\cos x = -\sqrt{1 - \sin^2 x} = -\sqrt{1 - \left(\dfrac{2}{3}\right)^2} = -\sqrt{\dfrac{5}{9}} = -\dfrac{\sqrt{5}}{3}$

Find tan x: $\tan x = \dfrac{\sin x}{\cos x} = \dfrac{2/3}{-\sqrt{5}/3} = -\dfrac{2}{\sqrt{5}}$

Find sec x: $\sec x = \dfrac{1}{\cos x} = \dfrac{1}{-\sqrt{5}/3} = -\dfrac{3}{\sqrt{5}}$

Find cot x: $\cot x = \dfrac{1}{\tan x} = \dfrac{1}{-2/\sqrt{5}} = -\dfrac{\sqrt{5}}{2}$

29. (A) From the identities for negatives,
$$\sin(-x) = -\sin x$$
Hence
$$\sin(-x) = -0.4350$$

(B) From the Pythagorean identity, $\cos^2 x + \sin^2 x = 1$, thus
$$\cos^2 x = 1 - \sin^2 x$$
Hence
$$(\cos x)^2 = 1 - 0.1892 = 0.8108$$

31. $\csc(-y)\cos(-y) = \dfrac{1}{\sin(-y)}\cos(-y)$ Reciprocal identity

$$= -\dfrac{1}{\sin y}\cos y$$ Identities for negatives

$$= -\dfrac{\cos y}{\sin y}$$ Algebra

$$= -\cot y$$ Quotient identity

33. $\cot x \cos x + \sin x = \dfrac{\cos x}{\sin x}\cos x + \sin x$ Quotient identity

$$= \dfrac{\cos^2 x}{\sin x} + \dfrac{\sin x}{1}$$ Algebra

$$= \dfrac{\cos^2 x}{\sin x} + \dfrac{\sin^2 x}{\sin x}$$ Algebra

$$= \dfrac{\cos^2 x + \sin^2 x}{\sin x}$$ Algebra

$$= \dfrac{1}{\sin x}$$ Pythagorean identity

$$= \csc x$$ Reciprocal identity

Key algebraic steps: $\dfrac{a}{b}a + b = \dfrac{a}{b}\cdot\dfrac{a}{1} + \dfrac{b}{1} = \dfrac{a^2}{b} + \dfrac{b}{1} = \dfrac{a^2}{b} + \dfrac{b^2}{b} = \dfrac{a^2 + b^2}{b}$

35. $\dfrac{\cot(-\theta)}{\csc\theta} + \cos\theta = \dfrac{\dfrac{\cos(-\theta)}{\sin(-\theta)}}{\dfrac{1}{\sin\theta}} + \cos\theta$ Quotient and reciprocal identities

$$= \dfrac{\dfrac{\cos\theta}{-\sin\theta}}{\dfrac{1}{\sin\theta}} + \cos\theta$$ Identities for negatives

$$= \dfrac{\cos\theta}{-\sin\theta}\cdot\dfrac{\sin\theta}{1} + \cos\theta$$ Algebra

$$= -\cos\theta + \cos\theta$$ Algebra
$$= 0$$ Algebra

37. $\dfrac{\cot x}{\tan x} + 1 = \cot x \div \tan x + 1$ Algebra

$\qquad\qquad\quad = \cot x \div \dfrac{1}{\cot x} + 1$ Reciprocal identity

$\qquad\qquad\quad = \cot x \cdot \dfrac{\cot x}{1} + 1$ Algebra

$\qquad\qquad\quad = \cot^2 x + 1$ Algebra

$\qquad\qquad\quad = \csc^2 x$ Pythagorean identity

39. $\sec w \csc w - \sec w \sin w = \dfrac{1}{\cos w} \cdot \dfrac{1}{\sin w} - \dfrac{1}{\cos w} \cdot \sin w$ Reciprocal identities

$\qquad\qquad\qquad\qquad\quad = \dfrac{1}{\cos w \sin w} - \dfrac{\sin w}{\cos w}$ Algebra

$\qquad\qquad\qquad\qquad\quad = \dfrac{1}{\cos w \sin w} - \dfrac{\sin^2 w}{\cos w \sin w}$ Algebra

$\qquad\qquad\qquad\qquad\quad = \dfrac{1 - \sin^2 w}{\cos w \sin w}$ Algebra

$\qquad\qquad\qquad\qquad\quad = \dfrac{\cos^2 w}{\cos w \sin w}$ Pythagorean identity (solved for $1 - \sin^2 x = \cos^2 x$)

$\qquad\qquad\qquad\qquad\quad = \dfrac{\cos w}{\sin w}$ Algebra

$\qquad\qquad\qquad\qquad\quad = \cot w$ Quotient identity

41. (A) By the Pythagorean identity, $\sin^2 \alpha + \cos^2 \alpha = 1$ for any α. Therefore,
$\sin^2 \dfrac{x}{2} + \cos^2 \dfrac{x}{2} = 1$ (this is independent of x).

(B) By the Pythagorean identity, $\csc^2 \alpha = 1 + \cot^2 \alpha$ for any α. Therefore,
$\csc^2 (2x) - \cos^2 (2x) = 1 + \cot^2 (2x) - \cot^2 (2x) = 1$ (this is independent of x).

43. $\sqrt{1 - \cos^2 x} = \sqrt{\sin^2 x}$ by the Pythagorean identity. Therefore, $\sqrt{1 - \cos^2 x} = \sin x$ when, and only when, $\sqrt{\sin^2 x} = \sin x$. The latter statement is true whenever $\sin x$ is positive, that is, when x is in quadrant I or II.

45. $\sqrt{1 - \sin^2 x} = \sqrt{\cos^2 x}$ by the Pythagorean identity. Therefore, $\sqrt{1 - \sin^2 x} = -\cos x$ when, and only when, $\sqrt{\cos^2 x} = -\cos x$. The latter statement is true whenever $\cos x$ is negative, that is, when x is in quadrant II or III.

47. $\sqrt{1 - \sin^2 x} = \sqrt{\cos^2 x}$ by the Pythagorean identity. $\sqrt{\cos^2 x} = |\cos x|$ is always true.
Hence, $\sqrt{1 - \sin^2 x} = |\cos x|$ in all quadrants.

49. $\dfrac{\sin x}{\sqrt{1 - \sin^2 x}} = \dfrac{\sin x}{\sqrt{\cos^2 x}}$ by the Pythagorean identity. $\tan x = \dfrac{\sin x}{\cos x}$ by the quotient identity.

Therefore, $\dfrac{\sin x}{\sqrt{1 - \sin^2 x}} = \dfrac{\sin x}{\sqrt{\cos^2 x}} = \dfrac{\sin x}{\cos x} = \tan x$ will be true whenever the middle two quantities

are equal, that is, when, and only when, $\sqrt{\cos^2 x} = \cos x$. This statement is true whenever $\cos x$ is
positive, that is, when x is in quadrant I or IV.

51. $\sqrt{a^2 - u^2} = \sqrt{a^2 - (a \sin x)^2}$ using the given substitution

$= \sqrt{a^2 - a^2 \sin^2 x}$ Algebra

$= \sqrt{a^2(1 - \sin^2 x)}$ Algebra

$= \sqrt{a^2 \cos^2 x}$ Pythagorean identity

$= |a| \, |\cos x|$ Algebra

$= a \cos x$ since $a > 0$ and x is in quadrant I or IV

$\left(\text{given } -\dfrac{\pi}{2} < x < \dfrac{\pi}{2} \right)$, thus, $\cos x > 0$.

53. $\sqrt{a^2 + u^2} = \sqrt{a^2 + (a \tan x)^2}$ using the given substitution

$= \sqrt{a^2 + a^2 \tan^2 x}$ Algebra

$= \sqrt{a^2(1 + \tan^2 x)}$ Algebra

$= \sqrt{a^2 \sec^2 x}$ Pythagorean identity

$= |a| \, |\sec x|$ Algebra

$= a \sec x$ since $a > 0$ and x is in quadrant I

$\left(\text{given } 0 < x < \dfrac{\pi}{2} \right)$, thus, $\sec x > 0$.

55.   (A) Following the hint, we write $x = 5 \cos t$ and $y = 2 \sin t$ in the form $\dfrac{x}{5} = \cos t$ and $\dfrac{y}{2} = \sin t$.

Then, $\left(\dfrac{x}{5}\right)^2 + \left(\dfrac{y}{2}\right)^2 = (\cos t)^2 + (\sin t)^2 = 1$ by the Pythagorean identity. Thus, $\dfrac{x^2}{25} + \dfrac{y^2}{4} = 1$.

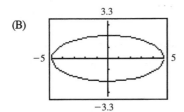

(B)

EXERCISE 4.2 Verifying Trigonometric Identities

1. $\cos x \sec x = \cos x \dfrac{1}{\cos x}$ Reciprocal identity

           $= 1$ Algebra

3. $\tan x \cos x = \dfrac{\sin x}{\cos x} \cos x$ Quotient identity

           $= \sin x$ Algebra

5. $\tan x = \dfrac{\sin x}{\cos x}$ Quotient identity

           $= \sin x \cdot \dfrac{1}{\cos x}$ Algebra

           $= \sin x \sec x$ Reciprocal identity

7. $\csc(-x) = \dfrac{1}{\sin(-x)}$ Reciprocal identity

           $= \dfrac{1}{-\sin x}$ Identities for negatives

           $= -\dfrac{1}{\sin x}$ Algebra

           $= -\csc x$ Reciprocal identity

9. $\dfrac{\sin \alpha}{\cos \alpha \tan \alpha} = \dfrac{\sin \alpha}{\cos \alpha \dfrac{\sin \alpha}{\cos \alpha}}$ Quotient identity

           $= \dfrac{\sin \alpha}{\sin \alpha}$ Algebra

           $= 1$ Algebra

11. $\dfrac{\cos \beta \sec \beta}{\tan \beta} = \dfrac{\cos \beta \dfrac{1}{\cos \beta}}{\tan \beta}$ Reciprocal identity

           $= \dfrac{1}{\tan \beta}$ Algebra

           $= \cot \beta$ Reciprocal identity

13. $\sec\theta(\sin\theta + \cos\theta)$ $= \sec\theta\sin\theta + \sec\theta\cos\theta$ Algebra

$= \dfrac{1}{\cos\theta}\sin\theta + \dfrac{1}{\cos\theta}\cos\theta$ Reciprocal identity

$= \dfrac{\sin\theta}{\cos\theta} + 1$ Algebra

$= \tan\theta + 1$ Quotient identity

15. $\dfrac{\cos^2 t - \sin^2 t}{\sin t \cos t} = \dfrac{\cos^2 t}{\sin t \cos t} - \dfrac{\sin^2 t}{\sin t \cos t}$ Algebra

$= \dfrac{\cos t}{\sin t} - \dfrac{\sin t}{\cos t}$ Algebra

$= \cot t - \tan t$ Quotient identity

Key algebraic steps:$\dfrac{b^2 - a^2}{ab} = \dfrac{b^2}{ab} - \dfrac{a^2}{ab} = \dfrac{b}{a} - \dfrac{a}{b}$

17. $\dfrac{\cos\beta}{\cot\beta} + \dfrac{\sin\beta}{\tan\beta}$ $= \cos\beta \div \cot\beta + \sin\beta \div \tan\beta$ Algebra

$= \cos\beta \div \dfrac{\cos\beta}{\sin\beta} + \sin\beta \div \dfrac{\sin\beta}{\cos\beta}$ Quotient identity

$= \dfrac{\cos\beta}{1}\cdot\dfrac{\sin\beta}{\cos\beta} + \dfrac{\sin\beta}{1}\cdot\dfrac{\cos\beta}{\sin\beta}$ Algebra

$= \sin\beta + \cos\beta$ Algebra

19. $\sec^2\theta - \tan^2\theta$ $= \tan^2\theta + 1 - \tan^2\theta$ Pythagorean identity

$= 1$ Algebra

21. $\sin^2 x(1 + \cot^2 x)$ $= \sin^2 x \csc^2 x$ Pythagorean identity

$= \sin^2 x \left(\dfrac{1}{\sin x}\right)^2$ Reciprocal identity

$= \dfrac{\sin^2 x}{1}\cdot\dfrac{1}{\sin^2 x}$ Algebra

$= 1$ Algebra

23. $(\csc\alpha + 1)(\csc\alpha - 1)$ $= \csc^2\alpha - 1$ Algebra

$= 1 + \cot^2\alpha - 1$ Pythagorean identity

$= \cot^2\alpha$ Algebra

Key algebraic step: $(x + 1)(x - 1) = x^2 - 1$

25.
$$\frac{\sin t}{\csc t} + \frac{\cos t}{\sec t} = \frac{\sin t}{1/\sin t} + \frac{\cos t}{1/\cos t} \qquad \text{Reciprocal identities}$$
$$= \sin t \div \frac{1}{\sin t} + \cos t \div \frac{1}{\cos t} \qquad \text{Algebra}$$
$$= \sin t \cdot \frac{\sin t}{1} + \cos t \cdot \frac{\cos t}{1} \qquad \text{Algebra}$$
$$= \sin^2 t + \cos^2 t \qquad \text{Algebra}$$
$$= 1 \qquad \text{Pythagorean identity}$$

27. To solve a conditional equation is to find, using equation solving strategies, all replacements of the variable that make the statement true. To verify an identity is to show that one side is equivalent to the other for all replacements of the variable for which both sides are defined. This is done by transforming one side, through a logical sequence of steps, into the other side.

29.
$$\frac{1 - (\cos\theta - \sin\theta)^2}{\cos\theta} = \frac{1 - (\cos^2\theta - 2\sin\theta\cos\theta + \sin^2\theta)}{\cos\theta} \qquad \text{Algebra}$$
$$= \frac{1 - \cos^2\theta + 2\sin\theta\cos\theta - \sin^2\theta}{\cos\theta} \qquad \text{Algebra}$$
$$= \frac{\sin^2\theta + 2\sin\theta\cos\theta - \sin^2\theta}{\cos\theta} \qquad \text{Pythagorean identity}$$
$$= \frac{2\sin\theta\cos\theta}{\cos\theta} \qquad \text{Algebra}$$
$$= 2\sin\theta \qquad \text{Algebra}$$

Key algebraic steps: $1 - (b-a)^2 = 1 - (b^2 - 2ab + a^2) = 1 - b^2 + 2ab - a^2$
$$\frac{a^2 + 2ab - a^2}{b} = \frac{2ab}{b} = 2a$$

31.
$$\frac{\tan w + 1}{\sec w} = \frac{\tan w}{\sec w} + \frac{1}{\sec w} \qquad \text{Algebra}$$
$$= \frac{\sin w/\cos w}{1/\cos w} + \cos w \qquad \text{Quotient and reciprocal identities}$$
$$= \frac{\sin w}{\cos w} \cdot \frac{\cos w}{1} + \cos w \qquad \text{Algebra}$$
$$= \sin w + \cos w \qquad \text{Algebra}$$

Key algebraic steps: $\dfrac{a/b}{1/b} + b = \dfrac{a}{b} \div \dfrac{1}{b} + b = \dfrac{a}{b} \cdot \dfrac{b}{1} + b = a + b$

33.
$$\frac{1}{1 - \cos^2\theta} = \frac{1}{\sin^2\theta} \qquad \text{Pythagorean identity}$$
$$= \left(\frac{1}{\sin\theta}\right)^2 \qquad \text{Algebra}$$
$$= \csc^2\theta \qquad \text{Reciprocal identity}$$
$$= 1 + \cot^2\theta \qquad \text{Pythagorean identity}$$

35. $\dfrac{\sin^2 \beta}{1 - \cos \beta} = \dfrac{1 - \cos^2 \beta}{1 - \cos \beta}$ Pythagorean identity

$= \dfrac{(1 - \cos \beta)(1 + \cos \beta)}{1 - \cos \beta}$ Algebra

$= 1 + \cos \beta$ Algebra

Key algebraic steps: $\dfrac{1 - b^2}{1 - b} = \dfrac{(1 - b)(1 + b)}{1 - b} = 1 + b$

37. $\dfrac{2 - \cos^2 \theta}{\sin \theta} = \dfrac{2 - (1 - \sin^2 \theta)}{\sin \theta}$ Pythagorean identity

$= \dfrac{2 - 1 + \sin^2 \theta}{\sin \theta}$ Algebra

$= \dfrac{1 + \sin^2 \theta}{\sin \theta}$ Algebra

$= \dfrac{1}{\sin \theta} + \dfrac{\sin^2 \theta}{\sin \theta}$ Algebra

$= \dfrac{1}{\sin \theta} + \sin \theta$ Algebra

$= \csc \theta + \sin \theta$ Reciprocal identity

Key algebraic steps: $\dfrac{2 - (1 - a^2)}{a} = \dfrac{2 - 1 + a^2}{a} = \dfrac{1 + a^2}{a} = \dfrac{1}{a} + \dfrac{a^2}{a} = \dfrac{1}{a} + a$

39. $\tan x + \cot x = \dfrac{\sin x}{\cos x} + \dfrac{\cos x}{\sin x}$ Quotient identity

$= \dfrac{\sin^2 x}{\sin x \cos x} + \dfrac{\cos^2 x}{\sin x \cos x}$ Algebra

$= \dfrac{\sin^2 x + \cos^2 x}{\sin x \cos x}$ Algebra

$= \dfrac{1}{\sin x \cos x}$ Pythagorean identity

$= \dfrac{1}{\sin x} \cdot \dfrac{1}{\cos x}$ Algebra

$= \sec x \csc x$ Reciprocal identities

Key algebraic steps: $\dfrac{a}{b} + \dfrac{b}{a} = \dfrac{a^2}{ab} + \dfrac{b^2}{ab} = \dfrac{a^2 + b^2}{ab}$

41. $\dfrac{1 - \csc x}{1 + \csc x} = \dfrac{1 - \dfrac{1}{\sin x}}{1 + \dfrac{1}{\sin x}}$ Reciprocal identity

$\qquad = \dfrac{\sin x \cdot 1 - \sin x \cdot \dfrac{1}{\sin x}}{\sin x \cdot 1 + \sin x \cdot \dfrac{1}{\sin x}}$ Algebra

$\qquad = \dfrac{\sin x - 1}{\sin x + 1}$ Algebra

43. $\csc^2 \alpha - \cos^2 \alpha - \sin^2 \alpha = \csc^2 \alpha - (\cos^2 \alpha + \sin^2 \alpha)$ Algebra

$\qquad\qquad\qquad\qquad\quad = \csc^2 \alpha - 1$ Pythagorean identity

$\qquad\qquad\qquad\qquad\quad = \cot^2 \alpha$ Pythagorean identity

45. $(\sin x + \cos x)^2 - 1 = \sin^2 x + 2 \sin x \cos x + \cos^2 x - 1$ Algebra

$\qquad\qquad\qquad\quad = 2 \sin x \cos x + \sin^2 x + \cos^2 x - 1$ Algebra

$\qquad\qquad\qquad\quad = 2 \sin x \cos x + 1 - 1$ Pythagorean identity

$\qquad\qquad\qquad\quad = 2 \sin x \cos x$ Algebra

47. $(\sin u - \cos u)^2 + (\sin u + \cos u)^2$

$\qquad\qquad = \sin^2 u - 2 \sin u \cos u + \cos^2 u + \sin^2 u + 2 \sin u \cos u + \cos^2 u$ Algebra

$\qquad\qquad = 2 \sin^2 u + 2 \cos^2 u$ Algebra

$\qquad\qquad = 2(\sin^2 u + \cos^2 u)$ Algebra

$\qquad\qquad = 2 \cdot 1 \text{ or } 2$ Pythagorean identity

Key algebraic steps: $(a - b)^2 + (a + b)^2 = a^2 - 2ab + b^2 + a^2 + 2ab + b^2 = 2a^2 + 2b^2 = 2(a^2 + b^2)$

49. $\sin^4 x - \cos^4 x = (\sin^2 x)^2 - (\cos^2 x)^2$ Algebra

$\qquad\qquad\quad = (\sin^2 x - \cos^2 x)(\sin^2 x + \cos^2 x)$ Algebra

$\qquad\qquad\quad = (\sin^2 x - \cos^2 x)(1)$ Pythagorean identity

$\qquad\qquad\quad = \sin^2 x - \cos^2 x$ Algebra

$\qquad\qquad\quad = (1 - \cos^2 x) - \cos^2 x$ Pythagorean identity

$\qquad\qquad\quad = 1 - 2 \cos^2 x$ Algebra

Key algebraic steps: $a^4 - b^4 = (a^2)^2 - (b^2)^2 = (a^2 - b^2)(a^2 + b^2)$

51.
$$\frac{\sin \alpha}{1 - \cos \alpha} - \frac{1 + \cos \alpha}{\sin \alpha} = \frac{\sin \alpha \cdot \sin \alpha}{(1 - \cos \alpha)\sin \alpha} - \frac{(1 - \cos \alpha)(1 + \cos \alpha)}{(1 - \cos \alpha)\sin \alpha} \quad \text{Algebra}$$

$$= \frac{\sin \alpha \sin \alpha - (1 - \cos \alpha)(1 + \cos \alpha)}{(1 - \cos \alpha)\sin \alpha} \quad \text{Algebra}$$

$$= \frac{\sin^2 \alpha - (1 - \cos^2 \alpha)}{(1 - \cos \alpha)\sin \alpha} \quad \text{Algebra}$$

$$= \frac{\sin^2 \alpha - \sin^2 \alpha}{(1 - \cos \alpha)\sin \alpha} \quad \text{Pythagorean identity}$$

$$= \frac{0}{(1 - \cos \alpha)\sin \alpha} \quad \text{Algebra}$$

$$= 0 \quad \text{Algebra}$$

Key algebraic steps: $\dfrac{a}{1 - b} - \dfrac{1 + b}{a} = \dfrac{a \cdot a}{a(1 - b)} - \dfrac{(1 - b)(1 + b)}{a(1 - b)} = \dfrac{a^2 - (1 - b^2)}{a(1 - b)}$

53.
$$\frac{\cos^2 n - 3 \cos n + 2}{\sin^2 n} = \frac{(\cos n - 2)(\cos n - 1)}{\sin^2 n} \quad \text{Algebra}$$

$$= \frac{(\cos n - 2)(\cos n - 1)}{1 - \cos^2 n} \quad \text{Pythagorean identity}$$

$$= \frac{(\cos n - 2)(\cos n - 1)}{(1 - \cos n)(1 + \cos n)} \quad \text{Algebra}$$

$$= \frac{(\cos n - 2)(-1)}{1 + \cos n} \quad \text{Algebra}$$

$$= \frac{2 - \cos n}{1 + \cos n} \quad \text{Algebra}$$

55.
$$\frac{1 - \cot^2 x}{\tan^2 x - 1} = \frac{1 - \cot^2 x}{\dfrac{1}{\cot^2 x} - 1} \quad \text{Reciprocal identity}$$

$$= \frac{\cot^2 x (1 - \cot^2 x)}{\cot^2 x \cdot \dfrac{1}{\cot^2 x} - \cot^2 x \cdot 1} \quad \text{Algebra}$$

$$= \frac{\cot^2 x (1 - \cot^2 x)}{1 - \cot^2 x} \quad \text{Algebra}$$

$$= \cot^2 x \quad \text{Algebra}$$

57. $\sec^2 x + \csc^2 x = \dfrac{1}{\cos^2 x} + \dfrac{1}{\sin^2 x}$ Reciprocal identities

$$= \dfrac{\sin^2 x}{\cos^2 x \sin^2 x} + \dfrac{\cos^2 x}{\cos^2 x \sin^2 x}$$ Algebra

$$= \dfrac{\sin^2 x + \cos^2 x}{\cos^2 x \sin^2 x}$$ Algebra

$$= \dfrac{1}{\cos^2 x \sin^2 x}$$ Pythagorean identity

$$= \dfrac{1}{\cos^2 x} \cdot \dfrac{1}{\sin^2 x}$$ Algebra

$$= \sec^2 x \csc^2 x$$ Reciprocal identities

59. $\dfrac{1 + \sin t}{\cos t} = \dfrac{(1 + \sin t)\cos t}{\cos t \cos t}$ Algebra

$$= \dfrac{(1 + \sin t)\cos t}{\cos^2 t}$$ Algebra

$$= \dfrac{(1 + \sin t)\cos t}{1 - \sin^2 t}$$ Pythagorean identity

$$= \dfrac{(1 + \sin t)\cos t}{(1 + \sin t)(1 - \sin t)}$$ Algebra

$$= \dfrac{\cos t}{1 - \sin t}$$ Algebra

61. (A)
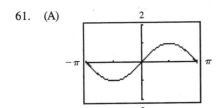

The graphs appear almost identical, but if the trace feature of the calculator is used to move from one graph to the other, different y values will arise for the same value of x. Although close, the graphs are not the same, and the equation is not an identity over the interval.

(B)
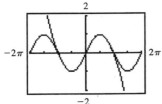

Outside the interval $[-\pi, \pi]$ the graphs differ widely.

63.   Graph both sides of the equation in the same viewing window.

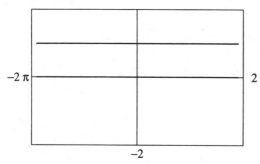

$$\frac{\cos x}{\sin(-x)\cot(-x)} = 1 \text{ appears to be an identity, which we verify.}$$

$$\frac{\cos x}{\sin(-x)\cot(-x)} = \frac{\cos x}{\sin(-x)\dfrac{\cos(-x)}{\sin(-x)}} \qquad \text{Quotient identity}$$

$$= \frac{\cos x}{\cos(-x)} \qquad \text{Algebra}$$

$$= \frac{\cos x}{\cos x} \qquad \text{Identities for negatives}$$

$$= 1 \qquad \text{Algebra}$$

65.   Graph both sides of the equation in the same viewing window.

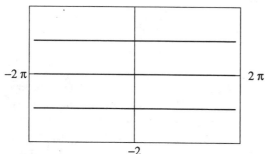

$$\frac{\cos(-x)}{\sin x \cot(-x)} = 1 \text{ is not an identity, since the graphs do not match. Try } x = \frac{\pi}{4}.$$

Left side: $\dfrac{\cos(-\pi/4)}{\sin(\pi/4)\cot(-\pi/4)} = \dfrac{\dfrac{1}{\sqrt{2}}}{\dfrac{1}{\sqrt{2}}(-1)} = -1$

Right side: 1

This verifies that the equation is not an identity.

67. Graph both sides of the equation in the same viewing window.

$$\frac{\cos x}{\sin x + 1} - \frac{\cos x}{\sin x - 1} = 2 \csc x$$

is not an identity, since the graphs do not match.

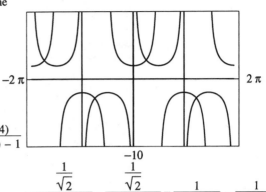

Try $x = -\frac{\pi}{4}$.

Left side: $\dfrac{\cos(-\pi/4)}{\sin(-\pi/4) + 1} - \dfrac{\cos(-\pi/4)}{\sin(-\pi/4) - 1}$

$$= \frac{\dfrac{1}{\sqrt{2}}}{-\dfrac{1}{\sqrt{2}} + 1} - \frac{\dfrac{1}{\sqrt{2}}}{-\dfrac{1}{\sqrt{2}} - 1} = \frac{1}{-1 + \sqrt{2}} - \frac{1}{-1 - \sqrt{2}}$$

$$= \frac{(-1 - \sqrt{2}) - (-1 + \sqrt{2})}{(-1 + \sqrt{2})(-1 - \sqrt{2})} = \frac{-2\sqrt{2}}{-1} = 2\sqrt{2}$$

Right side: $2 \csc\left(-\dfrac{\pi}{4}\right) = 2(-\sqrt{2}) = -2\sqrt{2}$. This verifies that the equation is not an identity.

69. Graph both sides of the equation in the same viewing window.

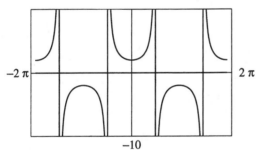

$\dfrac{\cos x}{1 - \sin x} + \dfrac{\cos x}{1 + \sin x} = 2 \sec x$ appears to be an identity, which we verify.

$$\frac{\cos x}{1 - \sin x} + \frac{\cos x}{1 + \sin x} = \frac{\cos x (1 + \sin x)}{(1 - \sin x)(1 + \sin x)} + \frac{\cos x (1 - \sin x)}{(1 + \sin x)(1 - \sin x)} \qquad \text{Algebra}$$

$$= \frac{\cos x (1 + \sin x) + \cos x (1 - \sin x)}{(1 - \sin x)(1 + \sin x)} \qquad \text{Algebra}$$

$$= \frac{\cos x + \cos x \sin x + \cos x - \cos x \sin x}{1 - \sin^2 x} \qquad \text{Algebra}$$

$$= \frac{2 \cos x}{1 - \sin^2 x} \qquad \text{Algebra}$$

$$= \frac{2 \cos x}{\cos^2 x} \qquad \text{Pythagorean identity}$$

$$= \frac{2}{\cos x} \qquad \text{Algebra}$$

$$= 2 \sec x \qquad \text{Reciprocal identity}$$

71. $\dfrac{\sin x}{1 - \cos x} - \cot x \ = \dfrac{\sin x}{1 - \cos x} - \dfrac{\cos x}{\sin x}$ 　　　Quotient identity

$= \dfrac{(\sin x)(\sin x)}{\sin x\,(1 - \cos x)} - \dfrac{(1 - \cos x)(\cos x)}{(1 - \cos x)(\sin x)}$ 　　Algebra

$= \dfrac{(\sin x)(\sin x) - (1 - \cos x)(\cos x)}{(1 - \cos x)\sin x}$ 　　Algebra

$= \dfrac{\sin^2 x - \cos x + \cos^2 x}{(1 - \cos x)\sin x}$ 　　Algebra

$= \dfrac{\sin^2 x + \cos^2 x - \cos x}{(1 - \cos x)\sin x}$ 　　Algebra

$= \dfrac{1 - \cos x}{(1 - \cos x)\sin x}$ 　　Pythagorean identity

$= \dfrac{1}{\sin x}$ 　　Algebra

$= \csc x$ 　　Reciprocal identity

Key algebraic steps: $\dfrac{a}{1 - b} - \dfrac{b}{a} \ = \dfrac{aa}{a(1 - b)} - \dfrac{b(1 - b)}{a(1 - b)} = \dfrac{a^2 - b(1 - b)}{a(1 - b)}$

$= \dfrac{a^2 - b + b^2}{a(1 - b)} = \dfrac{a^2 + b^2 - b}{a(1 - b)}$

73. $\dfrac{\cot \beta}{\csc \beta + 1} = \dfrac{\cot \beta\,(\csc \beta - 1)}{(\csc \beta + 1)(\csc \beta - 1)}$ 　　Algebra

$= \dfrac{\cot \beta\,(\csc \beta - 1)}{\csc^2 \beta - 1}$ 　　Algebra

$= \dfrac{\cot \beta\,(\csc \beta - 1)}{1 + \cot^2 \beta - 1}$ 　　Pythagorean identity

$= \dfrac{\cot \beta\,(\csc \beta - 1)}{\cot^2 \beta}$ 　　Algebra

$= \dfrac{\csc \beta - 1}{\cot \beta}$ 　　Algebra

75. $\dfrac{3 \cos^2 m + 5 \sin m - 5}{\cos^2 m} = \dfrac{3(1 - \sin^2 m) + 5 \sin m - 5}{1 - \sin^2 m}$ 　　Pythagorean identity

$= \dfrac{3 - 3 \sin^2 m + 5 \sin m - 5}{1 - \sin^2 m}$ 　　Algebra

$= \dfrac{-3 \sin^2 m + 5 \sin m - 2}{1 - \sin^2 m}$ 　　Algebra

$$= \frac{(-\sin m + 1)(3 \sin m - 2)}{(1 - \sin m)(1 + \sin m)} \qquad \text{Algebra}$$

$$= \frac{(1 - \sin m)(3 \sin m - 2)}{(1 - \sin m)(1 + \sin m)} \qquad \text{Algebra}$$

$$= \frac{3 \sin m - 2}{1 + \sin m} \qquad \text{Algebra}$$

77. In this problem, it is more straightforward to start with the right-hand side of the identity to be verified. The student can confirm that the steps would be valid if reversed.

$$\frac{\tan x + \tan y}{1 - \tan x \tan y} = \frac{\dfrac{\sin x}{\cos x} + \dfrac{\sin y}{\cos y}}{1 - \dfrac{\sin x \sin y}{\cos x \cos y}} \qquad \text{Quotient identity}$$

$$= \frac{\cos x \cos y \left(\dfrac{\sin x}{\cos x} + \dfrac{\sin y}{\cos y}\right)}{\cos x \cos y \left(1 - \dfrac{\sin x \sin y}{\cos x \cos y}\right)} \qquad \text{Algebra}$$

$$= \frac{\cos x \cos y \dfrac{\sin x}{\cos x} + \cos x \cos y \dfrac{\sin y}{\cos y}}{\cos x \cos y - \cos x \cos y \dfrac{\sin x \sin y}{\cos x \cos y}} \qquad \text{Algebra}$$

$$= \frac{\sin x \cos y + \cos x \sin y}{\cos x \cos y - \sin x \sin y} \qquad \text{Algebra}$$

EXERCISE 4.3 Sum, Difference, and Cofunction Identities

1. Use the sum identity for cosine, replacing y with 2π.
$$\cos(x + y) = \cos x \cos y - \sin x \sin y$$
$$\cos(x + 2\pi) = \cos x \cos 2\pi - \sin x \sin 2\pi = \cos x(1) - \sin x(0) = \cos x$$

3. Use the reciprocal identity together with the sum identity for tangent, replacing y with π.
$$\cot(x + \pi) = \frac{1}{\tan(x + \pi)} = \frac{1}{\dfrac{\tan x + \tan \pi}{1 - \tan x \tan \pi}} = \frac{1}{\dfrac{\tan x + 0}{1 - \tan x \cdot 0}} = \frac{1}{\tan x} = \cot x$$

5. Use the sum identity for sine, replacing y with $2k\pi$.
$$\sin(x + y) = \sin x \cos y + \cos x \sin y$$
$$\sin(x + 2k\pi) = \sin x \cos 2k\pi + \cos x \sin 2k\pi = \sin x(1) + \cos x(0) = \sin x$$

7. Use the sum identity for tangent, replacing y with $k\pi$.

$$\tan(x + y) = \frac{\tan x + \tan y}{1 - \tan x \tan y}$$

$$\tan(x + k\pi) = \frac{\tan x + \tan k\pi}{1 - \tan x \tan k\pi} = \frac{\tan x + 0}{1 - \tan x \cdot 0} = \tan x$$

9. $\tan\left(\dfrac{\pi}{2} - 2\right) = \dfrac{\sin\left(\dfrac{\pi}{2} - x\right)}{\cos\left(\dfrac{\pi}{2} - x\right)}$ Quotient identity

$\qquad\qquad = \dfrac{\cos x}{\sin x}$ Cofunction identities

$\qquad\qquad = \cot x$ Quotient identity

11. $\sec\left(\dfrac{\pi}{2} - x\right) = \dfrac{1}{\cos\left(\dfrac{\pi}{2} - x\right)}$ Reciprocal identity

$\qquad\qquad = \dfrac{1}{\sin x}$ Cofunction identity

$\qquad\qquad = \csc x$ Reciprocal identity

13. Use the difference identity for sine, replacing y with $45°$.

$$\sin(x - y) = \sin x \cos y - \cos x \sin y$$

$$\sin(x - 45°) = \sin x \cos 45° - \cos x \sin 45° = \sin x \frac{\sqrt{2}}{2} - \cos x \frac{\sqrt{2}}{2} = \frac{\sqrt{2}}{2}(\sin x - \cos x)$$

15. Use the sum identity for cosine, replacing y with $180°$.

$$\cos(x + y) = \cos x \cos y - \sin x \sin y$$

$$\cos(x + 180°) = \cos x \cos 180° - \sin x \sin 180° = \cos x(-1) - \sin x(0) = -\cos x$$

17. Use the difference identity for tangent, replacing x with $\dfrac{\pi}{4}$ and y with x.

$$\tan(x - y) = \frac{\tan x - \tan y}{1 + \tan x \tan y}$$

$$\tan\left(\frac{\pi}{4} - x\right) = \frac{\tan \dfrac{\pi}{4} - \tan x}{1 + \tan \dfrac{\pi}{4} \tan x} = \frac{1 - \tan x}{1 + 1 \cdot \tan x} = \frac{1 - \tan x}{1 + \tan x}$$

19. Since we can write $75° = 45° + 30°$, the sum of two special angles, we can use the sum identity for sine with $x = 45°$ and $y = 30°$.

$$\sin(x + y) = \sin x \cos y + \cos x \sin y$$

$$\sin(45° + 30°) = \sin 45° \cos 30° + \cos 45° \sin 30°$$

$$= \frac{1}{\sqrt{2}} \cdot \frac{\sqrt{3}}{2} + \frac{1}{\sqrt{2}} \cdot \frac{1}{2} = \frac{\sqrt{3}}{2\sqrt{2}} + \frac{1}{2\sqrt{2}} = \frac{\sqrt{3} + 1}{2\sqrt{2}}$$

21. Using the hint, we write $\dfrac{\pi}{12} = \dfrac{\pi}{4} - \dfrac{\pi}{6}$, the difference of two special angles, and use the difference identity for cosine with $x = \dfrac{\pi}{4}$ and $y = \dfrac{\pi}{6}$.

$$\cos(x - y) = \cos x \cos y + \sin x \sin y$$

$$\cos\left(\dfrac{\pi}{4} - \dfrac{\pi}{6}\right) = \cos\dfrac{\pi}{4}\cos\dfrac{\pi}{6} + \sin\dfrac{\pi}{4}\sin\dfrac{\pi}{6} = \dfrac{1}{\sqrt{2}} \cdot \dfrac{\sqrt{3}}{2} + \dfrac{1}{\sqrt{2}} \cdot \dfrac{1}{2} = \dfrac{\sqrt{3}}{2\sqrt{2}} + \dfrac{1}{2\sqrt{2}} = \dfrac{\sqrt{3} + 1}{2\sqrt{2}}$$

23. Since we can write $60° = 22° + 38°$, we can use the sum identity for sine with $x = 22°$ and $y = 38°$.

$$\sin x \cos y + \cos x \sin y = \sin(x + y)$$

$$\sin 22° \cos 38° + \cos 22° \sin 38° = \sin(22° + 38°) = \sin 60° = \dfrac{\sqrt{3}}{2}$$

25. Since we can write $60° = 110° - 50°$, we can use the difference identity for tangent with $x = 110°$ and $y = 50°$.

$$\dfrac{\tan x - \tan y}{1 + \tan x \tan y} = \tan(x - y) \quad \text{and} \quad \dfrac{\tan 110° - \tan 50°}{1 + \tan 110° \tan 50°} = \tan(110° - 50°) = \tan 60° = \sqrt{3}$$

27. To find $\sin(x - y)$, we start with the difference identity for sine:

$$\sin(x - y) = \sin x \cos y - \cos x \sin y.$$

We know $\sin x$ and $\cos y$, but not $\cos x$ and $\sin y$. We find the latter two values by using reference triangles and the Pythagorean theorem:

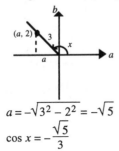

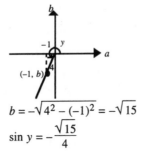

$$a = -\sqrt{3^2 - 2^2} = -\sqrt{5}$$

$$\cos x = -\dfrac{\sqrt{5}}{3}$$

$$b = -\sqrt{4^2 - (-1)^2} = -\sqrt{15}$$

$$\sin y = -\dfrac{\sqrt{15}}{4}$$

Thus, $\sin(x - y) = \sin x \cos y - \cos x \sin y$

$$= \dfrac{2}{3}\left(-\dfrac{1}{4}\right) - \left(-\dfrac{\sqrt{5}}{3}\right)\left(-\dfrac{\sqrt{15}}{4}\right) = \dfrac{-2}{12} - \dfrac{\sqrt{75}}{12} = \dfrac{-2 - \sqrt{75}}{12}$$

To find $\tan(x + y)$, we start with the sum identity for tangents: $\tan(x + y) = \dfrac{\tan x + \tan y}{1 - \tan x \tan y}$.

Since $\sin x = \dfrac{2}{3}$ and $\cos x = -\dfrac{\sqrt{5}}{3}$, we know $\tan x = \dfrac{\sin x}{\cos x} = \dfrac{2/3}{-\sqrt{5}/3} = -\dfrac{2}{\sqrt{5}}$. Since $\sin y = -\dfrac{\sqrt{15}}{4}$

and $\cos y = -\dfrac{1}{4}$, we know $\tan y = \dfrac{\sin y}{\cos y} = \dfrac{-\sqrt{15}/4}{-1/4} = \sqrt{15}$. Thus,

$$\tan(x + y) = \frac{\tan x + \tan y}{1 - \tan x \tan y} = \frac{-\dfrac{2}{\sqrt{5}} + \sqrt{15}}{1 - \left(-\dfrac{2}{\sqrt{5}}\right)\sqrt{15}} = \frac{\sqrt{5}\left(-\dfrac{2}{\sqrt{5}}\right) + \sqrt{5}\sqrt{15}}{\sqrt{5} - \sqrt{5}\left(-\dfrac{2}{\sqrt{5}}\right)\sqrt{15}} = \frac{-2 + \sqrt{75}}{\sqrt{5} + 2\sqrt{15}}$$

29. To find $\sin(x - y)$, we start with the difference identity for sine:

$\sin(x - y) = \sin x \cos y - \cos x \sin y$. We know $\cos x$ (and $\tan y$), but not $\sin x$, $\cos y$, or $\sin y$. We find the latter three by using reference triangles and the Pythagorean theorem:

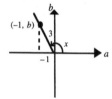   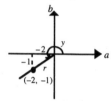

$b = \sqrt{3^2 - (-1)^2} = \sqrt{8}$; $\sin x = \dfrac{\sqrt{8}}{3}$ $r = \sqrt{(-2)^2 + (-1)^2} = \sqrt{5}$; $\sin y = \dfrac{-1}{\sqrt{5}}$; $\cos y = \dfrac{-2}{\sqrt{5}}$

Thus, $\sin(x - y) = \sin x \cos y - \cos x \sin y$

$$= \left(\frac{\sqrt{8}}{3}\right)\left(-\frac{2}{\sqrt{5}}\right) - \left(-\frac{1}{3}\right)\left(-\frac{1}{\sqrt{5}}\right) = \frac{-2\sqrt{8}}{3\sqrt{5}} - \frac{1}{3\sqrt{5}} = \frac{-2\sqrt{8} - 1}{3\sqrt{5}}$$

To find $\tan(x + y)$, we start with the sum identity for tangent: $\tan(x + y) = \dfrac{\tan x + \tan y}{1 - \tan x \tan y}$.

Since $\sin x = \dfrac{\sqrt{8}}{3}$, and $\cos x = -\dfrac{1}{3}$, we know $\tan x = \dfrac{\sin x}{\cos x} = \dfrac{\sqrt{8}/3}{-1/3} = -\sqrt{8}$. We also know that

$\tan y = \dfrac{1}{2}$. Thus,

$$\tan(x + y) = \frac{\tan x + \tan y}{1 - \tan x \tan y} = \frac{-\sqrt{8} + \dfrac{1}{2}}{1 - (-\sqrt{8})\left(\dfrac{1}{2}\right)}$$

$$= \frac{-2\sqrt{8} + 2\left(\dfrac{1}{2}\right)}{2 - 2(-\sqrt{8})\left(\dfrac{1}{2}\right)} = \frac{-2\sqrt{8} + 1}{2 + \sqrt{8}} = \frac{1 - 4\sqrt{2}}{2 + 2\sqrt{2}}$$

31. $\sin 2x = \sin(x + x)$ Algebra

$= \sin x \cos x + \cos x \sin x$ Sum identity for sine

$= \sin x \cos x + \sin x \cos x$ Algebra

$= 2 \sin x \cos x$ Algebra

33. $\cot(x - y) = \dfrac{\cos(x - y)}{\sin(x - y)}$ Quotient identity

$= \dfrac{\cos x \cos y + \sin x \sin y}{\sin x \cos y - \cos x \sin y}$ Difference identities for sine and cosine

$= \dfrac{\dfrac{\cos x \cos y}{\sin x \sin y} + \dfrac{\sin x \sin y}{\sin x \sin y}}{\dfrac{\sin x \cos y}{\sin x \sin y} - \dfrac{\cos x \sin y}{\sin x \sin y}}$ Algebra

$= \dfrac{\dfrac{\cos x \cos y}{\sin x \sin y} + 1}{\dfrac{\cos y}{\sin y} - \dfrac{\cos x}{\sin x}}$ Algebra

$= \dfrac{\cot x \cot y + 1}{\cot y - \cot x}$ Quotient identity

35. $\cot 2x = \dfrac{\cos 2x}{\sin 2x}$ Quotient identity

$= \dfrac{\cos(x + x)}{\sin(x + x)}$ Algebra

$= \dfrac{\cos x \cos x - \sin x \sin x}{\sin x \cos x + \cos x \sin x}$ Sum identities for sine and cosine

$= \dfrac{\cos x \cos x - \sin x \sin x}{2 \sin x \cos x}$ Algebra

$= \dfrac{\dfrac{\cos x \cos x}{\sin x \sin x} - \dfrac{\sin x \sin x}{\sin x \sin x}}{\dfrac{2 \sin x \cos x}{\sin x \sin x}}$ Algebra

$= \dfrac{\dfrac{\cos x \cos x}{\sin x \sin x} - 1}{2 \dfrac{\cos x}{\sin x}}$ Algebra

$= \dfrac{\cot x \cot x - 1}{2 \cot x}$ Quotient identity

$= \dfrac{\cot^2 x - 1}{2 \cot x}$ Algebra

37. $\dfrac{\tan\alpha + \tan\beta}{\tan\alpha - \tan\beta} = \dfrac{\dfrac{\sin\alpha}{\cos\alpha} + \dfrac{\sin\beta}{\cos\beta}}{\dfrac{\sin\alpha}{\cos\alpha} - \dfrac{\sin\beta}{\cos\beta}}$ 　　　　Quotient identity

$= \dfrac{\dfrac{\cos\alpha\cos\beta}{1}\cdot\dfrac{\sin\alpha}{\cos\alpha} + \dfrac{\cos\alpha\cos\beta}{1}\cdot\dfrac{\sin\beta}{\cos\beta}}{\dfrac{\cos\alpha\cos\beta}{1}\cdot\dfrac{\sin\alpha}{\cos\alpha} - \dfrac{\cos\alpha\cos\beta}{1}\cdot\dfrac{\sin\beta}{\cos\beta}}$ 　　Algebra

$= \dfrac{\sin\alpha\cos\beta + \cos\alpha\sin\beta}{\sin\alpha\cos\beta - \cos\alpha\sin\beta}$ 　　　　Algebra

$= \dfrac{\sin(\alpha+\beta)}{\sin(\alpha-\beta)}$ 　　　　Sum identity for sine

39. $\dfrac{\sin(x-y)}{\cos x\cos y} = \dfrac{\sin x\cos y - \cos x\sin y}{\cos x\cos y}$ 　　　　Difference identity for sine

$= \dfrac{\sin x\cos y}{\cos x\cos y} - \dfrac{\cos x\sin y}{\cos x\cos y}$ 　　　　Algebra

$= \dfrac{\sin x}{\cos x} - \dfrac{\sin y}{\cos y}$ 　　　　Algebra

$= \tan x - \tan y$ 　　　　Quotient identities

41. $\tan(x+y) = \dfrac{\sin(x+y)}{\cos(x+y)}$ 　　　　Quotient identity

$= \dfrac{\sin x\cos y + \cos x\sin y}{\cos x\cos y - \sin x\sin y}$ 　　　　Sum identities for sine and cosine

$= \dfrac{\dfrac{\sin x\cos y}{\sin x\sin y} + \dfrac{\cos x\sin y}{\sin x\sin y}}{\dfrac{\cos x\cos y}{\sin x\sin y} - \dfrac{\sin x\sin y}{\sin x\sin y}}$ 　　　　Algebra

$= \dfrac{\dfrac{\cos y}{\sin y} + \dfrac{\cos x}{\sin x}}{\dfrac{\cos x\cos y}{\sin x\sin y} - 1}$ 　　　　Algebra

$= \dfrac{\cot y + \cot x}{\cot y\cot x - 1}$ 　　　　Quotient identities

$= \dfrac{\cot x + \cot y}{\cot x\cot y - 1}$ 　　　　Algebra

43. $\dfrac{\sin(x+h) - \sin x}{h} = \dfrac{\sin x \cos h + \cos x \sin h - \sin x}{h}$ Sum identity for sine

$$= \dfrac{\sin x \cos h - \sin x + \cos x \sin h}{h}$$ Algebra

$$= \dfrac{\sin x(\cos h - 1) + \cos x \sin h}{h}$$ Algebra

$$= \sin x\dfrac{\cos h - 1}{h} + \cos x\dfrac{\sin h}{h}$$ Algebra

45. To show that $\csc(x - y) = \csc x - \csc y$ is not an identity, it is enough to find one pair of values x and y for which both sides are defined and the left side is not equal to the right side. For example, let $x = \dfrac{\pi}{2}$ and $y = \dfrac{\pi}{3}$, then

$$\csc(x - y) = \csc\left(\dfrac{\pi}{2} + \dfrac{\pi}{3}\right) = \csc\dfrac{\pi}{6} = 2$$

$$\csc x - \csc y = \csc\dfrac{\pi}{2} - \csc\dfrac{\pi}{3} = 1 - \dfrac{2}{\sqrt{3}}$$

47. Graph y1 = $\sin(x - 2)$ and y2 = $\sin x - \sin 2$ in the same viewing window and observe that the graphs are not the same:

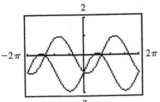

49. $\cos\left(x + \dfrac{5\pi}{6}\right) = \cos x \cos\dfrac{5\pi}{6} - \sin x \sin\dfrac{5\pi}{6}$ Sum identity for cosine

$$= \cos x\left(-\dfrac{\sqrt{3}}{2}\right) - \sin x\left(\dfrac{1}{2}\right)$$ Known values

$$= -\dfrac{\sqrt{3}}{2}\cos x - \dfrac{1}{2}\sin x$$

Graph y1 = $\cos\left(x + \dfrac{5\pi}{6}\right)$ and

y2 = $-\dfrac{\sqrt{3}}{2}\cos x - \dfrac{1}{2}\sin x$.

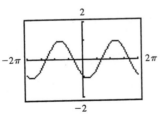

51. $\tan\left(x - \dfrac{\pi}{4}\right) = \dfrac{\tan x - \tan \dfrac{\pi}{4}}{1 + \tan x \tan \dfrac{\pi}{4}}$ Differerence identity for tangent

$= \dfrac{\tan x - 1}{1 + \tan x \cdot 1}$ Known values

$= \dfrac{\tan x - 1}{1 + \tan x}$

Graph y1 $= \tan\left(x - \dfrac{\pi}{4}\right)$ and

y2 $= \dfrac{\tan x - 1}{1 + \tan x}$.

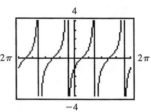

53. $\sin 3x \cos x - \cos 3x \sin x = \sin(3x - x)$ Difference identity for sine

$= \sin 2x$ Algebra

Graph y1 $= \sin 3x \cos x - \cos 3x \sin x$ and y2 $= \sin 2x$.

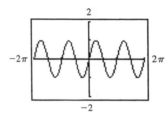

55. $\sin \dfrac{\pi x}{4} \cos \dfrac{3\pi x}{4} + \cos \dfrac{\pi x}{4} \sin \dfrac{3\pi x}{4} = \sin\left(\dfrac{\pi x}{4} + \dfrac{3\pi x}{4}\right)$ Sum identity for sine

$= \sin \pi x$ Algebra

Graph y1 $= \sin \dfrac{\pi x}{4} \cos \dfrac{3\pi x}{4} + \cos \dfrac{\pi x}{4} \sin \dfrac{3\pi x}{4}$ and y2 $= \sin \pi x$.

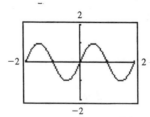

57. $\sin(x + y + z) = \sin[(x + y) + z]$ Algebra

$= \sin(x + y) \cos z + \cos(x + y) \sin z$ Sum identity for sine

$= (\sin x \cos y + \cos x \sin y) \cos z + (\cos x \cos y - \sin x \sin y) \sin z$ Sum identities for sine and cosine

$= \sin x \cos y \cos z + \cos x \sin y \cos z + \cos x \cos y \sin z - \sin x \sin y \sin z$ Algebra

59. $\tan(\theta_2 - \theta_1) = \dfrac{\tan\theta_2 - \tan\theta_1}{1 + \tan\theta_2 \tan\theta_1}$ Difference identity for tangent

$= \dfrac{m_2 - m_1}{1 + m_2 m_1}$ Given $m_1 = \tan\theta_1$ and $m_2 = \tan\theta_2$

$= \dfrac{m_2 - m_1}{1 + m_1 m_2}$ Algebra

61. Following the hint, label the text diagram as follows:

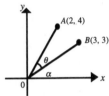

Then $\tan\theta = \dfrac{y_B}{x_B} = \dfrac{3}{3} = 1$ $\tan(\alpha + \theta) = \dfrac{y_A}{x_A} = \dfrac{4}{2} = 2$

$\tan\theta = \tan(\alpha + \theta - \alpha) = \tan[(\alpha + \theta) - \alpha]$

$= \dfrac{\tan(\alpha + \theta) - \tan\alpha}{1 + \tan(\alpha + \theta)\tan\alpha}$

$= \dfrac{2 - 1}{1 + 2 \cdot 1}$

$= \dfrac{1}{3}$

Therefore, $\theta = \tan^{-1}\dfrac{1}{3} = 0.322$ rad.

63. (A) In right triangle ABE, we have (1) $\cot\alpha = \dfrac{AB}{AE} = \dfrac{AB}{h}$

In right triangle BCD, we have (2) $\cot\alpha = \dfrac{BC}{CD} = \dfrac{BC}{H}$

In right triangle $EE'D$, we have (3) $\tan\beta = \dfrac{E'D}{EE'} = \dfrac{H - h}{AC} = \dfrac{H - h}{AB + BC}$

From (3), $H - h = (AB + BC)\tan\beta$. From (1) and (2), $AB = h\cot\alpha$ and $BC = H\cot\alpha$. Hence, substituting, we have (4) $H - h = (h\cot\alpha + H\cot\alpha)\tan\beta$,

or $= (h + H)\cot\alpha\tan\beta$

Solving (4) for H, we have $H - h = h\cot\alpha\tan\beta + H\cot\alpha\tan\beta$

$H - H\cot\alpha\tan\beta = h + h\cot\alpha\tan\beta$

$H(1 - \cot\alpha\tan\beta) = h(1 + \cot\alpha\tan\beta)$

$H = h\left(\dfrac{1 + \cot\alpha\tan\beta}{1 - \cot\alpha\tan\beta}\right)$

(B) Start with $H = h\left(\dfrac{1 + \cot \alpha \tan \beta}{1 - \cot \alpha \tan \beta}\right)$

Apply the quotient identities: $H = h\left(\dfrac{1 + \dfrac{\cos \alpha}{\sin \alpha} \dfrac{\sin \beta}{\cos \beta}}{1 - \dfrac{\cos \alpha}{\sin \alpha} \dfrac{\sin \beta}{\cos \beta}}\right)$

Reduce the complex fraction to a simple one: $H = h\left(\dfrac{\sin \alpha \cos \beta + \cos \alpha \sin \beta}{\sin \alpha \cos \beta - \cos \alpha \sin \beta}\right)$

Apply the sum and difference identities for sine: $H = h\left(\dfrac{\sin(\alpha + \beta)}{\sin(\alpha - \beta)}\right)$

(C) Substituting the given values, we have: $H = (5.50 \text{ ft})\dfrac{\sin(45.00° + 44.92°)}{\sin(45.00° - 44.92°)} = 3{,}940 \text{ ft}$

EXERCISE 4.4 Double-Angle and Half-Angle Identities

1. $\sin 2x = \sin(2 \cdot 60°) = \sin 120° = \dfrac{\sqrt{3}}{2}$; $2 \sin x \cos x = 2 \sin 60° \cos 60° = 2 \cdot \dfrac{\sqrt{3}}{2} \cdot \dfrac{1}{2} = \dfrac{\sqrt{3}}{2}$

3. $\tan 2x = \tan(2 \cdot 60°) = \tan 120° = -\sqrt{3}$;

$\dfrac{2 \tan x}{1 - \tan^2 x} = \dfrac{2 \tan 60°}{1 - (\tan^2 60°)} = \dfrac{2 \cdot \sqrt{3}}{1 - (\sqrt{3})^2} = \dfrac{2 \cdot \sqrt{3}}{1 - 3} = \dfrac{2\sqrt{3}}{-2} = -\sqrt{3}$

5. $\sin 105° = \sin \dfrac{210°}{2} = \sqrt{\dfrac{1 - \cos 210°}{2}}$

The positive sign is used since 105° is in the second quadrant and sine is positive there. We note that the reference triangle for 210° is a 30°–60° triangle in the third quadrant. Thus,

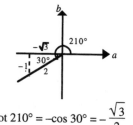

$\cot 210° = -\cos 30° = -\dfrac{\sqrt{3}}{2}$

$\sin 105° = \sqrt{\dfrac{1 - (-\sqrt{3}/2)}{2}} = \sqrt{\dfrac{2 + \sqrt{3}}{4}} = \dfrac{\sqrt{\sqrt{2} + \sqrt{3}}}{2}$

7. $\tan 15° = \tan \dfrac{30°}{2} = \dfrac{1 - \cos 30°}{\sin 30°} = \dfrac{1 - \sqrt{3}/2}{1/2} = \dfrac{2\left(1 - \dfrac{\sqrt{3}}{2}\right)}{2\left(\dfrac{1}{2}\right)} = \dfrac{2 - \sqrt{3}}{1} = 2 - \sqrt{3}$

9.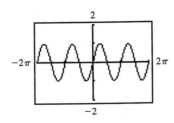

The graphs are identical, as can be seen from the trace function, or the double-angle identity for sine.

11.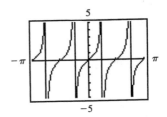

The graphs are identical, as can be seen from the trace function, or the double-angle identity for tangent.

13. $(\tan x)(1 + \cos 2x) = (\tan x)(1 + 2\cos^2 x - 1)$ Double-angle identity

$= (\tan x)\, 2\cos^2 x$ Algebra

$= \dfrac{\sin x}{\cos x}\, 2\cos^2 x$ Quotient identity

$= 2\sin x \cos x$ Algebra

$= \sin 2x$ Double-angle identity

15. $2\sin^2\dfrac{x}{2} = 2\left(\pm\sqrt{\dfrac{1-\cos x}{2}}\right)^2$ Half-angle identity

$= 2\left(\dfrac{1-\cos x}{2}\right)$ Algebra

$= \dfrac{1-\cos x}{1}$ Algebra

$= \dfrac{1-\cos x}{1}\cdot\dfrac{1+\cos x}{1+\cos x}$ Algebra

$= \dfrac{1-\cos^2 x}{1+\cos x}$ Algebra

$= \dfrac{\sin^2 x}{1+\cos x}$ Pythagorean identity

17. $(\sin\theta - \cos\theta)^2 = \sin^2\theta - 2\sin\theta\cos\theta + \cos^2\theta$ Algebra

$= \sin^2\theta + \cos^2\theta - 2\sin\theta\cos\theta$ Algebra

$= 1 - 2\sin\theta\cos\theta$ Pythagorean identity

$= 1 - \sin 2\theta$ Double-angle identity

19. $\cos^2\dfrac{w}{2} = \left(\pm\sqrt{\dfrac{1+\cos w}{2}}\right)^2$ Half-angle identity

$= \dfrac{1+\cos w}{2}$ Algebra

21. $\cot \dfrac{\alpha}{2} = \dfrac{1}{\tan \dfrac{\alpha}{2}}$ Reciprocal identity

$= \dfrac{1}{\dfrac{\sin \alpha}{1 + \cos \alpha}}$ Half-angle identity

$= \dfrac{1 + \cos \alpha}{\sin \alpha}$ Algebra

23. $\dfrac{\cos 2t}{1 - \sin 2t} = \dfrac{\cos^2 t - \sin^2 t}{1 - 2 \sin t \cos t}$ Double-angle identity

$= \dfrac{\cos^2 t - \sin^2 t}{\cos^2 t + \sin^2 t - 2 \sin t \cos t}$ Pythagorean identity

$= \dfrac{\cos^2 t - \sin^2 t}{\cos^2 t - 2 \sin t \cos t + \sin^2 t}$ Algebra

$= \dfrac{(\cos t - \sin t)(\cos t + \sin t)}{(\cos t - \sin t)(\cos t - \sin t)}$ Algebra

$= \dfrac{\cos t + \sin t}{\cos t - \sin t}$ Algebra

$= \dfrac{\dfrac{\cos t}{\cos t} + \dfrac{\sin t}{\cos t}}{\dfrac{\cos t}{\cos t} - \dfrac{\sin t}{\cos t}}$ Algebra

$= \dfrac{1 + \tan t}{1 - \tan t}$ Quotient identity

25. $\tan 2x = \tan(x + x)$ Algebra

$= \dfrac{\tan x + \tan x}{1 - \tan x \tan x}$ Sum identity for tangent

$= \dfrac{2 \tan x}{1 - \tan^2 x}$ Algebra

27. $\tan \dfrac{x}{2} = \dfrac{\sin \dfrac{x}{2}}{\cos \dfrac{x}{2}}$ Quotient identity

$= \dfrac{\pm\sqrt{\dfrac{1 - \cos x}{2}}}{\pm\sqrt{\dfrac{1 + \cos x}{2}}}$ Half-angle identities

$= \pm\sqrt{\dfrac{1 - \cos x}{1 + \cos x}}$ Algebra

$$\left| \tan \frac{x}{2} \right| = \sqrt{\frac{1 + \cos x}{1 + \cos x}} \qquad \text{Algebra}$$

$$= \sqrt{\frac{1 - \cos x}{1 + \cos x} \cdot \frac{1 + \cos x}{1 + \cos x}} \qquad \text{Algebra}$$

$$= \sqrt{\frac{1 - \cos^2 x}{(1 + \cos x)^2}} \qquad \text{Algebra}$$

$$= \sqrt{\frac{\sin^2 x}{(1 + \cos x)^2}} \qquad \text{Pythagorean identity}$$

$$= \left| \frac{\sin x}{1 + \cos x} \right| \qquad \text{Algebra}$$

Since $1 + \cos x \geq 0$ and $\sin x$ has the same sign as $\tan \frac{x}{2}$, we may drop the absolute value signs to obtain $\tan \frac{x}{2} = \dfrac{\sin x}{1 + \cos x}$

To show that $\sin x$ has the same sign as $\tan \frac{x}{2}$, we note the following cases:

If $0 < x < \pi$, $\sin x > 0$, then $0 < \frac{x}{2} < \frac{\pi}{2}$, $\tan \frac{x}{2} > 0$.

If $\pi < x < 2\pi$, $\sin x < 0$, then $\frac{\pi}{2} < \frac{x}{2} < \pi$, $\tan \frac{x}{2} < 0$.

The truth of the statement for other values of x follows since $\sin(x + 2k\pi) = \sin x$ and $\tan \dfrac{x + 2k\pi}{2}$ $= \tan \frac{x}{2}$ by the periodic properties of sine and tangent.

(Note: if $x = 0$ both sides of the proposed identity are 0, if $x = \pi$ both sides are meaningless.)

Alternative proof:

$$\tan \frac{x}{2} = \frac{\sin \frac{x}{2}}{\cos \frac{x}{2}} \qquad \text{Quotient identity}$$

$$= \frac{2 \sin \frac{x}{2} \cos \frac{x}{2}}{2 \cos \frac{x}{2} \cos \frac{x}{2}} \qquad \text{Algebra}$$

$$= \frac{2 \sin \frac{x}{2} \cos \frac{x}{2}}{2 \cos^2 \frac{x}{2}} \qquad \text{Algebra}$$

$$= \frac{2 \sin \frac{x}{2} \cos \frac{x}{2}}{1 + 2 \cos^2 \frac{x}{2} - 1} \qquad \text{Algebra}$$

$$= \frac{\sin 2\left(\frac{x}{2}\right)}{1 + \cos 2\left(\frac{x}{2}\right)}$$ Double-angle identities

$$= \frac{\sin x}{1 + \cos x}$$ Algebra

29. $(\sec 2x)(2 - \sec^2 x) = \frac{1}{\cos 2x}\left(2 - \frac{1}{\cos^2 x}\right)$ Reciprocal identity

$$= \frac{1}{\cos 2x}\left(\frac{2}{1} - \frac{1}{\cos^2 x}\right)$$ Algebra

$$= \frac{1}{\cos 2x}\left(\frac{2 \cos^2 x}{\cos^2 x} - \frac{1}{\cos^2 x}\right)$$ Algebra

$$= \frac{1}{\cos 2x}\left(\frac{2 \cos^2 x - 1}{\cos^2 x}\right)$$ Algebra

$$= \frac{1}{\cos 2x} \frac{\cos 2x}{\cos^2 x}$$ Double-angle identity

$$= \frac{1}{\cos^2 x}$$ Algebra

$$= \sec^2 x$$ Reciprocal identity

31. First draw a reference triangle in the second quadrant and find $\sin x$ and $\tan x$: $b = \sqrt{5^2 - (-4)^2} = 3$ $\sin x = \frac{3}{5}$; $\cos x = -\frac{4}{5}$; $\tan x = -\frac{3}{4}$. Now use the double-angle identities.

$$\sin 2x = 2 \sin x \cos x = 2\left(\frac{3}{5}\right)\left(-\frac{4}{5}\right) = -\frac{24}{25}$$

$$\cos 2x = 2 \cos^2 x - 1 = 2\left(-\frac{4}{5}\right)^2 - 1 = \frac{32}{25} - 1 = \frac{7}{25}$$

$$\tan 2x = \frac{2 \tan x}{1 - \tan^2 x} = \frac{2\left(-\frac{3}{4}\right)}{1 - \left(-\frac{3}{4}\right)^2} = \frac{-\frac{3}{2}}{1 - \left(\frac{9}{16}\right)} = -\frac{24}{7}$$

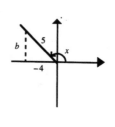

33. First draw a reference triangle in the fourth quadrant and find sin x, cos x, and tan x:

    $r = \sqrt{5^2 + (-12)^2} = 13$

    $\sin x = -\dfrac{12}{13}$; $\cos x = \dfrac{5}{13}$; $\tan x = -\dfrac{12}{5}$. Now use the double-angle identities.

    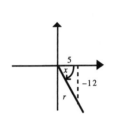

    $\sin 2x = 2 \sin x \cos x = 2\left(-\dfrac{12}{13}\right)\left(\dfrac{5}{13}\right) = -\dfrac{120}{169}$

    $\cos 2x = 1 - 2\sin^2 x = 1 - 2\left(-\dfrac{12}{13}\right)^2 = 1 - 2\left(\dfrac{144}{169}\right)$

    $\qquad = \dfrac{169}{169} - \dfrac{288}{169} = -\dfrac{119}{169}$

    $\tan 2x = \dfrac{2 \tan x}{1 - \tan^2 x} = \dfrac{2(-12/5)}{1 - (-12/5)^2} = \dfrac{-24/5}{1 - (144/25)} = \dfrac{25(-24/5)}{25 - 144} = \dfrac{120}{119}$

35. We are given cos x. We can find $\sin \dfrac{x}{2}$ and $\cos \dfrac{x}{2}$ from the half-angle identities, after determining their sign, as follows:

    If $0° < x < 90°$, then $0° < \dfrac{x}{2} < 45°$. Thus, $\dfrac{x}{2}$ is in the first quadrant, where sine and cosine are positive. Using half-angle identities, we obtain:

    $\sin \dfrac{x}{2} = \sqrt{\dfrac{1 - \cos x}{2}} = \sqrt{\dfrac{1 - \dfrac{1}{3}}{2}} = \sqrt{\dfrac{1}{3}} \text{ or } \dfrac{\sqrt{3}}{3}$

    $\cos \dfrac{x}{2} = \sqrt{\dfrac{1 + \cos x}{2}} = \sqrt{\dfrac{1 + \dfrac{1}{3}}{2}} = \sqrt{\dfrac{2}{3}} \text{ or } \dfrac{\sqrt{6}}{3}$

37. Draw a reference triangle in the third quadrant and find cos x. $a = -\sqrt{3^2 - (-1)^2} = -2\sqrt{2}$.

    $\cos x = \dfrac{-2\sqrt{2}}{3}$. If $\pi < x < \dfrac{3\pi}{2}$, then $\dfrac{\pi}{2} < \dfrac{x}{2} < \dfrac{3\pi}{4}$. Thus, $\dfrac{x}{2}$ is in the second quadrant, where sine is positive and cosine is negative. Using half-angle identities, we obtain:

    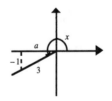

    $\sin \dfrac{x}{2} = \sqrt{\dfrac{1 - \cos x}{2}} = \sqrt{\dfrac{1 - (-2\sqrt{2}/3)}{2}} = \sqrt{\dfrac{3 + 2\sqrt{2}}{6}}$

    $\cos \dfrac{x}{2} = -\sqrt{\dfrac{1 + \cos x}{2}} = -\sqrt{\dfrac{1 + (-2\sqrt{2}/3)}{2}} = -\sqrt{\dfrac{3 - 2\sqrt{2}}{6}}$

39. (A) Since θ is a first quadrant angle and sec 2θ is negative for 2θ in the second quadrant and not for 2θ in the first, 2θ is a second quadrant angle.

    (B) Construct a reference triangle for 2θ in the second quadrant with $a = -4$ and $r = 5$. Use the Pythagorean theorem to find $b = 3$. Thus, $\sin 2\theta = \dfrac{3}{5}$ and $\cos 2\theta = -\dfrac{4}{5}$.

    (C) The double angle identities $\cos 2\theta = 1 - 2\sin^2 \theta$ and $\cos 2\theta = 2\cos^2 \theta - 1$.

(D) Use the identities in part (C) in the form

$$\sin \theta = \sqrt{\frac{1 - \cos 2\theta}{2}} \text{ and } \cos \theta = \sqrt{\frac{1 + \cos 2\theta}{2}}$$

The positive radicals are used because θ is in quadrant one.

(E) $\sin \theta = \sqrt{\dfrac{1 - \cos 2\theta}{2}} = \sqrt{\dfrac{1 - (-4/5)}{2}} = \sqrt{\dfrac{9}{10}} = \dfrac{3}{\sqrt{10}}$ or $\dfrac{3\sqrt{10}}{10}$

$\cos \theta = \sqrt{\dfrac{1 + \cos 2\theta}{2}} = \sqrt{\dfrac{1 + (-4/5)}{2}} = \sqrt{\dfrac{1}{10}} = \dfrac{1}{\sqrt{10}}$ or $\dfrac{\sqrt{10}}{10}$

41. (A)

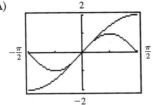

(B)

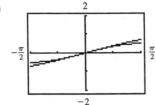

In each case, as x gets closer to zero, the two curves get closer together; the approximation improves.

43. $0 \le x \le 2\pi$

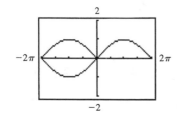

45. $-2\pi \le x \le -\pi/2,\ \pi/2 \le x \le 2\pi$

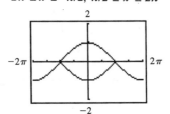

47. To obtain $\sin x$ and $\cos x$ from $\sin 2x$, we use the half-angle identities with x replaced by $2x$. Thus,

$\sin \dfrac{x}{2} = \pm\sqrt{\dfrac{1 - \cos x}{2}}$ becomes $\sin \dfrac{2x}{2} = \pm\sqrt{\dfrac{1 - \cos 2x}{2}}$ or $\sin x = \pm\sqrt{\dfrac{1 - \cos 2x}{2}}$

$\cos \dfrac{x}{2} = \pm\sqrt{\dfrac{1 + \cos x}{2}}$ becomes $\cos \dfrac{2x}{2} = \pm\sqrt{\dfrac{1 + \cos 2x}{2}}$ or $\cos x = \pm\sqrt{\dfrac{1 + \cos 2x}{2}}$

To obtain $\cos 2x$ from $\sin 2x$, we draw a reference triangle for $2x$ in the first quadrant.

$a = \sqrt{5^2 - 3^2} = 4,$

$\cos 2x = \dfrac{4}{5}.$

Since $0 < x < \dfrac{\pi}{4}$, $\sin x$ and $\cos x$ are positive.

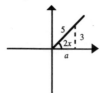

Thus,

$$\sin x = \sqrt{\frac{1 - \cos 2x}{2}} = \sqrt{\frac{1 - (4/5)}{2}} = \sqrt{\frac{1}{10}} \text{ or } \frac{1}{\sqrt{10}}$$

$$\cos x = \sqrt{\frac{1 + \cos 2x}{2}} = \sqrt{\frac{1 + (4/5)}{2}} = \sqrt{\frac{9}{10}} \text{ or } \frac{3}{\sqrt{10}};$$

$$\tan x = \frac{\sin x}{\cos x} = \frac{\dfrac{1}{\sqrt{10}}}{\dfrac{3}{\sqrt{10}}} = \frac{1}{3}$$

49. In Problem 47, we derived the identities:

$$\sin x = \pm\sqrt{\frac{1 - \cos 2x}{2}} \text{ and } \cos x = \pm\sqrt{\frac{1 + \cos 2x}{2}}$$

To obtain $\cos 2x$ from $\sec 2x$, we use the reciprocal identity:

$$\sec 2x = \frac{1}{\cos 2x} \text{ and } \cos 2x = \frac{1}{\sec 2x} = \frac{1}{(-5/3)} = -\frac{3}{5}$$

Since $-\dfrac{\pi}{2} < x < 0$, x is in the fourth quadrant, where sine is negative and cosine is positive. Thus,

$$\sin x = -\sqrt{\frac{1 - \cos 2x}{2}} = -\sqrt{\frac{1 - (-3/5)}{2}} = -\sqrt{\frac{4}{5}} \text{ or } -\frac{2}{\sqrt{5}}$$

$$\cos x = \sqrt{\frac{1 + \cos 2x}{2}} = \sqrt{\frac{1 + (-3/5)}{2}} = \sqrt{\frac{1}{5}} \text{ or } \frac{1}{\sqrt{5}}; \quad \tan x = \frac{\sin x}{\cos x} = \frac{-\dfrac{2}{\sqrt{5}}}{\dfrac{1}{\sqrt{5}}} = -2$$

51. $\sin 3x = \sin(2x + x)$ Algebra

     $= \sin 2x \cos x + \cos 2x \sin x$ Sum identity

     $= 2 \sin x \cos x \cos x + (1 - 2 \sin^2 x) \sin x$ Double-angle identities

     $= 2 \sin x \cos^2 x + \sin x - 2 \sin^3 x$ Algebra

     $= 2 \sin x(1 - \sin^2 x) + \sin x - 2 \sin^3 x$ Pythagorean identity

     $= 2 \sin x - 2 \sin^3 x + \sin x - 2 \sin^3 x$ Algebra

     $= 3 \sin x - 4 \sin^3 x$ Algebra

53. $\sin 4x = \sin 2(2x)$ Algebra

     $= 2 \sin 2x \cos 2x$ Double-angle identity

     $= 2(2 \sin x \cos x)(1 - 2 \sin^2 x)$ Double-angle identity

     $= \cos x(4 \sin x)(1 - 2 \sin^2 x)$ Algebra

     $= \cos x(4 \sin x - 8 \sin^3 x)$ Algebra

55. $\tan 3x = \tan(2x + x)$ Algebra

$$= \frac{\tan 2x + \tan x}{1 - \tan 2x \tan x}$$ Sum identity

$$= \frac{\dfrac{2 \tan x}{1 - \tan^2 x} + \tan x}{1 - \dfrac{2 \tan x}{1 - \tan^2 x} \tan x}$$ Double-angle identity

$$= \frac{(1 - \tan^2 x) \dfrac{2 \tan x}{1 - \tan^2 x} + (1 - \tan^2 x) \tan x}{(1 - \tan^2 x) \cdot 1 - (1 - \tan^2 x) \dfrac{2 \tan x}{1 - \tan^2 x} \tan x}$$ Algebra

$$= \frac{2 \tan x + (1 - \tan^2 x) \tan x}{1 - \tan^2 x - 2 \tan x \tan x}$$ Algebra

$$= \frac{2 \tan x + \tan x - \tan^3 x}{1 - \tan^2 x - 2 \tan^2 x}$$ Algebra

$$= \frac{3 \tan x - \tan^3 x}{1 - 3 \tan^2 x}$$ Algebra

57. The graph of $f(x)$ is shown in the figure. The graph appears to have vertical asymptotes $x = -2\pi$, $x = 0$, and $x = \pi$, x intercepts $-\pi$ and π, and period 2π. It appears that $g(x) = \cot \dfrac{x}{2}$ would be an appropriate choice. We verify $f(x) = g(x)$ as follows:

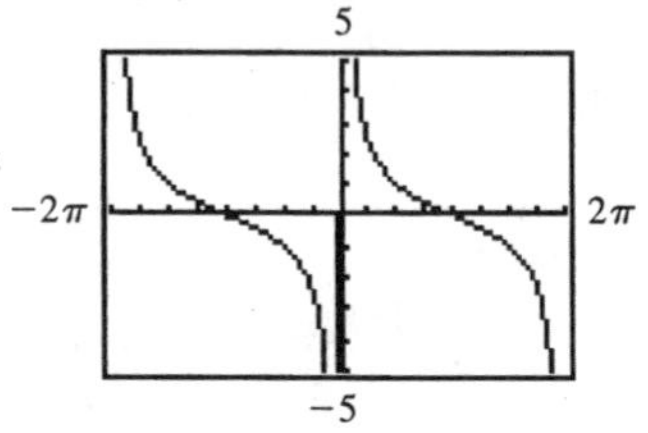

$f(x) = \csc x + \cot x$

$$= \frac{1}{\sin x} + \frac{\cos x}{\sin x}$$ Reciprocal and quotient identities

$$= \frac{1 + \cos x}{\sin x}$$ Algebra

$$= 1 + \frac{\sin x}{1 + \cos x}$$ Algebra

$$= 1 + \tan \frac{x}{2}$$ Half-angle identity

$$= \cot \frac{x}{2} = g(x)$$ Reciprocal identity

59. The graph of $f(x)$ is shown in the figure. The graph appears to have vertical asymptotes $x = -\pi, -\dfrac{\pi}{2}$, $0, \dfrac{\pi}{2}$, and π, and period π. It appears to have high and low points with y coordinate -1 and 1 respectively. It appears that $g(x) = \csc 2x$ would be an appropriate choice. We verify $f(x) = g(x)$ as follows:

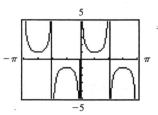

$$f(x) = \frac{\cot x}{1 + \cos 2x}$$

$$= \frac{\dfrac{\cos x}{\sin x}}{1 + \cos 2x} \qquad \text{Quotient identity}$$

$$= \frac{\dfrac{\cos x}{\sin x}}{1 + 2\cos^2 x - 1} \qquad \text{Double-angle identity}$$

$$= \frac{\cos x}{\sin x\,(2\cos^2 x)} \qquad \text{Algebra}$$

$$= \frac{1}{2\sin x \cos x} \qquad \text{Algebra}$$

$$= \frac{1}{\sin 2x} \qquad \text{Double-angle identity}$$

$$= \csc 2x = g(x) \qquad \text{Reciprocal identity}$$

61. The graph of $f(x)$ is shown in the figure. The graph appears to be a basic cosine curve with period 2π,

amplitude $= \frac{1}{2}(y\ \max - y\ \min) = \frac{1}{2}[1 - (-3)] = 2$, displaced downward by $|k| = 1$ unit. It appears that

$g(x) = 2\cos x - 1$ would be an appropriate choice. We verify $f(x) = g(x)$ as follows:

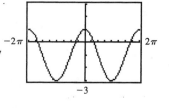

$$f(x) = \frac{1 + 2\cos 2x}{1 + 2\cos x}$$

$$= \frac{1 + 2(2\cos^2 x - 1)}{1 + 2\cos x} \qquad \text{Double-angle identity}$$

$$= \frac{1 + 4\cos^2 x - 2}{1 + 2\cos x} \qquad \text{Algebra}$$

$$= \frac{4\cos^2 x - 1}{2\cos x + 1} \qquad \text{Algebra}$$

$$= \frac{(2\cos x - 1)(2\cos x + 1)}{(2\cos x + 1)} \qquad \text{Algebra}$$

$$= 2\cos x - 1 = g(x) \qquad \text{Algebra}$$

63. For $n = 2$, the left side is $y1 = \frac{1}{2} + \cos x + \cos 2x$

and the right side is

$$y2 = \frac{\sin\left[\dfrac{2\cdot 2 + 1}{2}x\right]}{\sin\dfrac{1}{2}x} = \frac{\sin\dfrac{5}{2}x}{\sin\dfrac{x}{2}}.$$

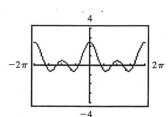

Since the identity holds, the graphs are identical.

65. (A) Since $2 \sin \theta \cos \theta = \sin 2\theta$ by the double-angle identity, we can write

$$d = \frac{2v_0^2 \sin \theta \cos \theta}{32 \text{ ft/sec}^2} = \frac{v_0^2 (2 \sin \theta \cos \theta)}{32 \text{ ft/sec}^2} = \frac{v_0^2 \sin 2\theta}{32 \text{ ft/sec}^2}$$

(B) Since v_0 is a given constant, d is maximum when $\sin 2\theta$ is maximum, and $\sin 2\theta$ is maximum when $2\theta = 90°$, that is, when $\theta = 45°$.

(C) Graph $d = \dfrac{100^2 \sin 2\theta}{32} = \dfrac{10,000 \sin 2\theta}{32}$.

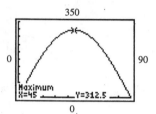

As θ increases from $0°$ to $90°$, d increases to a maximum of 312.5 feet when $\theta = 45°$, then decreases.

67. (A) Label the figure in the text as shown at the right. Then since triangle ABC is isosceles, the altitude BD bisects since AC and $AD = DC = \dfrac{b}{2}$.

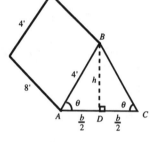

In triangle ABD,

$$\sin \theta = \frac{BD}{AB} = \frac{h}{4}, \text{ thus } h = 4 \sin \theta$$

$$\cos \theta = \frac{AD}{AB} = \frac{b/2}{4} = \frac{b}{8}, \text{ thus } b = 8 \cos \theta$$

Since the volume $V = 8 \cdot \dfrac{bh}{2}$, we have

$$V = 8 \cdot \frac{8 \cos \theta \cdot 4 \sin \theta}{2} = 128 \cos \theta \sin \theta$$

Hence, $V = 64 \cdot 2 \sin \theta \cos \theta = 64 \sin 2\theta$

(B) $\sin 2\theta$ has a maximum value of 1 when $2\theta = 90°$, that is, $\theta = 45°$. When $\theta = 45°$,
$V = 64 \sin 2(45°) = 64 \text{ ft}^3$

(C)

$\theta°$	30	35	40	45	50	55	60
V ft^3	55.4	60.1	63.0	64.0	63.0	60.1	55.4

(D)

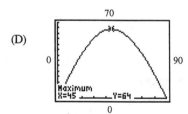

From the graphing utility, the maximum value of V is 64 when $\theta = 45°$.

69. We note that $\tan \theta = \dfrac{2}{x}$ and $\tan 2\theta = \dfrac{6}{x}$ (see figure). Using the hint, we have $\tan 2\theta = \dfrac{2 \tan \theta}{1 - \tan^2 \theta}$

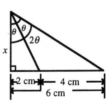

$$\frac{6}{x} = \frac{2\left(\dfrac{2}{x}\right)}{1 - \left(\dfrac{2}{x}\right)^2} = \frac{\dfrac{4}{x}}{1 - \left(\dfrac{4}{x^2}\right)} = \frac{x^2 \cdot \left(\dfrac{4}{x}\right)}{x^2 \cdot 1 - x^2 \cdot \left(\dfrac{4}{x^2}\right)} \qquad x \neq 0$$

$$\frac{6}{x} = \frac{4x}{x^2 - 4}$$

$$x(x^2 - 4) \cdot \frac{6}{x} = x(x^2 - 4) \cdot \frac{4x}{x^2 - 4} \qquad x \neq 2, -2$$

$$6(x^2 - 4) = x \cdot 4x$$
$$6x^2 - 24 = 4x^2$$
$$2x^2 = 24$$
$$x = \sqrt{12} \text{ or } 2\sqrt{3} \quad \text{(We discard the negative solution.)}$$

Then, $\dfrac{2}{x} = \tan \theta$, so $\tan \theta = \dfrac{2}{2\sqrt{3}} = \dfrac{1}{\sqrt{3}}$. To three decimal places $x \approx 3.464$ cm, $\theta = 30.000°$.

71. We label the figure as shown. From the Pythagorean theorem:

Since $AB = s$ and $MB = \dfrac{s}{2}$;

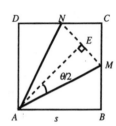

$$AM^2 = AB^2 + MB^2 = s^2 + \frac{s^2}{4} = \frac{5s^2}{4} \; ; \; AM = s\frac{\sqrt{5}}{2}$$

Since $CN = CM = \dfrac{s}{2}$;

$$NM^2 = CN^2 + CM^2 = \left(\frac{s}{2}\right)^2 + \left(\frac{s}{2}\right)^2 = \frac{2s^2}{4} \; ; \; NM = s\frac{\sqrt{2}}{2}$$

From the fact that $NA = MA$, thus triangle AMN is isosceles:

AE bisects NM, hence $ME = \dfrac{1}{2} MN = \dfrac{1}{2} \cdot s \dfrac{\sqrt{2}}{2} = s\dfrac{\sqrt{2}}{4}$.

From the definition of sine: $\sin \dfrac{\theta}{2} = \dfrac{ME}{MA} = s\dfrac{\sqrt{2}}{4} \div s\dfrac{\sqrt{5}}{2} = \dfrac{1}{2}\dfrac{\sqrt{2}}{\sqrt{5}}$

From the half-angle identity: $\sin \dfrac{\theta}{2} = \sqrt{\dfrac{1 - \cos \theta}{2}}$.

Hence, $\sqrt{\dfrac{1 - \cos \theta}{2}} = \dfrac{1}{2}\dfrac{\sqrt{2}}{\sqrt{5}}$; $\dfrac{1 - \cos \theta}{2} = \dfrac{1}{10}$; $1 - \cos \theta = \dfrac{1}{5}$; $\cos \theta = \dfrac{4}{5}$.

(The student may wish to compare with Exercise 1.4, Problem 37.)

EXERCISE 4.5   Product-Sum and Sum-Product Identities

1. $\cos x \cos y = \dfrac{1}{2}[\cos(x+y) + \cos(x-y)]$ Let $x = 7A$ and $y = 5A$.

$\cos 7A \cos 5A = \dfrac{1}{2}[\cos(7A + 5A) + \cos(7A - 5A)] = \dfrac{1}{2}(\cos 12A + \cos 2A) = \dfrac{1}{2}\cos 12A + \cos 2A$

3. $\cos x \sin y = \dfrac{1}{2}[\sin(x+y) - \sin(x-y)]$ Let $x = 2\theta$ and $y = 3\theta$.

$\cos 2\theta \sin 3\theta \;=\; \dfrac{1}{2}[\sin(2\theta + 3\theta) - \sin(2\theta - 3\theta)] = \dfrac{1}{2}[\sin 5\theta - \sin(-\theta)] = \dfrac{1}{2}(\sin 5\theta + \sin \theta)$

$\qquad\qquad\;\; = \dfrac{1}{2}\sin 5\theta + \dfrac{1}{2}\sin \theta$

5. $\cos x + \cos y = 2\cos\dfrac{x+y}{2}\cos\dfrac{x-y}{2}$ Let $x = 7\theta$ and $y = 5\theta$.

$\cos 7\theta + \cos 5\theta = 2\cos\dfrac{7\theta + 5\theta}{2}\cos\dfrac{7\theta - 5\theta}{2} = 2\cos 6\theta \cos \theta$

7. $\sin x - \sin y = 2\cos\dfrac{x+y}{2}\sin\dfrac{x-y}{2}$ Let $x = u$ and $y = 5u$.

$\sin u - \sin 5u = 2\cos\dfrac{u+5u}{2}\sin\dfrac{u-5u}{2} = 2\cos 3u \sin(-2u) = -2\cos 3u \sin 2u$

9. $\cos x \sin y = \dfrac{1}{2}[\sin(x+y) - \sin(x-y)]$ Let $x = 75°$ and $y = 15°$.

$\cos 75° \sin 15° \;=\; \dfrac{1}{2}[\sin(75° + 15°) - \sin(75° - 15°)] = \dfrac{1}{2}[\sin 90° - \sin 60°]$

$\qquad\qquad\;\; = \dfrac{1}{2}\left(1 - \dfrac{\sqrt{3}}{2}\right) = \dfrac{1}{2}\left(\dfrac{2 - \sqrt{3}}{2}\right) = \dfrac{2 - \sqrt{3}}{4}$

11. $\sin x \sin y = \dfrac{1}{2}[\cos(x-y) - \cos(x+y)]$ Let $x = 105°$ and $y = 165°$.

$\sin 105° \sin 165° \;=\; \dfrac{1}{2}[\cos(105° - 165°) - \cos(105° + 165°)]$

$\qquad\qquad\;\; = \dfrac{1}{2}[\cos(-60°) - \cos 270°] = \dfrac{1}{2}\left(\dfrac{1}{2} - 0\right) = \dfrac{1}{4}$

13. $\sin x + \sin y = 2\sin\dfrac{x+y}{2}\cos\dfrac{x-y}{2}$ Let $x = 195°$ and $y = 105°$.

$\sin 195° + \sin 105° \;=\; 2\sin\dfrac{195° - 105°}{2} = \cos\dfrac{195° - 105°}{2} = 2\sin 150° \cos 45°$

$\qquad\qquad\;\; = 2\left(\dfrac{1}{2}\right)\left(\dfrac{\sqrt{2}}{2}\right) = \dfrac{\sqrt{2}}{2}$

15. $\sin x - \sin y = 2 \cos \dfrac{x+y}{2} \sin \dfrac{x-y}{2}$ $\qquad$ Let $x = 75°$ and $y = 165°$.

$\sin 75° - \sin 165° = 2 \cos \dfrac{75° + 165°}{2} \sin \dfrac{75° - 165°}{2} = 2 \cos 120° \sin(-45°)$

$$= 2\left(-\dfrac{1}{2}\right)\left(-\dfrac{\sqrt{2}}{2}\right) = \dfrac{\sqrt{2}}{2}$$

17. $\qquad\qquad\quad \cos(x - y) = \cos x \cos y + \sin x \sin y$

$\underline{\qquad\qquad\quad \cos(x + y) = \cos x \cos y - \sin x \sin y}$

$\cos(x - y) - \cos(x + y) = 2 \sin x \sin y$ $\qquad$ subtracting the above

$\qquad\qquad \sin x \sin y = \dfrac{1}{2}[\cos(x - y) - \cos(x + y)]$

19. Let $x = u + v$ and $y = u - v$ and solve the resulting system for u and v in terms of x and y to obtain

$u = \dfrac{x+y}{2}$ and $v = \dfrac{x-y}{2}$. Substituting into the product-sum identity yields

$\qquad \cos \dfrac{x+y}{2} \cos \dfrac{x-y}{2} = \dfrac{1}{2}[\cos x + \cos y]$

or

$\qquad\qquad \cos x + \cos y = 2 \cos \dfrac{x+y}{2} \cos \dfrac{x-y}{2}$

21. $\dfrac{\cos t - \cos 3t}{\sin t + \sin 3t} = \dfrac{-2 \sin \dfrac{t + 3t}{2} \sin \dfrac{t - 3t}{2}}{2 \sin \dfrac{t + 3t}{2} \cos \dfrac{t - 3t}{2}}$ $\qquad$ Sum-product identities

$\qquad\qquad = \dfrac{-\sin 2t \sin(-t)}{\sin 2t \cos(-t)}$ $\qquad$ Algebra

$\qquad\qquad = \dfrac{\sin 2t \sin t}{\sin 2t \cos t}$ $\qquad$ Identities for negatives

$\qquad\qquad = \dfrac{\sin t}{\cos t}$ $\qquad$ Algebra

$\qquad\qquad = \tan t$ $\qquad$ Quotient identity

23. $\dfrac{\sin x + \sin y}{\cos x + \cos y} = \dfrac{2 \sin \dfrac{x + y}{2} \cos \dfrac{x - y}{2}}{2 \cos \dfrac{x + y}{2} \cos \dfrac{x - y}{2}}$ Sum-product identities

$$= \dfrac{\sin \dfrac{x + y}{2}}{\cos \dfrac{x + y}{2}}$$ Algebra

$$= \tan \dfrac{x + y}{2}$$ Quotient identity

25. $\dfrac{\cos x - \cos y}{\sin x + \sin y} = \dfrac{-2 \sin \dfrac{x + y}{2} \sin \dfrac{x - y}{2}}{2 \sin \dfrac{x + y}{2} \cos \dfrac{x - y}{2}}$ Sum-product identities

$$= \dfrac{-\sin \dfrac{x - y}{2}}{\cos \dfrac{x - y}{2}}$$ Algebra

$$= -\tan \dfrac{x - y}{2}$$ Quotient identity

27. $\dfrac{\sin x + \sin y}{\sin x - \sin y} = \dfrac{2 \sin \dfrac{x + y}{2} \cos \dfrac{x - y}{2}}{2 \cos \dfrac{x + y}{2} \sin \dfrac{x - y}{2}}$ Sum-product identities

$$= \dfrac{\sin \dfrac{x + y}{2} \cos \dfrac{x - y}{2}}{\cos \dfrac{x + y}{2} \sin \dfrac{x - y}{2}}$$ Algebra

$$= \tan \dfrac{x + y}{2} \cot \dfrac{x - y}{2}$$ Quotient identities

$$= \tan \dfrac{x + y}{2} \dfrac{1}{\tan \dfrac{x - y}{2}}$$ Reciprocal identity

$$= \dfrac{\tan \dfrac{x + y}{2}}{\tan \dfrac{x - y}{2}}$$ Algebra

29.
$$\cos u \cos v = \frac{1}{2}[\cos(u+v) + \cos(u-v)] \qquad \text{Let } u = 5x \text{ and } v = 3x.$$

$$\cos 5x \cos 3x = \frac{1}{2}[\cos(5x+3x) + \cos(5x-3x)] = \frac{1}{2}(\cos 8x + \cos 2x) = y2$$

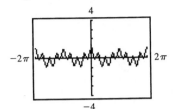

31.
$$\cos u \sin v = \frac{1}{2}[\sin(u+v) - \sin(u-v)] \qquad \text{Let } u = 1.9x \text{ and } v = 3.5x.$$

$$\cos 1.9x \sin 0.5x = \frac{1}{2}[\sin(1.9x+0.5x) - \sin(1.9x-0.5x)] = \frac{1}{2}(\sin 2.4x - \sin 1.4x) = y2$$

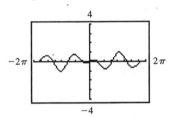

33.
$$\cos u + \cos v = 2\cos\frac{u+v}{2}\cos\frac{u-v}{2} \qquad \text{Let } u = 3x \text{ and } v = x.$$

$$\cos 3x + \cos x = 2\cos\frac{3x+x}{2}\cos\frac{3x-x}{2} = 2\cos 2x \cos x = y2$$

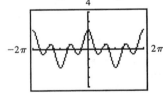

35.
$$\sin u - \sin y = 2\cos\frac{u+v}{2}\sin\frac{u-v}{2} \qquad \text{Let } u = 2.1x \text{ and } v = 0.5x.$$

$$\sin 2.1x - \sin 0.5x = 2\cos\frac{2.1x+0.5x}{2}\sin\frac{2.1x-0.5x}{2} = 2\cos 1.3x \sin 0.8x = y2$$

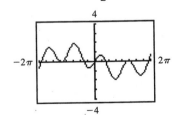

Chapter 4   Identities

37. $\sin x \sin y \sin z = \sin x \dfrac{1}{2}[\cos(y-z) - \cos(y+z)]$ Product-sum identity

$= \dfrac{1}{2}\sin x \cos(y-z) - \dfrac{1}{2}\sin x \cos(y+z)$ Algebra

$= \dfrac{1}{2}\left\{\dfrac{1}{2}[\sin(x+y-z) + \sin(x-\{y-z\})]\right\}$
$\quad - \dfrac{1}{2}\left\{\dfrac{1}{2}[\sin(x+y+z) + \sin(x-\{y+z\})]\right\}$ Product-sum identities

$= \dfrac{1}{4}\sin(x+y-z) + \dfrac{1}{4}\sin(x-y+z) - \dfrac{1}{4}\sin(x+y+z)$
$\quad - \dfrac{1}{4}\sin(x-y-z)$ Algebra

$= \dfrac{1}{4}[\sin(x+y-z) - \sin(x-y-z) + \sin(z+x-y)$
$\quad - \sin(x+y+z)]$ Algebra

$= \dfrac{1}{4}[\sin(x+y-z) + \sin\{-(x-y-z)\} + \sin(z+x-y)$
$\quad - \sin(x+y+z)]$ Identity for negatives

$= \dfrac{1}{4}[\sin(x+y-z) + \sin(y+z-x) + \sin(z+x-y)$
$\quad - \sin(x+y+z)]$ Algebra

39.   (A)

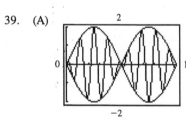

(B) $\cos u \sin v = \dfrac{1}{2}[\sin(u+v) - \sin(u-v)]$ Let $u = 16\pi x$ and $v = 2\pi x$.

$2\cos 16\pi x \sin 2\pi x = 2\left(\dfrac{1}{2}\right)[\sin(16\pi x + 2\pi x) - \sin(16\pi x - 2\pi x)]$

$= \sin(18\pi x) - \sin(14\pi x)$

The graph is the same as in part (A).

41.   (A)

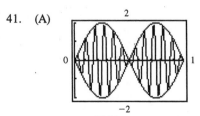

(B) $\sin u \sin v = \dfrac{1}{2}[\cos(u - v) - \cos(u + v)]$ Let $u = 24\pi x$ and $v = 2\pi x$.

$$2 \sin(24\pi x) \sin(2\pi x) = 2\left(\dfrac{1}{2}\right)[\cos(24\pi x - 2\pi x) - \cos(24\pi x + 2\pi x)]$$
$$= \cos(22\pi x) - \cos(26\pi x)$$

The graph is the same as in part (A).

43. The sum of the two tones is

$$y = k \sin 522\pi t + k \sin 512\pi t = k(\sin 522\pi t + \sin 512\pi t)$$
$$= k\left(2 \sin \dfrac{522\pi t + 512\pi t}{2} \cos \dfrac{522\pi t - 512\pi t}{2}\right) \quad \text{Sum-product identity}$$

This simplifies to $y = 2k \sin 517\pi t \cos 5\pi t$ \qquad Algebra

To find the beat frequency, we note

Period of first tone $= \dfrac{2\pi}{B_1} = \dfrac{2\pi}{522\pi} = \dfrac{1}{261}$

Frequency of first tone $= \dfrac{1}{\text{Period}} = \dfrac{1}{1/261} = 261$ Hz

Period of second tone $= \dfrac{2\pi}{B_2} = \dfrac{2\pi}{512\pi} = \dfrac{1}{256}$

Frequency of second tone $= \dfrac{1}{\text{Period}} = \dfrac{1}{1/256} = 256$ Hz

Beat frequency = Frequency of first tone – Frequency of second tone
$f_b = 261$ Hz $- 256$ Hz $= 5$ Hz

45. (A)

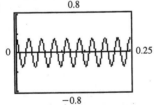

(B)

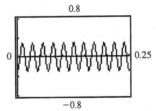

(C)

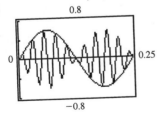

(D) $\cos x - \cos y = -2 \sin \dfrac{x+y}{2} \sin \dfrac{x-y}{2}$

Let $x = 72\pi t$ and $y = 88\pi t$

$0.3(\cos 72\pi t - \cos 88\pi t) = (0.3)(-2) \sin \dfrac{72\pi t + 88\pi t}{2} \sin \dfrac{72\pi t - 88\pi t}{2}$

$= -0.6 \sin 80\pi t \sin(-8\pi t)$

$= 0.6 \sin 80\pi t \sin 8\pi t$

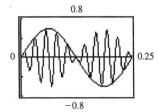

CHAPTER 4   REVIEW EXERCISE

1. Equation (1) is an identity, because it is true for all replacements of x by real numbers for which both sides are defined. Equation (2) is a conditional equation, because it is only true for $x = -2$ and $x = 3$, but not true for $x = 0$, for example.

2. $\csc x \sin x = \dfrac{1}{\sin x} \sin x$ Reciprocal identity

          $= 1$ Algebra

          $= \dfrac{1}{\cos x} \cos x$ Algebra

          $= \sec x \cos x$ Reciprocal identity

3. $\cot x \sin x = \dfrac{\cos x}{\sin x} \sin x$ Quotient identity

          $= \cos x$ Algebra

4. $\tan x = \dfrac{\sin x}{\cos x}$ Quotient identity

          $= \dfrac{-\sin(-x)}{\cos(-x)}$ Identities for negatives

          $= -\tan(-x)$ Quotient identity

5. $\dfrac{\sin^2 x}{\cos x} = \dfrac{1 - \cos^2 x}{\cos x}$ Pythagorean identity

          $= \dfrac{1}{\cos x} - \dfrac{\cos^2 x}{\cos x}$ Algebra

          $= \dfrac{1}{\cos x} - \cos x$ Algebra

          $= \sec x - \cos x$ Reciprocal identity

6. $\dfrac{\csc x}{\cos x} = \dfrac{\dfrac{1}{\sin x}}{\cos x}$ Reciprocal identity

$\qquad = \dfrac{1}{\sin x} \div \cos x$ Algebra

$\qquad = \dfrac{1}{\sin x} \cdot \dfrac{1}{\cos x}$ Algebra

$\qquad = \dfrac{1}{\sin x \cos x}$ Algebra

$\qquad = \dfrac{\sin^2 x + \cos^2 x}{\sin x \cos x}$ Pythagorean identity

$\qquad = \dfrac{\sin^2 x}{\sin x \cos x} + \dfrac{\cos^2 x}{\sin x \cos x}$ Algebra

$\qquad = \dfrac{\sin x}{\cos x} + \dfrac{\cos x}{\sin x}$ Algebra

$\qquad = \tan x + \cot x$ Quotient identities

7. $\cos^2 x\,(1 + \cot^2 x) = \cos^2 x \csc^2 x$ Pythagorean identity

$\qquad = \cos^2 x \dfrac{1}{\sin^2 x}$ Reciprocal identity

$\qquad = \dfrac{\cos^2 x}{\sin^2 x}$ Algebra

$\qquad = \cot^2 x$ Quotient identity

8. $\dfrac{\sin \alpha \csc \alpha}{\cot \alpha} = \dfrac{\sin \alpha \cdot \dfrac{1}{\sin \alpha}}{\cot \alpha}$ Reciprocal identity

$\qquad = \dfrac{1}{\cot \alpha}$ Algebra

$\qquad = \tan \alpha$ Reciprocal identity

9. $\dfrac{\sin^2 u - \cos^2 u}{\sin u \cos u} = \dfrac{\sin^2 u}{\sin u \cos u} - \dfrac{\cos^2 u}{\sin u \cos u}$ Algebra

$\qquad = \dfrac{\sin u}{\cos u} - \dfrac{\cos u}{\sin u}$ Algebra

$\qquad = \tan u - \cot u$ Quotient identities

10. $\dfrac{\sec \theta - \csc \theta}{\sec \theta \csc \theta} = \dfrac{\sec \theta}{\sec \theta \csc \theta} - \dfrac{\csc \theta}{\sec \theta \csc \theta}$ Algebra

$\qquad = \dfrac{1}{\csc \theta} - \dfrac{1}{\sec \theta}$ Algebra

$\qquad = \sin \theta - \cos \theta$ Reciprocal identities

Chapter 4   Identities

11. $\cos(x + y) = \cos x \cos y - \sin x \sin y$

$\quad\quad \cos(x + 2\pi) = \cos x \cos 2\pi - \sin x \sin 2\pi$

$\quad\quad\quad\quad\quad\quad = \cos x(1) - \sin x(0)$

$\quad\quad\quad\quad\quad\quad = \cos x$

12. $\sin(x + y) = \sin x \cos y + \cos x \sin y$

$\quad\quad \sin(x + y) = \sin x \cos \pi + \cos x \sin \pi$

$\quad\quad\quad\quad\quad\quad = \sin x(-1) + \cos x(0)$

$\quad\quad\quad\quad\quad\quad = -\sin x$

13. $\cos 2x = \cos 2(30°) = \cos 60° = \dfrac{1}{2}$

$\quad\quad 1 - 2 \sin^2 x = 1 - 2 \sin^2 (30°)$

$\quad\quad\quad\quad\quad\quad = 1 - 2(\sin 30°)^2$

$\quad\quad\quad\quad\quad\quad = 1 - 2\left(\dfrac{1}{2}\right)^2 = 1 - \dfrac{1}{2}$

$\quad\quad\quad\quad\quad\quad = \dfrac{1}{2}$

14. $\sin \dfrac{x}{2} = \sin \dfrac{\pi/2}{2} = \sin \dfrac{\pi}{4} = \dfrac{1}{\sqrt{2}}$

Since $\dfrac{\pi}{4}$ is in the first quadrant, the sign of the square root is chosen to be positive.

$$\sqrt{\dfrac{1 - \cos x}{2}} = \sqrt{\dfrac{1 - \cos \pi/2}{2}} = \sqrt{\dfrac{1 - 0}{2}} = \sqrt{\dfrac{1}{2}} = \dfrac{1}{\sqrt{2}}$$

15. $\sin x \sin y = \dfrac{1}{2}[\cos(x - y) - \cos(x + y)]$ Let $x = 8t$ and $y = 5t$

$\quad\quad \sin 8t \sin 5t = \dfrac{1}{2}[\cos(8t - 5t) - \cos(8t + 5t)] = \dfrac{1}{2}(\cos 3t - \cos 13t) = \dfrac{1}{2}\cos 3t - \dfrac{1}{2}\cos 13t$

16. $\sin x + \sin y = 2 \sin\dfrac{x + y}{2} \cos\dfrac{x - y}{2}$ Let $x = w$ and $y = 5w$

$\quad\quad \sin w + \sin 5w = 2 \sin\dfrac{w + 5w}{2} \cos\dfrac{w - 5w}{2} = 2 \sin 3w \cos(-2w) = 2 \sin 3w \cos 2w$

17. $\dfrac{1 - \cos^2 t}{\sin^3 t} = \dfrac{\sin^2 t}{\sin^3 t}$ Pythagorean identity

$\quad\quad\quad\quad = \dfrac{1}{\sin t}$ Algebra

$\quad\quad\quad\quad = \csc t$ Reciprocal identity

18. $$\frac{(\cos \alpha - 1)^2}{\sin^2 \alpha} = \frac{(\cos \alpha - 1)^2}{1 - \cos^2 \alpha} \qquad \text{Pythagorean identity}$$

$$= \frac{(\cos \alpha - 1)(\cos \alpha - 1)}{(1 - \cos \alpha)(1 + \cos \alpha)} \qquad \text{Algebra}$$

$$= \frac{(-1)(\cos \alpha - 1)}{1 + \cos \alpha} \qquad \text{Algebra}$$

$$= \frac{1 - \cos \alpha}{1 + \cos \alpha} \qquad \text{Algebra}$$

Key algebraic steps: $\dfrac{(b-1)^2}{1-b^2} = \dfrac{(b-1)(b-1)}{(1-b)(1+b)} = \dfrac{(-1)(b-1)}{1+b} = \dfrac{1-b}{1+b}$

19. $$\frac{1 - \tan^2 x}{1 - \tan^4 x} = \frac{1 - \tan^2 x}{(1)^2 - (\tan^2 x)^2} \qquad \text{Algebra}$$

$$= \frac{1 - \tan^2 x}{(1 - \tan^2 x)(1 + \tan^2 x)} \qquad \text{Algebra}$$

$$= \frac{1}{1 + \tan^2 x} \qquad \text{Algebra}$$

$$= \frac{1}{\sec^2 x} \qquad \text{Pythagorean identity}$$

$$= \left(\frac{1}{\sec x}\right)^2 \qquad \text{Algebra}$$

$$= \cos^2 x \qquad \text{Reciprocal identity}$$

Key algebraic steps: $\dfrac{1-c^2}{1-c^4} = \dfrac{1-c^2}{(1)^2-(c^2)^2} = \dfrac{1-c^2}{(1-c^2)(1+c^2)} = \dfrac{1}{1+c^2}$

20. $$\cot^2 x \cos^2 x = (\csc^2 x - 1)\cos^2 x \qquad \text{Pythagorean identity}$$

$$= \csc^2 x \cos^2 x - \cos^2 x \qquad \text{Algebra}$$

$$= \left(\frac{1}{\sin x}\right)^2 \cos^2 x - \cos^2 x \qquad \text{Reciprocal identity}$$

$$= \left(\frac{\cos x}{\sin x}\right)^2 - \cos^2 x \qquad \text{Algebra}$$

$$= \cot^2 x - \cos^2 x \qquad \text{Quotient identity}$$

21. The equation is not an identity. For example, let $x = \dfrac{\pi}{2}$. Then both sides are defined, however, the left side is $\sin \dfrac{\pi}{2}$ or 1 while the right side is 0, hence the equation is not true for all values of x for which both sides are defined.

22. Graph each side of the equation in the same viewing window and observe that the graphs are not the same, except where the graph of $y1 = \sin x$ crosses the x axis. (Note that the graph of $y2 = 0$ is the x axis.)

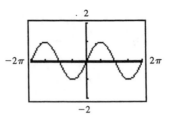

23. (A) From the identities for negatives,
$$\cos(-x) = \cos x$$
Hence
$$\cos(-x) = 0.9394$$

    (B) From the Pythagorean identity, $\cos^2 x + \sin^2 x = 1$, thus
$$\sin^2 x = 1 - \cos^2 x$$
Hence
$$(\sin x)^2 = 1 - 0.8824 = 0.1176$$

24. No. The equation is an identity for all real values of x, except for $x = k\pi$, k an integer (neither side of the equation is defined for these values).

25. Graph each side in the same viewing window and observe that the graphs are not the same.

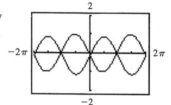

26. Find a value of x for which both sides are defined but are not equal. Try $x = 1$, for example: $\sin(1 - 3) = -0.9093$ and $\sin(1) - \sin(3) = 0.7004$.

27.
$$\frac{\sin x}{1 - \cos x} = \frac{\sin x}{1 - \cos x}\frac{1 + \cos x}{1 + \cos x} \qquad \text{Algebra}$$

$$= \frac{\sin x\,(1 + \cos x)}{1 - \cos^2 x} \qquad \text{Algebra}$$

$$= \frac{\sin x\,(1 + \cos x)}{\sin^2 x} \qquad \text{Pythagorean identity}$$

$$= \frac{\sin x}{\sin^2 x}(1 + \cos x) \qquad \text{Algebra}$$

$$= \frac{1}{\sin x}(1 + \cos x) \qquad \text{Algebra}$$

$$= \csc x\,(1 + \cos x) \qquad \text{Reciprocal identity}$$

162

28. $\dfrac{1 - \tan^2 x}{1 - \cot^2 x} = \dfrac{1 - \tan^2 x}{1 - \dfrac{1}{\tan^2 x}}$ Reciprocal identity

$= \dfrac{\tan^2 x\,(1 - \tan^2 x)}{\tan^2 x\left(1 - \dfrac{1}{\tan^2 x}\right)}$ Algebra

$= \dfrac{\tan^2 x\,(1 - \tan^2 x)}{\tan^2 x - 1}$ Algebra

$= \dfrac{-\tan^2 x\,(\tan^2 x - 1)}{\tan^2 x - 1}$ Algebra

$= -\tan^2 x$ Algebra

$= -(\sec^2 x - 1)$ Pythagorean identity

$= 1 - \sec^2 x$ Algebra

Key algebraic steps: $\dfrac{1 - a^2}{1 - \dfrac{1}{a^2}} = \dfrac{a^2(1 - a^2)}{a^2\left(1 - \dfrac{1}{a^2}\right)} = \dfrac{a^2(1 - a^2)}{a^2 - 1} = \dfrac{-a^2(a^2 - 1)}{a^2 - 1} = -a^2$

29. $\tan(x + \pi) = \dfrac{\tan x + \tan \pi}{1 - \tan x \tan \pi}$ Sum identity

$= \dfrac{\tan x + 0}{1 - \tan x \cdot 0}$ Known values

$= \tan x$ Algebra

30. $1 - (\cos \beta - \sin \beta)^2 = 1 - (\cos^2 \beta - 2 \sin \beta \cos \beta + \sin^2 \beta)$ Algebra

$= 1 - \cos^2 \beta + \sin^2 \beta - 2 \sin \beta \cos \beta)$ Algebra

$= 1 - (1 - 2 \sin \beta \cos \beta)$ Pythagorean identity

$= 1 - 1 + 2 \sin \beta \cos \beta$ Algebra

$= 2 \sin \beta \cos \beta$ Algebra

$= \sin 2\beta$ Double-angle identity

31. $\dfrac{\sin 2x}{\cot x} = \dfrac{2 \sin x \cos x}{\cot x}$ Double-angle identity

$= \dfrac{2 \sin x \cos x}{\dfrac{\cos x}{\sin x}}$ Quotient identity

$= 2 \sin x \cos x + \dfrac{\cos x}{\sin x}$ Algebra

$= 2 \sin x \cos x \cdot \dfrac{\sin x}{\cos x}$ Algebra

$= 2 \sin^2 x$ Algebra

$= 2 \sin^2 x - 1 + 1$ Algebra

$= 1 + (2 \sin^2 x - 1)$ Algebra

$= 1 - (1 - 2 \sin^2 x)$ Algebra

$= 1 - \cos 2x$ Double-angle identity

32. $\dfrac{2 \tan x}{1 + \tan^2 x} = \dfrac{2 \dfrac{\sin x}{\cos x}}{1 + \dfrac{\sin^2 x}{\cos^2 x}}$ Quotient identity

$$= \dfrac{\cos^2 x \cdot 2 \dfrac{\sin x}{\cos x}}{\cos^2 x \cdot 1 + \cos^2 x \cdot \dfrac{\sin^2 x}{\cos^2 x}}$$ Algebra

$$= \dfrac{2 \sin x \cos x}{\cos^2 x + \sin^2 x}$$ Algebra

$$= \dfrac{2 \sin x \cos x}{1}$$ Pythagorean identity

$$= 2 \sin x \cos x$$ Algebra

$$= \sin 2x$$ Double-angle identity

33. $2 \csc 2x = \dfrac{2}{\sin 2x}$ Reciprocal identity

$$= \dfrac{2}{2 \sin x \cos x}$$ Double-angle identity

$$= \dfrac{1}{\sin x \cos x}$$ Algebra

$$= \dfrac{\sin^2 x + \cos^2 x}{\sin x \cos x}$$ Pythagorean identity

$$= \dfrac{\sin^2 x}{\sin x \cos x} + \dfrac{\cos^2 x}{\sin x \cos x}$$ Algebra

$$= \dfrac{\sin x}{\cos x} + \dfrac{\cos x}{\sin x}$$ Algebra

$$= \tan x + \cot x$$ Quotient identities

34. $\dfrac{\cot \dfrac{x}{2}}{1 + \cos x} = \cot \dfrac{x}{2} \cdot \dfrac{1}{1 + \cos x}$ Algebra

$$= \dfrac{1}{\tan \dfrac{x}{2}} \cdot \dfrac{1}{1 + \cos x}$$ Reciprocal identity

$$= \dfrac{1}{\dfrac{\sin x}{1 + \cos x}} \cdot \dfrac{1}{1 + \cos x}$$ Half-angle identity

$$= \dfrac{1 + \cos x}{\sin x} \cdot \dfrac{1}{1 + \cos x}$$ Algebra

$$= \dfrac{1}{\sin x}$$ Algebra

$$= \csc x$$ Reciprocal identity

35. $\dfrac{\sin(x-y)}{\sin(x+y)} = \dfrac{\sin x \cos y - \cos x \sin y}{\sin x \cos y + \cos x \sin y}$ Sum and difference identities

$$= \dfrac{\dfrac{\sin x \cos y}{\cos x \cos y} - \dfrac{\cos x \sin y}{\cos x \cos y}}{\dfrac{\sin x \cos y}{\cos x \cos y} + \dfrac{\cos x \sin y}{\cos x \cos y}}$$ Algebra

$$= \dfrac{\dfrac{\sin x}{\cos x} - \dfrac{\sin y}{\cos y}}{\dfrac{\sin x}{\cos x} + \dfrac{\sin y}{\cos y}}$$ Algebra

$$= \dfrac{\tan x - \tan y}{\tan x + \tan y}$$ Quotient identity

36. $\csc 2x = \dfrac{1}{\sin 2x}$ Reciprocal identity

$$= \dfrac{1}{2 \sin x \cos x}$$ Double-angle identity

$$= \dfrac{\sin^2 x + \cos^2 x}{2 \sin x \cos x}$$ Pythagorean identity

$$= \dfrac{\sin^2 x}{2 \sin x \cos x} + \dfrac{\cos^2 x}{2 \sin x \cos x}$$ Algebra

$$= \dfrac{\sin x}{2 \cos x} + \dfrac{\cos x}{2 \sin x}$$ Algebra

$$= \dfrac{1}{2}\dfrac{\sin x}{\cos x} + \dfrac{1}{2}\dfrac{\cos x}{\sin x}$$ Algebra

$$= \dfrac{1}{2} \tan x + \dfrac{1}{2} \cot x$$ Quotient identities

$$= \dfrac{\tan x + \cot x}{2}$$ Algebra

37. $\dfrac{2 - \sec^2 x}{\sec^2 x} = \dfrac{1 - \dfrac{1}{\cos^2 x}}{\dfrac{1}{\cos^2 x}}$ Reciprocal identity

$$= \dfrac{\cos^2 x \cdot 2 - \cos^2 x \cdot \dfrac{1}{\cos^2 x}}{\cos^2 x \cdot \dfrac{1}{\cos^2 x}}$$ Algebra

$$= \dfrac{2 \cos^2 x - 1}{1}$$ Algebra

$$= 2 \cos^2 x - 1$$ Algebra

$$= \cos 2x$$ Double-angle identity

38. $\tan \dfrac{x}{2} = \dfrac{1 - \cos x}{\sin x}$ Half-angle identity

$$= \dfrac{\dfrac{1}{\cos x} - \dfrac{\cos x}{\cos x}}{\dfrac{\sin x}{\cos x}}$$ Algebra

$$= \dfrac{\sec x - \dfrac{\cos x}{\cos x}}{\dfrac{\sin x}{\cos x}}$$ Reciprocal identity

$$= \dfrac{\sec x - 1}{\dfrac{\sin x}{\cos x}}$$ Algebra

$$= \dfrac{\sec x - 1}{\tan x}$$ Quotient identity

39. $\dfrac{\sin t + \sin 5t}{\cos t + \cos 5t} = \dfrac{2 \sin \dfrac{t + 5t}{2} \cos \dfrac{t - 5t}{2}}{2 \cos \dfrac{t + 5t}{2} \cos \dfrac{t - 5t}{2}}$ Sum-product identities

$$= \dfrac{2 \sin 3t \cos(-2t)}{2 \cos 3t \cos(-2t)}$$ Algebra

$$= \dfrac{\sin 3t}{\cos 3t}$$ Algebra

$$= \tan 3t$$ Quotient identity

40. $\dfrac{\sin x + \sin y}{\cos x - \cos y} = \dfrac{2 \sin \dfrac{x + y}{2} \cos \dfrac{x - y}{2}}{-2 \sin \dfrac{x + y}{2} \sin \dfrac{x - y}{2}}$ Sum-product identities

$$= -\dfrac{\cos \dfrac{x - y}{2}}{\sin \dfrac{x - y}{2}}$$ Algebra

$$= -\cot \dfrac{x - y}{2}$$ Quotient identity

41. $\dfrac{\cos x - \cos y}{\cos x + \cos y} = \dfrac{-2 \sin \dfrac{x + y}{2} \sin \dfrac{x - y}{2}}{2 \cos \dfrac{x + y}{2} \cos \dfrac{x - y}{2}}$ Sum-product identities

$$= -\dfrac{\sin \dfrac{x + y}{2}}{\cos \dfrac{x + y}{2}} \dfrac{\sin \dfrac{x - y}{2}}{\cos \dfrac{x - y}{2}}$$ Algebra

$$= -\tan \dfrac{x + y}{2} \tan \dfrac{x - y}{2}$$ Quotient identity

42. $\sin x \sin y = \dfrac{1}{2}[\cos(x-y) - \cos(x+y)]$ Let $x = 165°$ and $y = 15°$.

$\sin 165° \sin 15° = \dfrac{1}{2}[\cos(165° - 15°) - \cos(165° + 15°)] = \dfrac{1}{2}[\cos 150° - \cos 180°]$

$= \dfrac{1}{2}\left[-\dfrac{\sqrt{3}}{2} - (-1)\right] = -\dfrac{\sqrt{3}}{4} + \dfrac{1}{2}$ or $\dfrac{1}{2} - \dfrac{\sqrt{3}}{4}$

43. $\cos x - \cos y = -2 \sin\dfrac{x+y}{2} \sin\dfrac{x-y}{2}$ Let $x = 165°$ and $y = 75°$.

$\cos 165° - \cos 75° = -2 \sin\dfrac{165° + 75°}{2} \sin\dfrac{165° - 75°}{2} = -2 \sin 120° \sin 45°$

$= -2\left(\dfrac{\sqrt{3}}{2}\right)\left(\dfrac{\sqrt{2}}{2}\right) = -\dfrac{\sqrt{6}}{2}$

44.   *Find sin x:* We start with the Pythagorean identity $\sin^2 x + \cos^2 x = 1$ and solve for $\sin x$:

$$\sin x = \pm\sqrt{1 - \cos^2 x}$$

Since $\cos x$ and $\tan x$ are negative, x is associated with the second quadrant, where $\sin x$ is positive; hence,

$$\sin x = \sqrt{1 - \cos^2 x} = \sqrt{1 - \left(-\dfrac{2}{3}\right)^2} = \sqrt{\dfrac{5}{9}} = \dfrac{\sqrt{5}}{3}$$

Find tan x: $\tan x = \dfrac{\sin x}{\cos x} = \dfrac{\dfrac{\sqrt{5}}{3}}{-\dfrac{2}{3}} = -\dfrac{\sqrt{5}}{2}$ *Find cot x:* $\cot x = \dfrac{1}{\tan x} = \dfrac{1}{-\dfrac{\sqrt{5}}{2}} = -\dfrac{2}{\sqrt{5}}$

Find sec x: $\sec x = \dfrac{1}{\cos x} = \dfrac{1}{-\dfrac{2}{3}} = -\dfrac{3}{2}$ *Find csc x:* $\csc x = \dfrac{1}{\sin x} = \dfrac{1}{\dfrac{\sqrt{5}}{3}} = \dfrac{3}{\sqrt{5}}$

45.   First draw a reference triangle in the first quadrant and find $\sin x$ and $\cos x$: $r = \sqrt{3^2 + 4^2} = 5$; $\sin x = \dfrac{4}{5}$, $\cos x = \dfrac{3}{5}$. Now use the double-angle identities:

$\sin 2x = 2 \sin x \cos x = 2\left(\dfrac{4}{5}\right)\left(\dfrac{3}{5}\right) = \dfrac{24}{25}$

$\cos 2x = 2 \cos^2 x - 1 = 2\left(\dfrac{3}{5}\right)^2 - 1 = \dfrac{18}{25} - 1 = -\dfrac{7}{25}$

$\tan 2x = \dfrac{2 \tan x}{1 - \tan^2 x} = \dfrac{2(4/3)}{1 - (4/3)^2} = \dfrac{(8/3)}{1 - (16/9)} = -\dfrac{24}{7}$

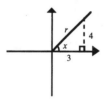

46.   We are given $\cos x$. We can find $\sin\dfrac{x}{2}$ and $\cos\dfrac{x}{2}$ from the half-angle identities, after determining their sign, as follows: if $-\pi < x < -\dfrac{\pi}{2}$, than $-\dfrac{\pi}{2} < \dfrac{x}{2} < -\dfrac{\pi}{4}$. Thus, $\dfrac{x}{2}$ is in the fourth quadrant, where sine is negative and cosine is positive. Using half-angle identities, we obtain:

$$\sin \frac{x}{2} = -\sqrt{\frac{1 - \cos x}{2}} = -\sqrt{\frac{1 - (-5/13)}{2}} = -\sqrt{\frac{9}{13}} \text{ or } -\frac{3}{\sqrt{13}}$$

$$\cos \frac{x}{2} = \sqrt{\frac{1 + \cos x}{2}} = \sqrt{\frac{1 + (-5/13)}{2}} = \sqrt{\frac{4}{13}} \text{ or } \frac{2}{\sqrt{13}}$$

$$\tan \frac{x}{2} = \frac{\sin\left(\frac{x}{2}\right)}{\cos\left(\frac{x}{2}\right)} = \frac{-\dfrac{3}{\sqrt{13}}}{\dfrac{2}{\sqrt{13}}} = -\frac{3}{2}$$

47. $$\tan\left(x + \frac{\pi}{4}\right) = \frac{\tan x + \tan \dfrac{\pi}{4}}{1 - \tan x \tan \dfrac{\pi}{4}} \qquad \text{Sum identity}$$

$$= \frac{\tan x + 1}{1 - \tan x \cdot 1} \qquad \text{Known values}$$

$$= \frac{\tan x + 1}{1 - \tan x} \qquad \text{Algebra}$$

Graph $y1 = \tan\left(x + \dfrac{\pi}{4}\right)$ and $y2 = \dfrac{\tan x + 1}{1 - \tan x}$.

Use the trace function to confirm that the graphs are identical.

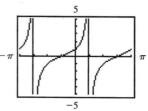

48. $$\cos 1.5x \cos 0.3x - \sin 1.5x \sin 0.3x = \cos(1.5x + 0.3x) \qquad \text{Sum identity}$$
$$= \cos(1.8x) \qquad \text{Algebra}$$

Graph $y1 = \cos 1.5x \cos 0.3x - \sin 1.5x \sin 0.3x$ and $y2 = \cos(1.8x)$. Use the trace function to confirm that the graphs are identical.

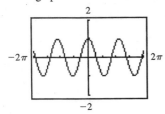

49.

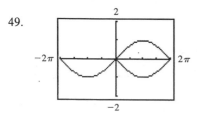

The two graphs coincide on the interval $-2\pi \leq x \leq 0$. $\sin \dfrac{x}{2} = -\sqrt{\dfrac{1 - \cos x}{2}}$ is an identity on this interval.

50. (A) Graph both sides of the equation in the same viewing window.

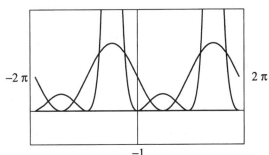

$$\frac{\sin^2 x}{1 + \sin x} = 1 - \sin x \text{ is not an identity, since the graphs do not match.}$$

Try $x = 0$

Left side: $\dfrac{\sin^2 0}{1 + \sin 0} = \dfrac{0}{1 + 0} = 0$

Right side: $1 - \sin 0 = 1 - 0 = 1$

This verifies that the equation is not an identity.

(B) Graph both sides of the equation in the same viewing window.

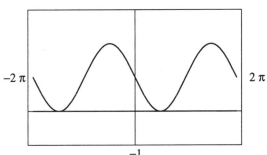

$\dfrac{\cos^2 x}{1 + \sin x} = 1 - \sin x$ appears to be an identity which we verify.

$$\dfrac{\cos^2 x}{1 + \sin x} = \dfrac{1 - \sin^2 x}{1 + \sin x} \qquad \text{Pythagorean identity}$$

$$= \dfrac{(1 + \sin x)(1 - \sin x)}{1 + \sin x} \qquad \text{Algebra}$$

$$= 1 - \sin x \qquad \text{Algebra}$$

51. To obtain $\sin x$ and $\cos x$ from $\sec 2x$, we use the half-angle identities with x replaced by $2x$. Thus,

$$\sin \frac{x}{2} = \pm\sqrt{\frac{1 - \cos x}{2}} \text{ becomes } \sin \frac{2x}{2} = \pm\sqrt{\frac{1 - \cos 2x}{2}} \text{ or } \sin x = \pm\sqrt{\frac{1 - \cos 2x}{2}}$$

$$\cos \frac{x}{2} = \pm\sqrt{\frac{1 + \cos x}{2}} \text{ becomes } \cos \frac{2x}{2} = \pm\sqrt{\frac{1 + \cos 2x}{2}} \text{ or } \cos x = \pm\sqrt{\frac{1 + \cos 2x}{2}}$$

To obtain $\cos 2x$ from $\sec 2x$, we use the reciprocal identity

$$\sec 2x = \frac{1}{\cos 2x} \qquad\qquad \cos 2x = \frac{1}{\sec 2x} = \frac{1}{-\frac{13}{12}} = -\frac{12}{13}$$

Since $-\frac{\pi}{2} < x < 0$, x is in the fourth quadrant, where sine is negative and cosine is positive. Thus,

$$\sin x = -\sqrt{\frac{1 - \cos 2x}{2}} = -\sqrt{\frac{1 - (-12/13)}{2}} = -\sqrt{\frac{25}{26}} \text{ or } -\frac{5}{\sqrt{26}}$$

$$\cos x = -\sqrt{\frac{1 + \cos 2x}{2}} = \sqrt{\frac{1 + (-12/13)}{2}} = \sqrt{\frac{1}{26}} \text{ or } \frac{1}{\sqrt{26}}$$

$$\tan x = \frac{\sin x}{\cos x} = \frac{-\dfrac{5}{\sqrt{26}}}{\dfrac{1}{\sqrt{26}}} = -5$$

52.
$$\frac{\cot x}{\csc x + 1} = \frac{(\csc x - 1)\cot x}{(\csc x - 1)(\csc x + 1)} \qquad\qquad \text{Algebra}$$

$$= \frac{(\csc x - 1)\cot x}{\csc^2 x - 1} \qquad\qquad \text{Algebra}$$

$$= \frac{(\csc x - 1)\cot x}{\cot^2 x} \qquad\qquad \text{Pythagorean identity}$$

$$= \frac{\csc x - 1}{\cot x} \qquad\qquad \text{Algebra}$$

53.
$$\cot 3x = \frac{1}{\tan 3x} \qquad\qquad\qquad\qquad \text{Reciprocal identity}$$

$$= 1 \div \tan 3x \qquad\qquad\qquad\qquad \text{Algebra}$$

$$= 1 \div \tan(2x + x) \qquad\qquad\qquad \text{Algebra}$$

$$= 1 \div \frac{\tan 2x + \tan x}{1 - \tan 2x \tan x} \qquad\qquad \text{Sum idenitty for tangent}$$

$$= \frac{1 - \tan 2x \tan x}{\tan 2x + \tan x} \qquad\qquad\qquad \text{Algebra}$$

$$= \frac{1 - \dfrac{2 \tan x}{1 - \tan^2 x} \cdot \tan x}{\dfrac{2 \tan x}{1 - \tan^2 x} + \tan x} \qquad\qquad \text{Double-angle identity}$$

$$= \frac{(1 - \tan^2 x) \cdot 1 - \dfrac{(1 - \tan^2 x)}{1} \cdot \dfrac{2 \tan x}{1 - \tan^2 x} \cdot \tan x}{\dfrac{2 \tan x}{1 - \tan^2 x} \cdot \dfrac{(1 - \tan^2 x)}{1} + \tan x \, (1 - \tan^2 x)} \qquad \text{Algebra}$$

$$= \frac{1 - \tan^2 x - 2 \tan x \cdot \tan x}{2 \tan x + \tan x - \tan^3 x} \qquad \text{Algebra}$$

$$= \frac{1 - \tan^2 x - 2 \tan^2 x}{3 \tan x - \tan^3 x} \qquad \text{Algebra}$$

$$= \frac{1 - 3 \tan^2 x}{3 \tan x - \tan^3 x} \qquad \text{Algebra}$$

$$= \frac{(-1)(1 - 3 \tan^2 x)}{(-1)(3 \tan x - \tan^3 x)} \qquad \text{Algebra}$$

$$= \frac{3 \tan^2 x - 1}{\tan^3 x - 3 \tan x} \qquad \text{Algebra}$$

54. The definitions of the circular functions involved a point (a, b) on a unit circle. Recall:

$$\sin x = b \qquad\qquad \cos x = a \qquad\qquad \tan x = \frac{b}{a} \qquad\qquad a \neq 0$$

Thus, $\tan x = \dfrac{b}{a} = \dfrac{\sin x}{\cos x}$

55. We are to show $\cos(x + 2k\pi) = \cos x$. But,

$\cos(x + 2k\pi) = \cos x \cos 2k\pi - \sin x \sin 2k\pi$ by the sum identity for cosine. A quick sketch shows $(a, b) = (1, 0)$, $r = 1$, $\cos 2k\pi = 1$, and $\sin 2k\pi = 0$. Hence,

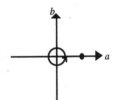

$$\cos(x + 2k\pi) = \cos x \cdot 1 - \sin x \cdot 0$$
$$\cos(x + 2k\pi) = \cos x$$

56. We are to show $\cot(x + k\pi) = \cot x$. But, $\cot(x + k\pi) = \dfrac{1}{\tan(x + k\pi)}$

$$= \frac{1}{\dfrac{\tan x + \tan k\pi}{1 - \tan x \tan k\pi}} \qquad \text{Sum identity}$$

$$= \frac{1 - \tan x \tan k\pi}{\tan x + \tan k\pi} \qquad \text{Algebra}$$

A quick sketch shows $(a, b) = (\pm 1, 0)$, $r = 1$.

$\tan k\pi = \dfrac{0}{\pm 1} = 0$. Hence, $\cot(x + k\pi) = \dfrac{1 - \tan x \cdot 0}{\tan x + 0}$ Known values

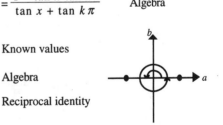

$$= \frac{1}{\tan x} \qquad \text{Algebra}$$

$$= \cot x \qquad \text{Reciprocal identity}$$

57. If we let $x + y = u$, $x - y = v$, and solve the system, we obtain $2x = u + v$, $2y = u - v$; that is,

$x = \dfrac{u + v}{2}$, $y = \dfrac{u - v}{2}$. Hence,

$$\sin x \sin y = \frac{1}{2}[\cos(x - y) - \cos(x + y)] \text{ becomes } \sin\frac{u + v}{2}\sin\frac{u - v}{2} = \frac{1}{2}[\cos v - \cos u]$$

upon substitution. Solving for the quantity in the brackets, we obtain

$$\frac{1}{2}[\cos v - \cos u] = \sin\frac{u + v}{2}\sin\frac{u - v}{2}; \quad \cos v - \cos u = 2\sin\frac{u + v}{2}\sin\frac{u - v}{2}$$

58. The graph of $f(x)$ is shown in the figure. The graph appears to be a basic cosine curve with period 2π, amplitude $= \dfrac{1}{2}(y\max - y\min) = \dfrac{1}{2}(6 - 2) = 2$, displaced upward by $k = 4$ units. It appears that $g(x) = 4 + 2\cos x$ would be an appropriate choice. We verify $f(x) = g(x)$ as follows:

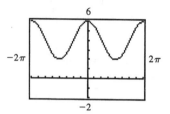

$$f(x) = \frac{3\sin^2 x}{1 - \cos x} + \frac{\tan^2 x \cos^2 x}{1 + \cos x}$$

$$= \frac{3\sin^2 x}{1 - \cos x} + \frac{\dfrac{\sin^2 x}{\cos^2 x}\cos^2 x}{1 + \cos x} \qquad \text{Quotient identity}$$

$$= \frac{3\sin^2 x}{1 - \cos x} + \frac{\sin^2 x}{1 + \cos x} \qquad \text{Algebra}$$

$$= \frac{3(1 - \cos^2 x)}{1 - \cos x} + \frac{1 - \cos^2 x}{1 + \cos x} \qquad \text{Pythagorean identity}$$

$$= \frac{3(1 + \cos x)(1 - \cos x)}{1 - \cos x} + \frac{(1 - \cos x)(1 + \cos x)}{1 + \cos x} \qquad \text{Algebra}$$

$$= 3 + 3\cos x + 1 - \cos x \qquad \text{Algebra}$$

$$= 4 + 2\cos x = g(x)$$

59. The graph of $f(x)$ is shown in the figure. The graph appears to have vertical asymptotes $x = -\dfrac{\pi}{4}$ and $x = \dfrac{\pi}{4}$, x intercepts $-\dfrac{\pi}{2}$, 0, and $\dfrac{\pi}{2}$, and period $\dfrac{\pi}{2}$. It appears that $g(x) = \tan 2x$ would be an appropriate choice. We verify $f(x) = g(x)$ as follows:

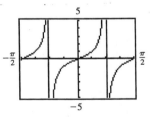

$$f(x) = \frac{\sin x}{\cos x - \sin x} + \frac{\sin x}{\cos x + \sin x}$$

$$= \frac{\sin x(\cos x + \sin x)}{(\cos x - \sin x)(\cos x + \sin x)} + \frac{\sin x(\cos x - \sin x)}{(\cos x + \sin x)(\cos x - \sin x)} \qquad \text{Algebra}$$

$$= \frac{\sin x(\cos x + \sin x) + \sin x(\cos x - \sin x)}{(\cos x - \sin x)(\cos x + \sin x)} \qquad \text{Algebra}$$

$$= \frac{\sin x \cos x + \sin^2 x + \sin x \cos x - \sin^2 x}{\cos^2 x - \sin^2 x} \qquad \text{Algebra}$$

$$= \frac{2 \sin x \cos x}{\cos^2 x - \sin^2 x} \qquad \text{Algebra}$$

$$= \frac{\sin 2x}{\cos 2x} \qquad \text{Double-angle identities}$$

$$= \tan 2x = g(x) \qquad \text{Quotient identity}$$

Key algebraic steps: $\dfrac{a}{b-a} + \dfrac{a}{b+a} = \dfrac{a(b+a)}{(b-a)(b+a)} + \dfrac{a(b-a)}{(b+a)(b-a)} = \dfrac{a(b+a) + a(b-a)}{(b-a)(b+a)}$

$$= \frac{ab + a^2 + ab - a^2}{b^2 - a^2} = \frac{2ab}{b^2 - a^2}$$

60. The graph of $f(x)$ is shown in the figure. The graph appears to be an upside down cosine curve with period π, amplitude $= \dfrac{1}{2}(y \max - y \min) = \dfrac{1}{2}(3-1) = 1$, displaced upward by $k = 2$ units. It appears that $g(x) = 2 - \cos 2x$ would be an appropriate choice. We verify $f(x) = g(x)$ as follows:

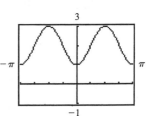

$f(x) = 3 \sin^2 x + \cos^2 x$

$\quad = 3 \sin^2 x + 1 - \sin^2 x \qquad$ Pythagorean identity

$\quad = 2 \sin^2 x + 1 \qquad$ Algebra

$\quad = 2 - (1 - 2\sin^2 x) \qquad$ Algebra

$\quad = 2 - \cos 2x \qquad$ Double-angle identity

61. The graph of $f(x)$ is shown in the figure. The graph appears to have vertical asymptotes $x = -\dfrac{3\pi}{4}, -\dfrac{\pi}{4}, \dfrac{\pi}{4},$ and $\dfrac{3\pi}{4}$, and period π. It appears to have high and low points with y coordinates -3 and -1, respectively. It appears that $g(x) = -2 + \sec 2x$ would be an appropriate choice. We verify $f(x) = g(x)$ as follows:

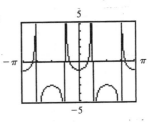

$f(x) = \dfrac{3 - 4\cos^2 x}{1 - 2\sin^2 x}$

$\quad = \dfrac{1 - (-2 + 4\cos^2 x)}{1 - 2\sin^2 x} \qquad$ Algebra

$\quad = \dfrac{1 - 2(2\cos^2 x - 1)}{1 - 2\sin^2 x} \qquad$ Algebra

$\quad = \dfrac{1 - 2\cos 2x}{\cos 2x} \qquad$ Double-angle identities

$\quad = \dfrac{1}{\cos 2x} - \dfrac{2\cos 2x}{\cos 2x} \qquad$ Algebra

$\quad = \sec 2x - 2 \text{ or } -2 + \sec 2x = g(x) \qquad$ Reciprocal identity, Algebra

62. The graph of $f(x)$ is shown in the figure. The graph appears to have vertical asymptotes $x = -2\pi$, $x = 0$, and $x = 2\pi$, and period 2π. Its x intercepts are difficult to determine, but since there appears to be symmetry with respect to the points where the curve crosses the line $y = 2$, it appears to be a cotangent curve displaced upward by $k = 2$ units. It appears that $g(x) = 2 + \cot\dfrac{x}{2}$ would be an appropriate choice. We verify

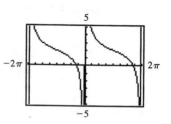

$f(x) = g(x)$ as follows:

$$
\begin{aligned}
f(x) &= \frac{2 + \sin x - 2\cos x}{1 - \cos x} \\
&= \frac{2 - 2\cos x + \sin x}{1 - \cos x} && \text{Algebra} \\
&= \frac{2 - 2\cos x}{1 - \cos x} + \frac{\sin x}{1 - \cos x} && \text{Algebra} \\
&= 2 + 1 \div \frac{1 - \cos x}{\sin x} && \text{Algebra} \\
&= 2 + 1 \div \tan\frac{x}{2} && \text{Half-angle identity} \\
&= 2 + \cot\frac{x}{2} && \text{Reciprocal identity}
\end{aligned}
$$

Key algebraic steps: $\dfrac{2 + a - 2b}{1 - b} = \dfrac{2 - 2b + a}{1 - b} = \dfrac{2 - 2b}{1 - b} + \dfrac{a}{1 - b} = 2 + 1 \div \dfrac{1 - b}{a}$

63. $-2\pi \le x < -\pi,\ 0 \le x < \pi$

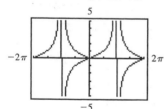

64. $-\pi < x \le 0,\ \pi < x \le 2\pi$

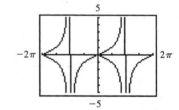

65.

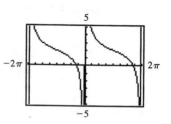

66. $\cos x \sin y = \frac{1}{2}[\sin(x+y) - \sin(x-y)]$

Let $x = 30\pi X$ and $y = 2\pi X$

$2 \cos 30\pi X \sin 2\pi X = 2\left(\frac{1}{2}\right)[\sin(30\pi X + 2\pi X) - \sin(30\pi X - 2\pi X)]$

$= \sin 32\pi X - \sin 28\pi X$

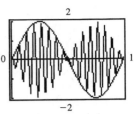

67. $\sqrt{u^2 - a^2} = \sqrt{(a \sec x)^2 - a^2}$ Using the given substitution

$= \sqrt{a^2 \sec^2 x - a^2}$ Algebra

$= \sqrt{a^2(\sec^2 x - 1)}$ Algebra

$= \sqrt{a^2 \tan^2 x}$ Pythagorean identity

$= |a| \, |\tan x|$ Algebra

$= a \tan x$ Since $a > 0$ and x is in quadrant I or IV

$\left(\text{given} -\frac{\pi}{2} < x < \frac{\pi}{2}\right)$, thus, $\tan x > 0$.

68. In Problem 59, Exercise 4.3, the formula

$$\tan(\theta_2 - \theta_1) = \frac{m_2 - m_1}{1 + m_1 m_2}$$

was derived. Since the given lines have slopes $4 = m_2$ and $\frac{1}{3} = m_1$, we can write

$$\tan(\theta_2 - \theta_1) = \frac{4 - \frac{1}{3}}{1 + \left(\frac{1}{3}\right)(4)} = \frac{3 \cdot 4 - 3 \cdot \frac{1}{3}}{3 + 3\left(\frac{1}{3}\right)4} = \frac{12 - 1}{3 + 4} = \frac{11}{7}$$

$$\theta_2 - \theta_1 = \tan^{-1}\left(\frac{11}{7}\right)$$

$$\approx 57.5°$$

69. We note that $\tan \theta = \frac{5}{8}$ and $\tan 2\theta = \frac{5 + x}{8}$ (see figure). Then,

$\tan 2\theta = \frac{2 \tan \theta}{1 - \tan^2 \theta}$, $\theta = \tan^{-1}\frac{5}{8} \approx 32.005°$

$$\frac{5 + x}{8} = \frac{2\left(\frac{5}{8}\right)}{1 - \left(\frac{5}{8}\right)^2} = \frac{\frac{5}{4}}{1 - \frac{25}{64}} = \frac{\frac{5}{4}}{\frac{39}{64}} = \frac{80}{39}$$

$$8 \cdot \frac{5 + x}{8} = 8 \cdot \frac{80}{39}$$

$$5 + x = \frac{640}{39}$$

$$x = \frac{445}{39} \approx 11.410$$

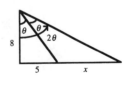

70. Following the hint, label the text diagram as follows:

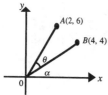

Then $\tan \alpha = \dfrac{y_B}{x_B} = \dfrac{4}{4} = 1$ $\tan(\alpha + \theta) = \dfrac{y_A}{x_A} = \dfrac{6}{2} = 3$

$$
\begin{aligned}
\tan \theta &= \tan(\alpha + \theta - \alpha) = \tan[(\alpha + \theta) - \alpha] \\
&= \frac{\tan(\alpha + \theta) - \tan \alpha}{1 + \tan(\alpha + \theta) \tan \alpha} \\
&= \frac{3 - 1}{1 + 3 \cdot 1} \\
&= \frac{1}{2}
\end{aligned}
$$

Therefore, $\theta = \tan^{-1} \dfrac{1}{2} = 0.464$ rad.

71. (A) Redrawing the front of the trim and labeling, we have:

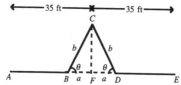

The length of this edge is given by

$$
\begin{aligned}
\ell = AB + BC + CD + DE &= AF - BF + BC + CD + FE - FD \\
&= 35 - a + b + b + 35 - a \\
&= 70 + 2b - 2a
\end{aligned}
$$

The length b of the entire trim is then $2\ell + 2 \cdot 50 = 2(70 + 2b - 2a) + 100$

Thus, $L = 140 + 4b - 4a + 100 = 240 + 4b - 4a$.

To express L in terms of θ, we note:

In triangle BCF, $\sin \theta = \dfrac{10}{b}$, hence $b \sin \theta = 10$ and $b = \dfrac{10}{\sin \theta}$

Also, $\cos \theta = \dfrac{a}{b}$, hence $a = b \cos \theta = \dfrac{10}{\sin \theta} \cos \theta = \dfrac{10 \cos \theta}{\sin \theta}$.

Hence, $L = 240 + 4b - 4a$

$$
\begin{aligned}
&= 240 + 4\left(\frac{10}{\sin \theta}\right) - 4\left(\frac{10 \cos \theta}{\sin \theta}\right) \\
&= 240 + 40\left(\frac{1}{\sin \theta} - \frac{\cos \theta}{\sin \theta}\right)
\end{aligned}
$$

$$= 240 + 40\frac{1 - \cos \theta}{\sin \theta}$$

$$= 240 + 40 \tan \frac{\theta}{2} \text{ as required}$$

(B) As θ varies from 30° to 60 °, $\frac{\theta}{2}$ varies from 15° to 30°; $\tan \frac{\theta}{2}$ and therefore L should increase steadily on this interval.

(C)

$\theta°$	30	35	40	45	50	55	60
L ft	250.7	252.6	254.6	256.6	258.7	260.8	263.1

(D) From the table, we find:
Min $L = 250.7$ ft; Max $L = 263.1$ ft

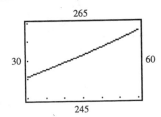

72. Sum $= 0.3 \cos 120\pi t - 0.3 \cos 140\pi t$.

$\cos x - \cos y = -2 \sin \frac{x + y}{2} \sin \frac{x - y}{2}$. Let $x = 120\pi t$ and $y = 140\pi t$.

$$0.3 \cos 120\pi t - 0.3 \cos 140\pi t = 0.3(\cos 120\pi t - \cos 140\pi t)$$

$$= 0.3(-2) \sin \frac{120\pi t + 140\pi t}{2} \sin \frac{120\pi t - 140\pi t}{2}$$

$$= -0.6 \sin 130\pi t \sin (-10\pi t)$$

$$= 0.6 \sin 130\pi t \sin 10\pi t$$

To find the beat frequency, we note:

Period of first tone $= \dfrac{2\pi}{B_1} = \dfrac{2\pi}{120\pi} = \dfrac{1}{60}$; Frequency of first tone $= \dfrac{1}{\text{Period}} = \dfrac{1}{\frac{1}{60}} = 60$ Hz

Period of second tone $= \dfrac{2\pi}{B_2} = \dfrac{2\pi}{140\pi} = \dfrac{1}{70}$; Frequency of first tone $= \dfrac{1}{\text{Period}} = \dfrac{1}{\frac{1}{70}} = 70$ Hz

Beat frequency = Frequency of second tone − Frequency of first tone
$f_b = 70$ Hz $- 60$ Hz $= 10$ Hz

73. (A)

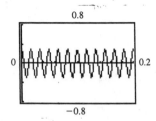

(B)

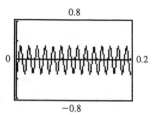

(C) y3 = 0.3 cos 120πt – 0.3 cos 140πt

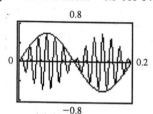

(D) y3 = 0.6 sin 130πt sin 10πt

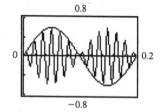

74. In the formula $M \sin Bt + N \sin Bt = A \sin(Bt + C)$, we have $M = -8$ and $N = -6$.

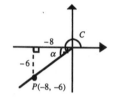

Locate $P(M, N) = P(-8, -6)$ to determine C: $r = \sqrt{(-8)^2 + (-6)^2} = 10$, $\sin C = \dfrac{-6}{10} = -0.6$, $\tan C = \dfrac{-6}{-8} = 0.75$. Find the reference angle α.

Then $C = \pi + \alpha$. $\tan \alpha = 0.75$, $\alpha = \tan^{-1}(0.75) = 0.6435$.
Thus, $C \approx 3.79$. We can now write:
$y = -8 \sin 3t - 6 \cos 3t = 10 \sin(3t + 3.79)$
Amplitude $= |10| = 10$

Period and Phase Shift:

$3t + 3.79 = 0$

$t = -1.26$

Period $= \dfrac{2\pi}{3}$

Frequency $= \dfrac{1}{\text{Period}} = \dfrac{1}{\dfrac{3}{2\pi}} = \dfrac{2\pi}{3}$

$3t + 3.79 = 2\pi$

$t = -1.26 + \dfrac{2\pi}{3}$

Phase Shift ≈ -1.26

75. The graph of $y = -8 \sin 3t - 6 \cos 3t$ is shown in the figure. We use the zoom feature or the built-in approximation routine to locate the t intercepts in this interval at $t = -1.26, -0.21, 0.83,$ and 1.88.
The phase shift for $y = 10 \sin(3t + 3.79)$ is -1.26.

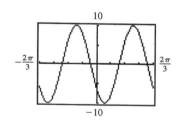

Chapter 5   Inverse Trigonometric Functions; Trigonometric Equations

EXERCISE 5.1   Inverse Sine, Cosine, and Tangent Functions

1. $y = \sin^{-1} 0$ is equivalent to $\sin y = 0$. No reference triangle can be drawn, but the only y between $-\dfrac{\pi}{2}$ and $\dfrac{\pi}{2}$ which has sine equal to 0 is $y = 0$. Thus, $\sin^{-1} 0 = 0$.

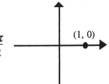

3. $y = \arccos \dfrac{\sqrt{3}}{2}$ is equivalent to $\cos y = \dfrac{\sqrt{3}}{2}$. What y between 0 and π has cosine equal to $\dfrac{\sqrt{3}}{2}$? y must be associated with a reference triangle in the first quadrant. Reference triangle is a special 30°–60° triangle.

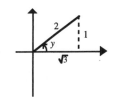

$y = \dfrac{\pi}{6}$ $\arccos \dfrac{\sqrt{3}}{2} = \dfrac{\pi}{6}$

5. $y = \tan^{-1} 1$ is equivalent to $\tan y = 1$. What y between $-\dfrac{\pi}{2}$ and $\dfrac{\pi}{2}$ has tangent equal to 1? y must be associated with a reference triangle in the first quadrant. Reference triangle is a special 45° triangle.

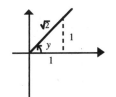

$y = \dfrac{\pi}{4}$ $\tan^{-1} 1 = \dfrac{\pi}{4}$

7. $y = \cos^{-1} \dfrac{1}{2}$ is equivalent to $\cos y = \dfrac{1}{2}$. What y between 0 and π has cosine equal to $\dfrac{1}{2}$? y must be associated with a reference triangle in the first quadrant. Reference triangle is a special 30°–60° triangle.

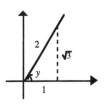

$y = \dfrac{\pi}{3}$ $\cos^{-1} \dfrac{1}{2} = \dfrac{\pi}{3}$

9.   Calculator in radian mode: $\cos^{-1} 0.4038 = 1.155$

11. Calculator in radian mode: $\tan^{-1} 43.09 = 1.548$

13. 1.131 is not in the domain of the inverse sine function. $-1 \le 1.131 \le 1$ is false. arcsin 1.131 is not defined.

15. $\sin^{-1} x = 37$ is equivalent to $\sin 37 = x$. Since the calculator is in degree mode, calculate $x = \sin 37° = 0.601815$.

17. $y = \arccos\left(-\dfrac{1}{2}\right)$ is equivalent to $\cos y = -\dfrac{1}{2}$. What y between

   0 and π has cosine equal to $-\dfrac{1}{2}$? y must be associated with a
   reference triangle in the second quadrant. Reference triangle is a
   special 30°–60° triangle.

   $y = \dfrac{2\pi}{3}$ $\arccos\left(-\dfrac{1}{2}\right) = \dfrac{2\pi}{3}$

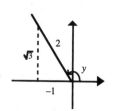

19. $y = \tan^{-1}(-1)$ is equivalent to $\tan y = -1$. What y between $-\dfrac{\pi}{2}$ and

   $\dfrac{\pi}{2}$ has tangent equal to -1? y must be negative and associated with a
   reference triangle in the fourth quadrant. Reference triangle is a
   special 45° triangle.

   $y = -\dfrac{\pi}{4}$ $\tan^{-1}(-1) = -\dfrac{\pi}{4}$

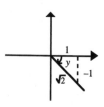

21. $y = \sin^{-1}\left(-\dfrac{\sqrt{3}}{2}\right)$ is equivalent to $\sin y = -\dfrac{\sqrt{3}}{2}$. What y between $-\dfrac{\pi}{2}$ and $\dfrac{\pi}{2}$ has sine equal to $-\dfrac{\sqrt{3}}{2}$?

   y must be negative and associated with a
   reference triangle in the fourth quadrant.
   Reference triangle is a special 30°–60°
   triangle.

   $y = -\dfrac{\pi}{3}$ $\sin^{-1}\left(\dfrac{\sqrt{3}}{2}\right) = -\dfrac{\pi}{3}$

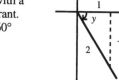

23. $y = \cos^{-1}\left(-\dfrac{\sqrt{3}}{2}\right)$ is equivalent to $\cos y = -\dfrac{\sqrt{3}}{2}$. What y between 0 and π has cosine equal to $-\dfrac{\sqrt{3}}{2}$?

   y must be associated with a reference triangle
   in the second quadrant. Reference triangle is a
   special 30°–60° triangle.

   $y = \dfrac{5\pi}{6}$ $\cos^{-1}\left(-\dfrac{\sqrt{3}}{2}\right) = \dfrac{5\pi}{6}$

25. $\sin[\sin^{-1}(-0.6)] = -0.6$ from the sine-inverse sine identity.

27. Let $y = \sin^{-1}\left(-\dfrac{\sqrt{2}}{2}\right)$, then $\sin y = -\dfrac{\sqrt{2}}{2}$, $-\dfrac{\pi}{2} \le y \le \dfrac{\pi}{2}$. Draw the reference triangle associated with y.

   Then $\cos y = \cos\left[\sin^{-1}\left(-\dfrac{\sqrt{2}}{2}\right)\right]$ can be determined directly from the

   triangle or by recognizing that $y = -\dfrac{\pi}{4}$.

   $\cos\left[\sin^{-1}\left(-\dfrac{\sqrt{2}}{2}\right)\right] = \cos y = \dfrac{\sqrt{2}}{2}$ or $= \cos\left(-\dfrac{\pi}{4}\right) = \dfrac{\sqrt{2}}{2}$

29. Calculator in radian mode: $\tan^{-1}(-4.038) = -1.328$

31. Calculator in radian mode: $\sec[\sin^{-1}(-0.0399)] = \dfrac{1}{\cos[\sin^{-1}(-0.0399)]} = 1.001$

33. Calculator in radian mode: $\sqrt{2} + \tan^{-1}\left(\sqrt[3]{5}\right) = 2.456$

35.

x	−1.0	−0.8	−0.6	−0.4	−0.2	0
$\sin^{-1}x$	−1.57	−0.92	−0.64	−0.41	−0.20	0

x	0.2	0.4	0.6	0.8	1.0
$\sin^{-1}x$	0.20	0.41	0.64	0.92	1.57

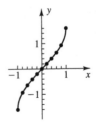

37. $\theta = \arccos\left(-\dfrac{1}{2}\right)$ is equivalent to $\cos\theta = -\dfrac{1}{2}$; $0° \le \theta \le 180°$. Thus, $\theta = 120°$.

39. $\theta = \tan^{-1}(-1)$ is equivalent to $\tan\theta = -1$; $-90° < \theta < 90°$. Thus, $\theta = -45°$.

41. $\theta = \sin^{-1}\left(-\dfrac{\sqrt{3}}{2}\right)$ is equivalent to $\sin\theta = -\dfrac{\sqrt{3}}{2}$; $-90° \le \theta \le 90°$. Thus, $\theta = -60°$

43. Calculator in degree mode: $\theta = \tan^{-1}3.0413 = 71.80°$

45. Calculator in degree mode: $\theta = \arcsin(-0.8107) = -54.16°$

47. Calculator in degree mode: $\theta = \arctan(-17.305) = -86.69°$

49. $\cos^{-1}[\cos(-0.3)] = 0.3$. This does not illustrate a cosine-inverse cosine identity because $\cos^{-1}(\cos x) = x$ only if $0 \le x \le \pi$.

51. (A)   (B) The domain of $\sin^{-1}$ is restricted to $-1 \le x \le 1$; hence no graph will appear for other x.

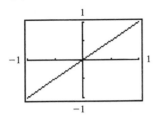

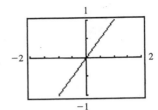

53. Since we recognize $\arccos\dfrac{1}{2} = \dfrac{\pi}{3}$ and $\arcsin(-1) = -\dfrac{\pi}{2}$, we can write

$$\sin\left[\arccos\dfrac{1}{2} + \arcsin(-1)\right] = \sin\left(\dfrac{\pi}{3} + \left(-\dfrac{\pi}{2}\right)\right) = \sin\dfrac{\pi}{3}\cos\left(-\dfrac{\pi}{2}\right) + \cos\dfrac{\pi}{3}\sin\left(-\dfrac{\pi}{2}\right)$$

$$= \dfrac{\sqrt{3}}{2}(0) + \left(\dfrac{1}{2}\right)(-1) = -\dfrac{1}{2}$$

55. Let $y = \sin^{-1}\left(-\frac{4}{5}\right)$. Then, $\sin y = -\frac{4}{5}$. We are asked to evaluate $\sin(2y)$, which is $2 \sin y \cos y$ by

the double-angle identity. Draw a reference triangle associated with y; then, $\cos y = \cos\left[\sin^{-1}\left(-\frac{4}{5}\right)\right]$
can be determined directly from the triangle.

$\sin y = \sin\left[\sin^{-1}\left(-\frac{4}{5}\right)\right] = -\frac{4}{5}$ by the sine-inverse sine identity.

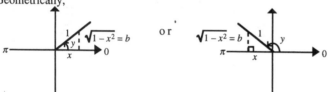

Note: $-\frac{\pi}{2} \le y \le \frac{\pi}{2}$ $a = \sqrt{5^2 - (-4)^2} = 3$ $\cos y = \frac{3}{5}$.

Thus, $\sin\left[2 \sin^{-1}\left(-\frac{4}{5}\right)\right] = 2 \sin\left[\sin^{-1}\left(-\frac{4}{5}\right)\right] \cos\left[\sin^{-1}\left(-\frac{4}{5}\right)\right] = 2\left(-\frac{4}{5}\right)\left(\frac{3}{5}\right) = -\frac{24}{25}$

57. Let $y = \cos^{-1} x$ $(-1 \le x \le 1$ corresponds to $0 \le y \le \pi)$
or, equivalently, $\cos y = x$ $0 \le y \le \pi$
Geometrically,

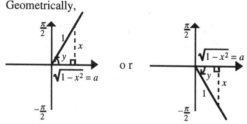

In either case, $b = \sqrt{1 - x^2}$. Thus, $\sin(\cos^{-1} x) = \sin y = \frac{b}{r} = \sqrt{1 - x^2}$

59. Let $y = \arcsin x$ $\left(-1 \le x \le 1$ corresponds to $-\frac{\pi}{2} \le y \le \frac{\pi}{2}\right)$

or, equivalently, $\sin y = x$ $-\frac{\pi}{2} \le y \le \frac{\pi}{2}$
Geometrically,

In either case, $a = \sqrt{1 - x^2}$. Thus, $\tan(\arcsin x) = \tan y = \frac{b}{a} = \frac{x}{\sqrt{1 - x^2}}$

61. Let $y = \tan^{-1}(-x)$.
This is equivalent to $-x = \tan y$ $\left(-\frac{\pi}{2} < y < \frac{\pi}{2}\right)$ by the definition of the inverse tangent function.
By algebra, this is equivalent to $x = -\tan y$, which in turn is equivalent to $x = \tan(-y)$ by the
identities for negatives. This equation is equivalent to $-y = \tan^{-1} x$ $\left(-\frac{\pi}{2} < -y < \frac{\pi}{2}\right)$ or
$y = -\tan^{-1} x$.

63. (A) $\cos^{-1} x$ has domain $-1 \le x \le 1$, therefore, $\cos^{-1}(2x - 3)$ has domain
$-1 \le 2x - 3 \le 1$ or
$2 \le 2x \le 4$ or
$1 \le x \le 2$

(B) The graph appears only for the domain values $1 \le x \le 2$.

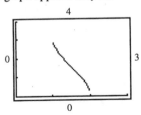

65. (A) To find the inverse function to $h(x)$, set $y = h(x)$, interchange x and y, then solve for y in terms of x.

$h(x) = y = 3 + 5 \sin(x - 1)$ $-\dfrac{\pi}{2} \le x \le 1 + \dfrac{\pi}{2}$

Inverse function:

$h(y) = x = 3 + 5 \sin(y - 1)$ $-\dfrac{\pi}{2} \le y \le 1 + \dfrac{\pi}{2}$

Solve for y in terms of x:
$x - 3 = 5 \sin(y - 1)$
$\dfrac{x - 3}{5} = \sin(y - 1)$
$y - 1 = \sin^{-1} \dfrac{x - 3}{5}$
$y = 1 + \sin^{-1}\left(\dfrac{x - 3}{5}\right)$

(B) $\sin^{-1} x$ has domain $-1 \le x \le 1$, therefore, $1 + \sin^{-1}\left(\dfrac{x - 3}{5}\right)$ has domain
$-1 \le \dfrac{x - 3}{5} \le 1$
$-5 \le x - 3 \le 5$
$-2 \le x \le 8$

67. (A)

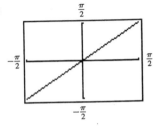

(B) The domain for sin x is $(-\infty, \infty)$ and the range is
$[-1, 1]$, which is the domain for $\sin^{-1} x$.
Thus, $y = \sin^{-1}(\sin x)$ has a graph over the interval
$(-\infty, \infty)$, but $\sin^{-1}(\sin x) = x$ only on the restricted

domain of cos x, $\left[-\dfrac{\pi}{2}, \dfrac{\pi}{2}\right]$.

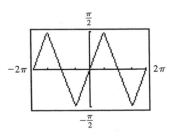

69. (A) If $\sin \dfrac{\theta}{2} = \dfrac{1}{M}$, then $\dfrac{\theta}{2} = \sin^{-1}\left(\dfrac{1}{M}\right)$ $-1 \le \dfrac{1}{M} \le 1$.

Thus, $\theta = 2 \sin^{-1}\left(\dfrac{1}{M}\right)$. Since M must be positive, $\dfrac{1}{M} \le 1$, or $M \ge 1$.

(B) Calculator in degree mode:

For $M = 1.7$, $\theta = 2 \sin^{-1}\left(\dfrac{1}{1.7}\right) = 72°$. For $M = 2.3$, $\theta = 2 \sin^{-1}\left(\dfrac{1}{2.3}\right) = 52°$

71. (A) By comparison of θ_1, θ_2, θ_3, drawn for various
values of x, it appears that θ increases with
increasing x, then decreases somewhat.

(B) Take an arbitrary value of θ (shown as θ_2 in the
diagram.)
Then $\theta = \angle CPA - \angle BPA$
$AB + BC + CD = 60 - 2 \cdot 5 = 50$ yd since this
distance is the width of the field minus two 5-yard
lengths.

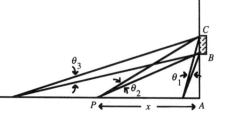

Let $y = AB = CD$, $BC = 8$ yd (width of the goal)

Then $2y + 8 = 50$, hence $y = 21$

$$\tan BPA = \dfrac{y}{x} = \dfrac{21}{x} \qquad \tan CPA = \dfrac{AB + BC}{x} = \dfrac{21 + 8}{x} = \dfrac{29}{x}$$

$$\tan \theta = \tan(\angle CPA - \angle BPA) = \dfrac{\tan \angle CPA - \tan \angle BPA}{1 + (\tan \angle CPA)(\tan \angle BPA)}$$

$$\tan \theta = \dfrac{\dfrac{29}{x} - \dfrac{21}{x}}{1 + \dfrac{29}{x} \cdot \dfrac{21}{x}} = \dfrac{\dfrac{8}{x}}{1 + \dfrac{609}{x^2}} = \dfrac{\dfrac{8}{x} \cdot x^2}{x^2 + \dfrac{609}{x^2} \cdot x^2} = \dfrac{8x}{x^2 + 609}$$

Hence $\theta = \tan^{-1} \dfrac{8x}{x^2 + 609}$

(C) From the table the Max $\theta = 9.21°$ when $x = 25$ yd.

x yd	10	15	20	25	30	35
$\theta°$	6.44	8.19	9.01	9.21	9.04	8.68

(D) 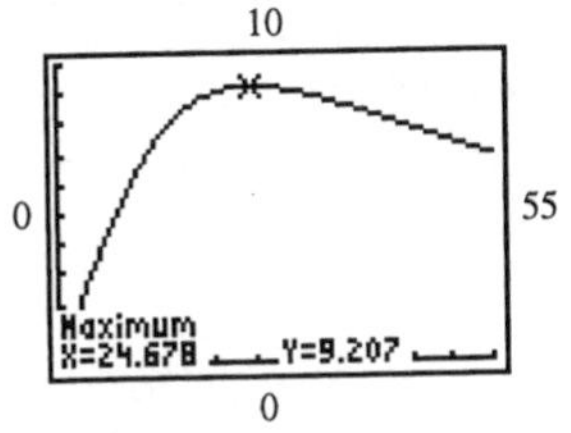

The angle θ increases rapidly until a maximum is reached then declines more slowly. The maximum $\theta = 9.21°$ when $x = 24.68$ yd, which is about 7 yd before from the left corner of the penalty area (shown in the text diagram.)

73. (A) The volume of the fuel is clearly given by Volume = (height)(cross-sectional area) with L = height. To determine the cross-sectional area (see figure), we reason that

Area of segment $AEBF$ = Area of sector $ACBF$ – Area of triangle ABC

Area of sector $ACBF$ = 2(area of sector CFB) $= 2\left(\frac{1}{2}R^2\theta\right) = R^2\theta$

Since triangle CEB is a right triangle, we have

$$\cos\theta = \frac{EC}{BC} = \frac{R-x}{R} \qquad \theta = \cos^{-1}\frac{R-x}{R}$$

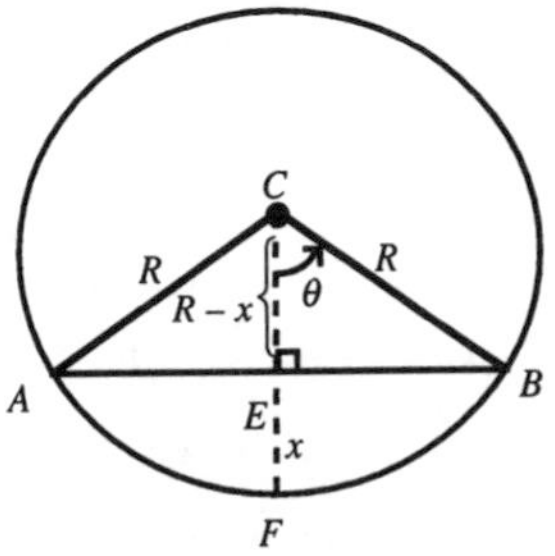

Therefore,

$$\text{Area of sector } ACBF = R^2\cos^{-1}\frac{R-x}{R}$$

Area of triangle $ABC = \frac{1}{2}$(base)(altitude) $= \frac{1}{2}(AB)(CE) = \frac{1}{2}(2 \cdot EB)(CE) = (EB)(R-x)$

By the Pythagorean theorem applied to triangle CEB, we have $EB^2 + (R-x)^2 = R^2$.
$EB = \sqrt{R^2 - (R-x)^2}$. Therefore, Area of triangle $ABC = (R-x)\sqrt{R^2-(R-x)^2}$.
Finally,

$$\text{Area of segment } AEBF = R^2\cos^{-1}\frac{R-x}{R} - (R-x)\sqrt{R^2-(R-x)^2}$$

$$\text{Volume} = \left[R^2\cos^{-1}\frac{R-x}{R} - (R-x)\sqrt{R^2-(R-x)^2}\right]L$$

(B) Substituting the given values, we have $R = 3$, $x = 2$, $R - x = 1$, $L = 30$

$$\begin{aligned}\text{Volume} &= [3^2\cos^{-1}\tfrac{1}{3} - 1\sqrt{3^2-1^2}\,]\,30 \\ &= [9\cos^{-1}\tfrac{1}{3} - \sqrt{8}\,]\,30 \qquad \text{(Calculator in radian mode)} \\ &= 248 \text{ ft}^3\end{aligned}$$

(C) The graphs of y1 and y2 are shown in the figure. We use the zoom feature or the built-in approximation routine to locate the x coordinate of the point of intersection at $x = 2.6$. The depth is 2.6 feet.

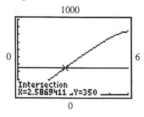

EXERCISE 5.2   Inverse Cotangent, Secant, and Cosecant Functions

1. $y = \cot^{-1} \sqrt{3}$ is equivalent to $\cot y = \sqrt{3}$ and $0 < y < \pi$.
   What number between 0 and π has cotangent equal to
   $\sqrt{3}$? y must be in the first quadrant. $\cot y = \sqrt{3} = \dfrac{\sqrt{3}}{1}$,
   $y = \dfrac{\pi}{6}$. Thus, $\cot^{-1} \sqrt{3} = \dfrac{\pi}{6}$

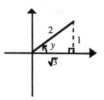

3. $y = \text{arccsc } 1$ is equivalent to $\csc y = 1$ and $-\dfrac{\pi}{2} \le y \le \dfrac{\pi}{2}$, $y \ne 0$.
   What number between $-\dfrac{\pi}{2}$ and $\dfrac{\pi}{2}$ has cosecant equal to 1? No
   reference triangle can be drawn, but from the diagram we see
   $(a, b) = (0, 1)$ $r = 1$ $\csc y = \dfrac{1}{1} = 1$ $y = \dfrac{\pi}{2}$
   Thus, $\text{arccsc } 1 = \dfrac{\pi}{2}$

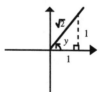

5. $y = \sec^{-1} \sqrt{2}$ is equivalent to $\sec y = \sqrt{2}$ and $0 \le y \le \pi$, $y \ne \dfrac{\pi}{2}$.
   What number between 0 and π has secant equal to $\sqrt{2}$? y must be
   in the first quadrant.
   $\sec y = \sqrt{2} = \dfrac{\sqrt{2}}{1}$ $y = \dfrac{\pi}{4}$
   Thus, $\sec^{-1} \sqrt{2} = \dfrac{\pi}{4}$

7. Let $y = \cot^{-1} 0$, then $\cot y = 0$, $0 < y < \pi$. No reference triangle
   can be drawn, but from the diagram we see
   $(a, b) = (0, 1)$ $r = 1$ $\cot y = \dfrac{0}{1} = 0$ $y = \dfrac{\pi}{2}$
   Thus, $\sin(\cot^{-1} 0) = \sin \dfrac{\pi}{2} = 1$

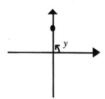

186

Chapter 5   Inverse Trigonometric Functions; Trigonometric Equations

9. Let $y = \csc^{-1}\dfrac{5}{4}$, the $\csc y = \dfrac{5}{4}$, $-\dfrac{\pi}{2} \le y \le \dfrac{\pi}{2}$, $y \ne 0$. y is in the first quadrant. Draw a reference triangle, find the third side, then determine $\tan y$ from the triangle.

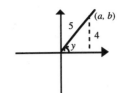

$a = \sqrt{5^2 - 4^2} = 3$

Thus, $\tan\left(\csc^{-1}\dfrac{5}{4}\right) = \tan y = \dfrac{b}{a} = \dfrac{4}{3}$

11. $y = \cot^{-1}(-1)$ is equivalent to $\cot y = -1$ and $0 < y < \pi$. What number between 0 and π has cotangent equal to -1? y must be positive and in the second quadrant.

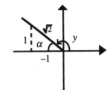

$\cot y = -1 = \dfrac{-1}{1}$ $\alpha = \dfrac{\pi}{4}$ $y = \dfrac{3\pi}{4}$

Thus, $\cot^{-1}(-1) = \dfrac{3\pi}{4}$

13. $y = \text{arcsec}(-2)$ is equivalent to $\sec y = -2$ and $0 \le y \le \pi$, $y \ne \dfrac{\pi}{2}$. What number between 0 and π has secant equal to -2? y must be positive and in the second quadrant.

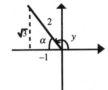

$\sec y = -2 = \dfrac{2}{-1}$ $\alpha = \dfrac{\pi}{3}$ $y = \dfrac{2\pi}{3}$

Thus, $\text{arcsec}(-2) = \dfrac{2\pi}{3}$

15. $y = \text{arccsc}(-2)$ is equivalent to $\csc y = -2$ and $-\dfrac{\pi}{2} \le y \le \dfrac{\pi}{2}$, $y \ne 0$. What number between $-\dfrac{\pi}{2}$ and $\dfrac{\pi}{2}$ has cosecant equal to -2? y must be negative and in the fourth quadrant.

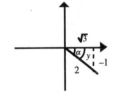

$\csc y = -2 = \dfrac{2}{-1}$ $\alpha = \dfrac{\pi}{6}$ $y = -\dfrac{\pi}{6}$

Thus, $\text{arccsc}(-2) = -\dfrac{\pi}{6}$

17. If $x = \dfrac{1}{2}$, neither $x \le -1$ nor $x \ge 1$ is a true statement. Thus, x is not in the domain of the inverse cosecant function.

$\csc^{-1}\dfrac{1}{2}$ is not defined.

19. Let $y = \csc^{-1}\left(-\dfrac{5}{3}\right)$; then, $\csc y = -\dfrac{5}{3}$, $-\dfrac{\pi}{2} \le y \le \dfrac{\pi}{2}$, $y \ne 0$. y is
negative and in the fourth quadrant. Draw a reference triangle, find
the third side, and then determine $\cos y$ from the triangle.

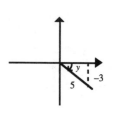

$$a = \sqrt{5^2 - (-3)^2} = 4 \quad \left(\csc y = -\dfrac{5}{3} = \dfrac{5}{-3}\right)$$

$$\text{Thus, } \cos\left[\csc^{-1}\left(-\dfrac{5}{3}\right)\right] = \cos y = \dfrac{a}{r} = \dfrac{4}{5}$$

21. Let $y = \sec^{-1}\left(-\dfrac{5}{4}\right)$; then, $\sec y = -\dfrac{5}{4}$, $0 \le y \le \pi$, $y \ne \dfrac{\pi}{2}$. y is
positive and in the second quadrant. Draw a reference triangle, find
the third side, and then determine $\cot y$ from the triangle.

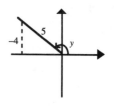

$$a = \sqrt{5^2 - (-4)^2} = 3 \quad \left(\sec y = -\dfrac{5}{4} = \dfrac{5}{-4}\right)$$

$$\text{Thus, } \cot\left[\sec^{-1}\left(-\dfrac{5}{4}\right)\right] = \cot y = \dfrac{a}{b} = -\dfrac{4}{3}$$

23. From the inverse trigonometric identities, we see $\sec^{-1} x = \cos^{-1}\dfrac{1}{x}$, $x \ge 1$ or $x \le -1$.

Thus, $\cos(\sec^{-1} x) = \cos\left(\cos^{-1}\dfrac{1}{x}\right) = \dfrac{1}{x}$ if $\dfrac{1}{x}$ is in the domain of the inverse cosine function.

Hence, $\cos[\sec^{-1}(-2)] = \cos\left[\cos^{-1}\left(\dfrac{1}{-2}\right)\right] = \dfrac{1}{-2}$ or $-\dfrac{1}{2}$, since $-\dfrac{1}{2}$ is in the domain of the inverse
cosine function.

25. We note: $\cot^{-1} x = \tan^{-1}\dfrac{1}{x}$ for $x > 0$.

$$\text{Thus,} \quad \cot(\cot^{-1} x) = \cot\left(\tan^{-1}\dfrac{1}{x}\right)$$

$$= \dfrac{1}{\tan\left(\tan^{-1}\dfrac{1}{x}\right)} \qquad \text{Reciprocal identity}$$

$$= \dfrac{1}{\dfrac{1}{x}} \qquad\qquad\qquad \text{Tangent-inverse tangent identity}$$

$$= x \qquad\qquad\qquad\qquad \text{Algebra}$$

Since $33.4 > 0$, the above reasoning applies; thus $\cot(\cot^{-1} 33.4) = 33.4$

27. We note: $\csc^{-1} x = \sin^{-1} \dfrac{1}{x}$ for $x \geq 1$ or $x \leq -1$.

Thus, $\csc(\csc^{-1} x) = \csc\left(\sin^{-1}\dfrac{1}{x}\right)$

$$= \dfrac{1}{\sin\left(\sin^{-1}\dfrac{1}{x}\right)}$$ Reciprocal identity

$$= \dfrac{1}{\dfrac{1}{x}}$$ $\begin{cases} \text{Sine-inverse sine identity. [If } x \geq 1 \text{ or } x \leq -1 \\ \text{then } 1/x \text{ will be in the domain of the inverse} \\ \text{sine function.]} \end{cases}$

$$= x$$ Algebra

Since $-4 \leq -1$, the above reasoning applies; thus $\csc[\csc^{-1}(-4)] = -4$

29. $\cot^{-1} 3.065 = \tan^{-1}\dfrac{1}{3.065}$ (Calculator in radian mode)

$= 0.315$

31. $\sec^{-1}(-1.963) = \cos^{-1}\dfrac{1}{-1.963}$ (Calculator in radian mode)

$= 2.105$

33. $\csc^{-1} 1.172 = \sin^{-1}\dfrac{1}{1.172}$ (Calculator in radian mode)

$= 1.022$

35. We note: $\cot^{-1} x = \pi + \tan^{-1}\dfrac{1}{x}$ if $x < 0$.

Thus, $\cot^{-1}(-5.104) = \pi + \tan^{-1}\dfrac{1}{-5.104}$ (Calculator in radian mode)

$= 2.948$

37. $\theta = \text{arcsec}(-2)$ is equivalent to $\sec\theta = -2$ $0 \leq \theta \leq 180°,\ \theta \neq 90°$

$\cos\theta = -\dfrac{1}{2}$ $0 \leq \theta \leq 180°$

Thus, $\theta = 120°$

39. $\theta = \cot^{-1}(-1)$ is equivalent to $\cot\theta = -1$ $0 < \theta < 180°$

$\tan\theta = \dfrac{1}{-1} = -1$ $0 < \theta < 180°$

Thus, $\theta = 135°$

41. $\theta = \csc^{-1}\left(-\dfrac{2}{\sqrt{3}}\right)$ is equivalent to $\csc\theta = -\dfrac{2}{\sqrt{3}}$ $-90° \leq \theta \leq 90°,\ \theta \neq 0°$

$\sin\theta = -\dfrac{\sqrt{3}}{2}$ $-90° \leq \theta \leq 90°$

Thus, $\theta = -60°$

43.   Calculator in degree mode: $\theta = \cot^{-1} 0.3288 = \tan^{-1} \dfrac{1}{0.3288} = 71.80°$

45.   Calculator in degree mode: $\theta = \operatorname{arccsc}(-1.2336) = \arcsin \dfrac{1}{-1.2336} = -54.16°$

47.   Calculator in degree mode: $\theta = \operatorname{arccot}(-0.0578) = 180° + \tan^{-1}\left(\dfrac{1}{-0.0578}\right) = 93.31°$

49.   Let $u = \csc^{-1}\left(-\dfrac{5}{3}\right)$ and $v = \tan^{-1}\dfrac{1}{4}$. Then we are asked to evaluate $\tan(u + v)$, which is

$\dfrac{\tan u + \tan v}{1 + \tan u \tan v}$ by the sum identity for tangent. We know $\tan v = \tan\left(\tan^{-1}\dfrac{1}{4}\right) = \dfrac{1}{4}$ by the

tangent-inverse tangent identity. It remains to find $\tan u = \tan\left[\csc^{-1}\left(-\dfrac{5}{3}\right)\right]$.

See sketch, Problem 19. $\tan\left[\csc^{-1}\left(-\dfrac{5}{3}\right)\right] = \dfrac{b}{a} = \dfrac{-3}{4}$ from the reference triangle. Hence,

$$\tan\left[\csc^{-1}\left(-\dfrac{5}{3}\right) + \tan^{-1}\dfrac{1}{4}\right] = \tan(u + v) = \dfrac{\tan u + \tan v}{1 - \tan u \tan v} = \dfrac{-\dfrac{3}{4} + \dfrac{1}{4}}{1 - \left(-\dfrac{3}{4}\right)\left(\dfrac{1}{4}\right)} = \dfrac{-\dfrac{2}{4}}{1 + \dfrac{3}{16}} = \dfrac{-8}{19}$$

51.   Let $y = \cot^{-1}\left(-\dfrac{3}{4}\right)$. Then we are asked to evaluate $\tan(2y)$, which is $\dfrac{2 \tan y}{1 - \tan^2 y}$ from the

double-angle identity. Draw a reference triangle associated with y. Then, $\tan y = \tan\left[\cot^{-1}\left(-\dfrac{3}{4}\right)\right]$
can be determined directly from the triangle.

y is positive and in the second quadrant. $\tan y = \dfrac{4}{-3} = -\dfrac{4}{3}$. Thus,

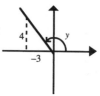

$$\tan\left[2 \cot^{-1}\left(-\dfrac{3}{4}\right)\right] = \tan 2y = \dfrac{2\left(-\dfrac{4}{3}\right)}{1 - \left(-\dfrac{4}{3}\right)^2} = \dfrac{-\dfrac{8}{3}}{1 - \left(\dfrac{16}{9}\right)} = \dfrac{24}{7}$$

53. Let $y = \cot^{-1} x$ $0 < y < \pi$ or, equivalently,
     $x = \cot y$ $0 < y < \pi$

Geometrically,

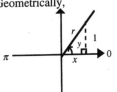   or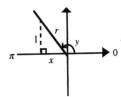

In either case, $r = \sqrt{x^2 + 1}$. Thus, $\sin(\cot^{-1} x) = \sin y = \dfrac{b}{R} = \dfrac{1}{\sqrt{x^2 + 1}}$.

55. Let $y = \sec^{-1} x$ $0 \le y \le \pi$ $y \ne \dfrac{\pi}{2}$ or, equivalently,

     $x = \sec y$ $0 \le y \le \pi$

Geometrically,

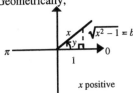   or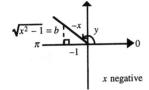

x positive   x negative

In either case, $b = \sqrt{x^2 - 1}$. Thus, if $x > 0$, $\csc(\sec^{-1} x) = \csc y = \dfrac{r}{b} = \dfrac{x}{\sqrt{x^2 - 1}}$.

If $x < 0$, $\csc(\sec^{-1} x) = \csc y = \dfrac{r}{b} = \dfrac{-x}{\sqrt{x^2 - 1}}$. A convenient notation for the quantity x if $x > 0$, $-x$

if $x < 0$, is $|x|$. Hence, we can write $\csc(\sec^{-1} x) = \dfrac{|x|}{\sqrt{x^2 - 1}}$.

57. Let $y = \cot^{-1} x$ $0 < y < \pi$ or, equivalently,
     $x = \cot y$ $0 < y < \pi$

Then, $\sin(2 \cot^{-1} x) = \sin(2y) = 2 \sin y \cos y$. From Problem 53, we have

$$\sin y = \sin(\cot^{-1} x) = \dfrac{1}{\sqrt{x^2 + 1}}.$$

From the figure in Problem 53, we also have $\cos y = \cos(\cot^{-1} x) = \dfrac{a}{r} = \dfrac{x}{\sqrt{x^2 + 1}}$. Thus,

$$\sin(2 \cot^{-1} x) = 2 \cdot \dfrac{1}{\sqrt{x^2 + 1}} \cdot \dfrac{x}{\sqrt{x^2 + 1}} = \dfrac{2x}{x^2 + 1}.$$

59. Let $y = \sec^{-1} x$ .... $x \le -1$ or $x \ge 1$

    Then, $\sec y = x$ .... $0 \le y \le \pi,\ y \ne \dfrac{\pi}{2}$ .... Definition of $\sec^{-1}$

    $\dfrac{1}{\cos y} = x$ .... $0 \le y \le \pi,\ y \ne \dfrac{\pi}{2}$ .... Reciprocal identity

    $\cos y = \dfrac{1}{x}$ .... $0 \le y \le \pi,\ y \ne \dfrac{\pi}{2}$ .... Algebra

    $y = \cos^{-1} \dfrac{1}{x}$ .... $0 \le y \le \pi,\ y \ne \dfrac{\pi}{2}$ .... Definition of $\cos^{-1}$

    Thus, $\sec^{-1} x = \cos^{-1} \dfrac{1}{x}$ for $x \le -1$ and $x \ge 1$

61.

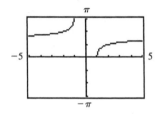

63.

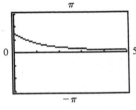 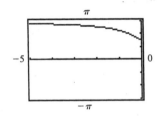

            $-5 \le x \le 0$ $0 \le x \le 5$

EXERCISE 5.3   Trigonometric Equations: An Algebraic Approach

1. $2 \cos x + 1 = 0$ $0 \le x < 2\pi$

    Solve for cos x: $2 \cos x = -1$

                     $\cos x = -\dfrac{1}{2}$

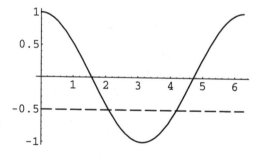

    Solve over $0 \le x < 2\pi$: Sketch a graph of $y = \cos x$ and $y = -\dfrac{1}{2}$ in the same coordinate system. We see that the solutions are in the second and third quadrants.

           $x = \dfrac{2\pi}{3},\ \dfrac{4\pi}{3}$

3. We have found all solutions of $2\cos x + 1 = 0$ over one period in Problem 1. Because the cosine function is periodic with period 2π, all solutions of $2\cos x + 1 = 0$ are given by

$$\frac{2\pi}{3} + 2k\pi, \quad \frac{4\pi}{3} + 2k\pi, \quad k \text{ any integer}$$

5. $\sqrt{2}\sin\theta - 1 = 0 \quad 0° \le \theta < 360°$

    Solve for sin θ: $\sqrt{2}\sin\theta = 1$

    $$\sin\theta = \frac{1}{\sqrt{2}}$$

    Solve over 0° ≤ θ < 360°: Sketch a graph of $y = \sin\theta$ and $y = \dfrac{1}{\sqrt{2}}$ in the same coordinate system. We see that the solutions are in the first and second quadrants.
    $\theta = 45°, 135°$

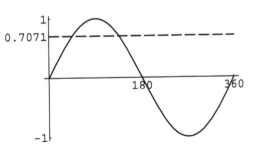

7. We have found all solutions of $\sqrt{2}\sin\theta - 1 = 0$ over one period in Problem 5. Because the sine function is periodic with period 360°, all solutions of $\sqrt{2}\sin\theta - 1 = 0$ are given by
    $45° + k(360°), 135° + k(360°), k$ any integer

9. $4\tan\theta + 15 = 0 \quad 0° \le \theta < 180°$

    Solve for tan θ: $\tan\theta = -\dfrac{15}{4}$

    Solve over 0° ≤ θ < 180°: Sketch a graph of $y = \tan\theta$ and $y = -\dfrac{15}{4}$ in the same coordinate system. We see that the solution is in the second quadrant.

    $$\theta = 180 + \tan^{-1}\left(-\frac{15}{4}\right) = 104.9314°$$

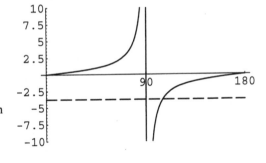

11. $5\cos x - 2 = 0 \quad 0 \le x < 2\pi$

    Solve for cos x: $\cos x = \dfrac{2}{5}$

    Solve over 0 ≤ x < 2π: Sketch a graph of $y = \cos x$ and $y = \dfrac{2}{5}$ in the same coordinate system. We see that the solutions are in the first and fourth quadrants.

    $$x = \cos^{-1}\frac{2}{5} = 1.1593$$
    $$x = 2\pi - 1.1593 = 5.1239$$

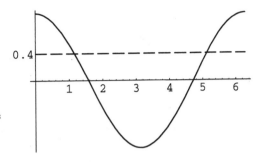

13. $5.0118 \sin x - 3.1105 = 0 \quad 0 \le x < 2\pi$

Solve for sin x: $\sin x = \dfrac{3.1105}{5.0118}$

$\sin x = 0.620635$

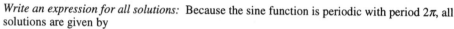

Solve over $0 \le x < 2\pi$: Sketch a graph of $y = \sin x$ and $y = 0.620635$ in the same coordinate system. We see that the solutions are in the first and second quadrants.

$x = \sin^{-1} 0.620635 = 0.6696$
$x = \pi - 0.6696 = 2.4720$

Write an expression for all solutions: Because the sine function is periodic with period 2π, all solutions are given by

$x = 0.6696 + 2k\pi, \quad x = 2.4720 + 2k\pi, \quad k$ any integer

15. $\cos x = \cot x \qquad 0 \le x < 2\pi$

Solve for sin x and/or cos x: $\cos x = \dfrac{\cos x}{\sin x}$ \qquad Use quotient identity

$\cos x \sin x = \cos x$ \qquad $(\sin x \ne 0)$

$\cos x \sin x - \cos x = 0$

$\cos x (\sin x - 1) = 0$

$\cos x = 0 \quad$ or $\quad \sin x = 1$
$x = \dfrac{\pi}{2}, \dfrac{3\pi}{2} \qquad x = \dfrac{\pi}{2}$

(The student should sketch graphs as appropriate to confirm these values.)

$x = \dfrac{\pi}{2}, \dfrac{3\pi}{2}$

17. $\cos^2 \theta = \dfrac{1}{2} \sin 2\theta$

Solve over $0° \le \theta < 360°$: $\cos^2 \theta = \dfrac{1}{2} \cdot 2 \sin \theta \cos \theta$ \qquad Use double-angle identity

$\cos^2 \theta = \sin \theta \cos \theta$

$\cos^2 \theta - \sin \theta \cos \theta = 0$

$\cos \theta (\cos \theta - \sin \theta) = 0$

$\cos \theta = 0 \quad$ or $\quad \cos \theta - \sin \theta = 0$
$\theta = 90°, 270° \qquad \cos \theta = \sin \theta$
$\qquad\qquad 1 = \dfrac{\sin \theta}{\cos \theta}$
$\qquad\qquad 1 = \tan \theta$
$\qquad\qquad \theta = 45°, 225°$

(The student should sketch graphs as appropriate to confirm these values.)

$\theta = 45°, 90°, 225°, 270°$

Write an expression for all solutions: Because the cosine function is periodic with period 360° and the tangent function is periodic with period 180°, all solutions are given by

$$45° + k180°, 90° + k360°, 225° + k180°, 270° + k360°, k \text{ any integer}$$

This can be written more compactly as $45° + k180°, 90° + k180°, k$ any integer.

19. $\tan \dfrac{x}{2} - 1 = 0 \qquad 0 \le x < 2\pi$

This is equivalent to

$\tan \dfrac{x}{2} - 1 = 0 \qquad 0 \le \dfrac{x}{2} < \pi$

Solve over $0 \le \dfrac{x}{2} < \pi$: $\tan \dfrac{x}{2} = 1$

$$\dfrac{x}{2} = \dfrac{\pi}{4}$$

$$x = \dfrac{\pi}{2}$$

21. $\sin^2 \theta + 2 \cos \theta = -2 \qquad 0° \le \theta < 360°$

Solve for sin θ and/or cos θ: $1 - \cos^2 \theta + 2 \cos \theta = -2$ Use Pythagorean identity

$$-\cos^2 \theta + 2 \cos \theta + 3 = 0$$

$$\cos^2 \theta - 2 \cos \theta - 3 = 0$$

$$(\cos \theta - 3)(\cos \theta + 1) = 0$$

$\cos \theta - 3 = 0 \quad$ or $\quad \cos \theta + 1 = 0$

$\cos \theta = 3 \qquad\qquad \cos \theta = -1$

No solution $\qquad\qquad\qquad \theta = 180°$

23. $\cos 2\theta + \sin^2 \theta = 0 \qquad 0° \le \theta < 360°$

Solve for sin θ and/or cos θ: $1 - 2 \sin^2 \theta + \sin^2 \theta = 0$ Use double-angle identity

$$1 - \sin^2 \theta = 0$$

$$\cos^2 \theta = 0$$

$$\cos \theta = 0$$

$$\theta = 90°, 270°$$

25. $4 \cos^2 2x - 4 \cos 2x + 1 = 0$ $\qquad$ $0 \le x < 2\pi$

Solve for cos 2x: $(2 \cos 2x - 1)^2 = 0$

$$2 \cos 2x - 1 = 0$$

$$\cos 2x = \frac{1}{2}$$

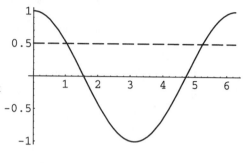

Solve over $0 \le x < 2\pi$: Sketch a graph of $y = \cos u$ and $y = \frac{1}{2}$ in the same coordinate system. We see that the solutions are in the first and fourth quadrants.

$$2x = \frac{\pi}{3} \quad \text{or} \quad 2x = \frac{5\pi}{3}$$

$$x = \frac{\pi}{6} \qquad\qquad x = \frac{5\pi}{6}$$

27. $4 \cos^2 \theta = 7 \cos \theta + 2$ $\qquad$ $0° \le \theta < 180°$

Solve for cos θ: $4 \cos^2 \theta - 7 \cos \theta - 2 = 0$

$$(4 \cos \theta + 1)(\cos \theta - 2) = 0$$

$$4 \cos \theta + 1 = 0 \quad \text{or} \quad \cos \theta - 2 = 0$$

$$\cos \theta = -\frac{1}{4} \qquad\qquad \cos \theta = 2$$

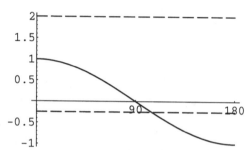

Solve over $0° \le \theta \le 180°$: Sketch a graph of $y = \cos \theta$, $y = -\frac{1}{4}$, and $y = 2$ in the same coordinate system.

$\cos \theta = 2$ $\qquad$ No solution (2 is not in the range of the cosine function.)

$\cos \theta = -\frac{1}{4}$ $\qquad$ From the graph, we see that the solution is in the second quadrant.

$$\theta = \cos^{-1}\left(-\frac{1}{4}\right) = 104.5°$$

29. $\cos 2x + 10 \cos x = 5$ $\qquad$ $0 \le x < 2\pi$

Solve for cos x:

$\qquad$ $2 \cos^2 x - 1 + 10 \cos x = 5$ $\qquad$ Use double-angle identity

$\qquad$ $2 \cos^2 x + 10 \cos x - 6 = 0$

$\qquad\qquad$ $\cos^2 x + 5 \cos x - 3 = 0$ $\qquad$ Use quadratic formula

$$\cos x = \frac{-5 \pm \sqrt{5^2 - 4(1)(-3)}}{2(1)}$$

$$\cos x = \frac{-5 \pm \sqrt{37}}{2}$$

$$\cos x = -5.54138 \quad \text{or} \quad \cos x = 0.54138$$

Solve over $0 \leq x < 2\pi$: Sketch a graph of

$y = \cos x$, $y = -5.54138$, and $y = 0.54138$

in the same coordinate system.

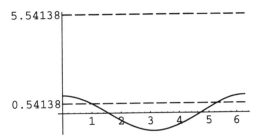

$\cos x = -5.54138$ No solution (-5.54138 is not in the domain of the cosine function.)

$\cos x = 0.54138$ From the graph, we see that the solutions are in the first and fourth quadrants.

$x = \cos^{-1}(0.54138) = 0.9987$

$x = 2\pi - 0.9987 = 5.284$

31. $\cos^2 x = 3 - 5 \cos x$
     $\cos^2 x + 5 \cos x - 3 = 0$

We have found all solutions of this equation over one period in Problem 29. Because the cosine function is periodic with period 2π, all solutions are given by

$0.9987 + 2k\pi$, $5.284 + 2k\pi$, k any integer

If 5.284 is replaced by $5.284 - 2\pi = -0.9987$, this answer is seen to be equivalent to the text answer,

$0.9987 + 2k\pi$, $-0.9987 + 2k\pi$, k any integer

33. $\cos^{-1}(-0.7334)$ has exactly one value, 2.3941; the equation $\cos x = -0.7334$ has infinitely many solutions, which are found by adding $2\pi k$, k any integer, to each solution in one period of $\cos x$, thus, $2.3941 + 2\pi k$.

35. $\sin x + \cos x = 1$ $0 \leq x < 2\pi$

$\pm\sqrt{1 - \cos^2 x} + \cos x = 1$

$\pm\sqrt{1 - \cos^2 x} = 1 - \cos x$

$1 - \cos^2 x = (1 - \cos x)^2$ Squaring both sides

$1 - \cos^2 x = 1 - 2 \cos x + \cos^2 x$

$0 = -2 \cos x + 2 \cos^2 x$

$0 = 2 \cos x (-1 + \cos x)$

$2 \cos x = 0$	$-1 + \cos x = 0$
$\cos x = 0$	$\cos x = 1$
$x = \dfrac{\pi}{2}, \dfrac{3\pi}{2}$	$x = 0$

In squaring both sides, we may have introduced extraneous solutions; hence, it is necessary to check solutions of these equations in the original equation.

Check:

$$x = \frac{\pi}{2}$$

$$\sin x + \cos x = 1$$

$$\sin \frac{\pi}{2} + \cos \frac{\pi}{2} = 1$$

$$1 + 0 = 1$$

A solution

$$x = \frac{3\pi}{2}$$

$$\sin x + \cos x = 1$$

$$\sin \frac{3\pi}{2} + \cos \frac{3\pi}{2} = 1$$

$$-1 + 0 \neq 1$$

Not a solution

$$x = 0$$

$$\sin x + \cos x = 1$$

$$\sin 0 + \cos 0 = 1$$

$$0 + 1 = 1$$

A solution

Solutions: $x = 0, \frac{\pi}{2}$

37. $\sec x + \tan x = 1 \quad 0 \leq x < 2\pi$

$$\pm\sqrt{1 + \tan^2 x} + \tan x = 1$$

$$\pm\sqrt{1 + \tan^2 x} = 1 - \tan x$$

$$1 + \tan^2 x = (1 - \tan x)^2 \qquad \text{Squaring both sides}$$

$$1 + \tan^2 x = 1 - 2\tan x + \tan^2 x$$

$$0 = -2\tan x$$

$$\tan x = 0$$

$$x = 0, \pi$$

In squaring both sides, we may have introduced extraneous solutions; hence, it is necessary to check solutions of these equations in the original equation.

$$x = 0$$

$$\sec x + \tan x = 1$$

$$\sec 0 + \tan 0 = 1$$

$$1 + 0 = 1$$

A solution

$$x = \pi$$

$$\sec x + \tan x = 1$$

$$\sec \pi + \tan \pi = 1$$

$$-1 + 0 \neq 1$$

Not a solution

Solution: $x = 0$

39. $I = 30 \sin 120\pi t \quad I = 25$

$$25 = 30 \sin 120\pi t$$

$$\sin 120\pi t = \frac{25}{30}$$

$$120\pi t = \sin^{-1}\frac{25}{30} \text{ will yield the least positive solution of the equation}$$

$$t = \frac{1}{120\pi} \sin^{-1}\frac{25}{30} = 0.002613 \text{ sec}$$

41.   Following the hint, we solve:

$$I \cos^2 \theta = 0.70I \qquad 0° \leq \theta \leq 180°$$

$$\cos^2 \theta = 0.70$$

$$\cos \theta = \pm\sqrt{0.70}$$

$$\theta = \cos^{-1}\sqrt{0.70} \text{ will yield the least positive solution of the equation}$$

$$\theta = 33.21°$$

43.   We are to solve $3.78 \times 10^7 = \dfrac{3.44 \times 10^7}{1 - 0.206 \cos \theta}$. For convenience, we can divide both sides of this

equation by 10^7:

$$3.78 = \frac{3.44}{1 - 0.206 \cos \theta}$$

$$3.78(1 - 0.206 \cos \theta) = 3.44$$

$$3.78 - (3.78)(0.206) \cos \theta = 3.44$$

$$-(3.78)(0.206) \cos \theta = 3.44 - 3.78$$

$$\cos \theta = \frac{3.44 - 3.78}{-(3.78)(0.206)}$$

$$\theta = \cos^{-1} \frac{3.44 - 3.78}{-(3.78)(0.206)} = 64.1°$$

45. $r = 2 \sin \theta \qquad 0° \le \theta \le 360°$
$r = 2(1 - \sin \theta)$

We solve this system of equations by equating the right sides:

$$2 \sin \theta = 2(1 - \sin \theta) = 2 - 2 \sin \theta$$

$$4 \sin \theta = 2$$

$$\sin \theta = \frac{1}{2}, \ \theta = 30° \text{ or } 150°$$

If we substitute these values of θ in either of the original equations, we obtain

$$r = 2 \sin 30° = 1 \qquad r = 2 \sin 150° = 1$$

Thus, the solutions of the system of equations are

$$(r, \theta) = (1, 30°) \text{ and } (r, \theta) = (1, 150°)$$

47. $xy = -2$

$$(u \cos \theta - v \sin \theta)(u \sin \theta + v \cos \theta) = -2 \qquad \text{Substitution}$$

$$u \cos \theta u \sin \theta + u \cos \theta v \cos \theta - v \sin \theta u \sin \theta - v \sin \theta v \cos \theta = -2 \qquad \text{Multiplication}$$

$$u^2 \cos \theta \sin \theta + uv \cos^2 \theta - uv \sin^2 \theta - v^2 \sin \theta \cos \theta = -2$$

$$u^2 \sin \theta \cos \theta + uv (\cos^2 \theta - \sin^2 \theta) - v^2 \sin \theta \cos \theta = -2$$

We are to find the least positive θ so that the coefficient of the uv term will be zero.

$$\cos^2 \theta - \sin^2 \theta = 0$$

$$\cos 2\theta = 0$$

$$2\theta = \cos^{-1} 0 \qquad \text{yields the least positive } \theta$$

$$2\theta = 90°$$

$$\theta = 45°$$

EXERCISE 5.4   Trigonometric Equations and Inequalities: A Graphing Utility Approach

1. Graph y1 = 2x and y2 = cos x in the same viewing window in a graphing utility and find the point(s) of intersection using an automatic intersection routine. The single intersection point is found to be 0.4502.

Check: 2(0.4502) = 0.9004
          cos(0.4502) = 0.9004

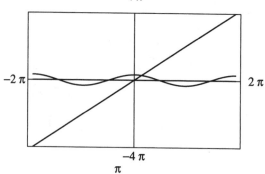

3. Graph y1 = 3x + 1 and y2 = tan 2x in the same viewing window in a graphing utility and find the point(s) of intersection using an automatic intersection routine. The single intersection point is found to be 0.6167.

Check: 3(0.6167) + 1 = 2.8501
          tan 2(0.6167) = 2.8505

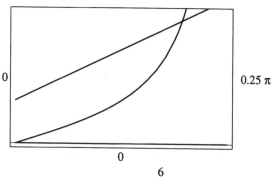

5. Graph y1 = cos 2x + 10 cos x and y2 = 5 in the same viewing window in a graphing utility and find the points of intersection using an automatic intersection routine. The intersection points are found to be 0.9987 and 5.2845.

Check: cos 2(0.9987) + 10 cos(0.9987) = 5.0002
          cos 2(5.2845) + 10 cos(5.2845) = 5.0003

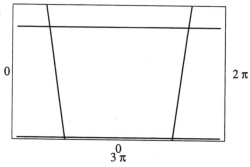

7. Graph y1 = cos² x and y2 = 3 − 5 cos x in the same viewing window, 0 ≤ x < 2π, in a graphing utility and find the points of intersection using an automatic intersection routine. The intersection points are found to be 0.9987 and 5.2845.

Check: cos² (0.9987) = 0.2931
          3 − 5 cos (0.9987) = 0.2930
          cos² (5.2845) = 0.2931
          3 − 5 cos(5.2845) = 0.2930

Because the cosine function is periodic with period 2π, all solutions are given by
    x = 0.9987 + 2kπ,   x = 5.2845 + 2kπ,   k any integer.

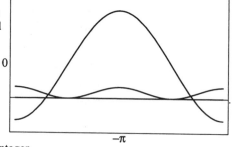

9. Graph $y1 = 2 \sin(x - 2)$ and $y2 = 3 - x^2$ in the same viewing window in a graphing utility. Finding the two points of intersection using an automatic intersection routine, we see that the graph of y1 is below the graph of y2 for the interval $(-1.5099, 1.8281)$.

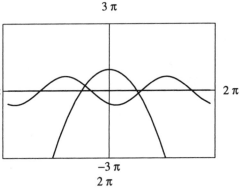

11. Graph $y1 = \sin(3 - 2x)$ and $y2 = 1 - 0.4x$ in the same viewing window in a graphing utility. Finding the points of intersection using an automatic intersection routine, we see that the graph of y1 is not below the graph of y2 for the intervals $[0.4204, 1.2346]$ and $[2.9752, \infty)$.

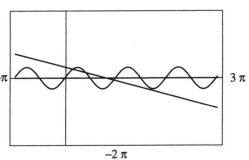

13. $\sin^2 x - 2 \sin x + 1$ is greater than or equal to 0 for all real x because $\sin^2 x - 2 \sin x + 1 = (\sin x - 1)^2$, and the latter is greater than or equal to 0 for all real x.

15. Graph $y1 = 2 \cos \dfrac{1}{x}$ and $y2 = 950x - 4$ in the same viewing window in a graphing utility. Find the points of intersection using an automatic intersection routine. The intersection points are found to be 0.006104 and 0.006137.

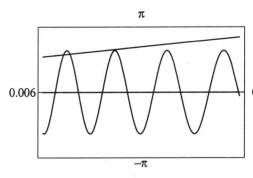

17. Graph $y1 = \cos(\sin x)$ and $y2 = \sin(\cos x)$ in the same viewing window in a graphing utility. Since y1 is above y2 for the entire width of the window, we conclude that on the interval $[-2\pi, 2\pi]$, the inequality $\cos(\sin x) > \sin(\cos x)$ holds everywhere.

19. $\sqrt{3\sin(2x) - 2\cos\left(\dfrac{x}{2}\right) + x}$ is a real

    number as long as $3\sin 2x - 2\cos\dfrac{x}{2} + x \geq 0$.

    Graph $y1 = 3\sin 2x - 2\cos\dfrac{x}{2} + x$ and find the

    zeros of y1 using an automatic intersection
    routine. The zeros are found to be 0.2974,
    1.6073, and 2.7097. The value 3.1416 is the
    right-hand endpoint of the stated interval and the
    desired inequality holds in the intervals
    [0.2974, 1.6073] and [2.7097, 3.1416].

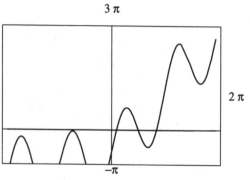

Check: $\sqrt{3\sin[2(0.2974)] - 2\cos\left(\dfrac{0.2974}{2}\right) + 0.2974} = 0.0022$

    The remaining checking is left to the student.

21. We use the given formula $A = \dfrac{1}{2}r^2(\theta - \sin\theta)$ with

    $r = 10$ and $A = 40$.
    $$40 = \frac{1}{2}(10)^2(\theta - \sin\theta) = 50(\theta - \sin\theta)$$
    $$\theta - \sin\theta = 0.8$$
    We graph $y1 = x - \sin x$ and $y2 = 0.8$ on the interval
    from 0 to π. From the figure, we see that y1 and y2
    intersect once on the given interval. Using the
    automatic intersection routine, the solution is found to
    be 1.78 radians.

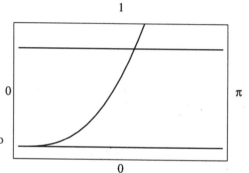

23. The doorway consists of a rectangle of area $4 \times 8 = 32$ square feet
    plus a segment $ABCDA$ as shown. To find the area of the
    segment, we need to find θ and r. In triangle OCD, we have
    $\sin\dfrac{\theta}{2} = \dfrac{2}{r}$. Since $s = r\theta$, we can also write $r = \dfrac{s}{\theta} = \dfrac{5}{\theta}$.
    Therefore, θ must satisfy the relation
    $$\sin\frac{\theta}{2} = 2 \div r = 2 \div \frac{5}{\theta} = \frac{2\theta}{5}; \quad \sin\frac{\theta}{2} = \frac{2\theta}{5}$$

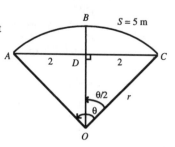

We graph $y1 = \sin\dfrac{x}{2}$ and $y2 = \dfrac{2x}{5}$. From the figure, we see that y1 and y2 intersect once for positive x on the interval 0 to π. Using the automatic intersection routine, the solution is found to be $\theta = 2.262205$ radians. Then,

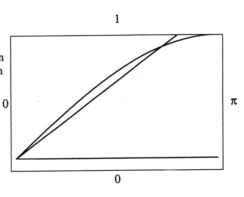

$$r = \frac{5}{\theta} = 2.21023 \text{ ft. Finally,}$$

$$A = \frac{1}{2} r^2 (\theta - \sin\theta)$$

$$= \frac{1}{2} (2.21023)^2 (2.262205 - \sin 2.262205)$$

$$= 3.64 \text{ sq. ft.}$$

Hence, the area of the doorway $= 32 + 3.64 = 35.64$ square feet.

25. (A) Analyzing triangle OAB; we see the following:

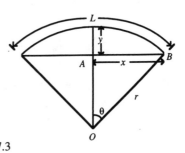

$$OA = r - y$$

$$OA^2 + AB^2 = r^2$$

$$(r - y)^2 + x^2 = r^2$$

$$r^2 - 2ry + y^2 + x^2 = r^2$$

$$y^2 + x^2 = 2ry$$

$$r = \frac{y^2 + x^2}{2y} = \frac{2.5^2 + 5.5^2}{2(2.5)} = 7.3$$

$$\sin\theta = \frac{x}{r}$$

$$\theta = \sin^{-1}\frac{x}{r} = \sin^{-1}\frac{5.5}{7.3} = 0.85325 \text{ rad.}$$

$$L = 2\theta \cdot r = 2(0.85325)(7.3) = 12.4575 \text{ mm}$$

(B) From part (A), we can write $r = \dfrac{y^2 + x^2}{2y}$, $\theta = \sin^{-1}\dfrac{x}{r} = \sin^{-1}\dfrac{2xy}{x^2 + y^2}$,

$$L = 2r\theta = \frac{y^2 + x^2}{y} \sin^{-1}\frac{2xy}{x^2 + y^2}$$

We want to find y if $L = 12.4575$ and $x = 5.4$. Substituting, we find that y must satisfy

$$12.4575 = \frac{y^2 + 29.16}{y} \sin^{-1}\frac{10.8y}{29.16 + y^2}$$

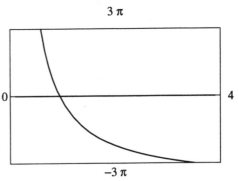

We graph

$$y1 = \frac{y^2 + 29.16}{y} \sin^{-1}\frac{10.8}{29.16 + y^2} - 12.4575.$$

From the figure, we see that y1 has one zero on the interval from 0 to 4. Using the automatic intersection routine, the solution is found to be $y = 2.6495$ mm.

CHAPTER 5   REVIEW EXERCISE

1. $y = \cos^{-1}\dfrac{\sqrt{3}}{2}$ is equivalent to $\cos y = \dfrac{\sqrt{3}}{2}$. What y between 0

and π has cosine equal to $\dfrac{\sqrt{3}}{2}$? y must be associated with a
reference triangle in the first quadrant. Reference triangle is a
special 30°–60° triangle.

$$y = \frac{\pi}{6} \qquad \cos^{-1}\frac{\sqrt{3}}{2} = \frac{\pi}{6}$$

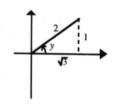

2. $y = \arcsin\dfrac{1}{2}$ is equivalent to $\sin y = \dfrac{1}{2}$. What y between $-\dfrac{\pi}{2}$ and $\dfrac{\pi}{2}$ has sine equal to $\dfrac{1}{2}$? y must be

associated with a reference triangle in the first quadrant. See graph in Problem 1. $\arcsin\dfrac{1}{2} = \dfrac{\pi}{6}$

3. $y = \sin^{-1}\left(-\dfrac{\sqrt{3}}{2}\right)$ is equivalent to $\sin y = -\dfrac{\sqrt{3}}{2}$. What y between

$-\dfrac{\pi}{2}$ and $\dfrac{\pi}{2}$ has sine equal to $-\dfrac{\sqrt{3}}{2}$? y must be in the fourth
quadrant. Reference triangle is a special 30°–60° triangle.

$$y = -\frac{\pi}{3} \qquad \sin^{-1}\left(-\frac{\sqrt{3}}{2}\right) = -\frac{\pi}{3}$$

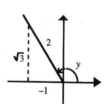

4. $y = \arccos\left(-\dfrac{1}{2}\right)$ is equivalent to $\cos y = -\dfrac{1}{2}$. What y between 0

and π has cosine equal to $-\dfrac{1}{2}$? y must be associated with a
reference triangle in the second quadrant. Reference triangle is a
special 30°–60° triangle.

$$y = \frac{2\pi}{3} \qquad \arccos\left(-\frac{1}{2}\right) = \frac{2\pi}{3}$$

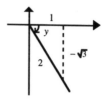

5. $y = \arctan 1$ is equivalent to $\tan y = 1$. What y between $-\dfrac{\pi}{2}$ and $\dfrac{\pi}{2}$
has tangent equal to 1? y must be associated with a reference
triangle in the first quadrant. Reference triangle is a special 45°
triangle.

$$y = \frac{\pi}{4} \qquad \arctan 1 = \frac{\pi}{4}$$

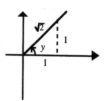

6. $y = \tan^{-1}(-\sqrt{3})$ is equivalent to $\tan y = -\sqrt{3}$. What y between $-\dfrac{\pi}{2}$ and $\dfrac{\pi}{2}$ has tangent equal to $-\sqrt{3}$?
y must be associated with a reference triangle in the fourth quadrant. See graph in Problem 3.

$$\tan^{-1}(-\sqrt{3}) = -\frac{\pi}{3}$$

7. $y = \cot^{-1}\left(-\dfrac{1}{\sqrt{3}}\right)$ is equivalent to $\cot y = -\dfrac{1}{\sqrt{3}}$. What y between 0 and π has cotangent equal

to $-\dfrac{1}{\sqrt{3}}$? y must be associated with a reference triangle in the second quadrant. See graph in

Problem 4.

$$y = \frac{2\pi}{3} \qquad \cot^{-1}\left(-\frac{1}{\sqrt{3}}\right) = \frac{2\pi}{3}$$

8. $y = \operatorname{arcsec}(-2)$ is equivalent to $\sec y = -2$. What y between 0 and π has secant equal to -2?
    y must be associated with a reference triangle in the second quadrant. See graph in Problem 4.

$$y = \frac{2\pi}{3} \qquad \sec^{-1}(-2) = \frac{2\pi}{3}$$

9. $y = \operatorname{arccsc}(-1)$ is equivalent to $\csc y = -1$. No reference triangle can be

    drawn, but the only y between $-\dfrac{\pi}{2}$ and $\dfrac{\pi}{2}$ that has cosecant equal to -1 is

    $y = -\dfrac{\pi}{2}$. Thus, $\operatorname{arccsc}(-1) = -\dfrac{\pi}{2}$.

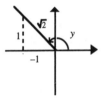

$(0, -1)$

10. $y = \sec^{-1}\left(-\sqrt{2}\right)$ is equivalent to $\sec y = -\sqrt{2}$. What y between 0

    and π has secant equal to $-\sqrt{2}$? y must be associated with a
    reference triangle in the second quadrant. Reference triangle is a
    special 45° triangle.

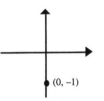

$$y = \frac{3\pi}{4} \qquad \sec^{-1}\left(-\sqrt{2}\right) = \frac{3\pi}{4}$$

11. Calculator in radian mode: $\sin^{-1}(-0.8277) = -0.9750$

12. -1.328 is not in the domain of the inverse cosine function. $-1 \le -1.328 \le 1$ is false.
    $\arccos(-1.328)$ is not defined.

13. Calculator in radian mode: $\tan^{-1}(75.14) = 1.557$

14. Calculator in radian mode: $\cot^{-1} 5.632 = \tan^{-1}\dfrac{1}{5.632} = 0.1757$

15. Calculator in degree mode: $\theta = \cos^{-1} 0.3456 = 69.78°$

16. Calculator in degree mode: $\theta = \arctan(-12.45) = -85.41°$

17. Calculator in degree mode: $\theta = \sin^{-1}(0.0025) = 0.14°$

18. $2 \cos x - \sqrt{3} = 0 \quad 0 \le x < 2\pi$

Solve for cos x: $2 \cos x = \sqrt{3}$

$$\cos x = \frac{\sqrt{3}}{2}$$

Solve over $0 \le x < 2\pi$: Sketch a graph of $y = \cos x$ and $y = \dfrac{\sqrt{3}}{2}$ in the same coordinate system. We see that the solutions are in the first and fourth quadrants.

$$x = \frac{\pi}{6}, \frac{11\pi}{6}$$

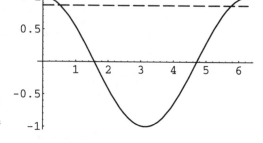

19. $2 \sin^2 \theta = \sin \theta \qquad 0° \le \theta < 360°$

Solve for sin θ: $2 \sin^2 \theta - \sin \theta = 0$

$$\sin \theta (2 \sin \theta - 1) = 0$$

$\sin \theta = 0 \qquad\qquad 2 \sin \theta - 1 = 0$

$\theta = 0°, 180° \qquad\qquad \sin \theta = \dfrac{1}{2}$

$$\theta = 30°, 150°$$

Solutions: $\theta = 0°, 30°, 150°, 180°$

(The student should sketch graphs as appropriate to confirm these values.)

20. $4 \cos^2 x - 3 = 0 \qquad 0 \le x < 2\pi$

Solve for cos x: $\cos^2 x = \dfrac{3}{4}$

$$\cos x = \pm \sqrt{\frac{3}{4}}$$

$$\cos x = \pm \frac{\sqrt{3}}{2}$$

$\cos x = \pm\dfrac{\sqrt{3}}{2} \qquad\qquad \cos x = -\dfrac{\sqrt{3}}{2}$

$x = \dfrac{\pi}{6}, \dfrac{11\pi}{6} \qquad\qquad x = \dfrac{5\pi}{6}, \dfrac{7\pi}{6}$

Solutions: $x = \dfrac{\pi}{6}, \dfrac{5\pi}{6}, \dfrac{7\pi}{6}, \dfrac{11\pi}{6}$

(The student should sketch graphs as appropriate to confirm these values.)

21. $2 \cos^2 \theta + 3 \cos \theta + 1 = 0 \quad 0° \le \theta < 360°$

Solve for cos θ: $(2 \cos \theta + 1)(\cos \theta + 1) = 0$

$2 \cos \theta + 1 = 0 \quad$ or $\quad \cos \theta + 1 = 0$

$\cos \theta = -\dfrac{1}{2} \qquad\qquad \cos \theta = -1$

$\theta = 120°, 240° \qquad\qquad \theta = 180°$

Solutions: $\theta = 120°, 180°, 240°$

(The student should sketch graphs as appropriate to confirm these values.)

22. $\sqrt{2} \sin 4x - 1 = 0$ $0 \le x < \dfrac{\pi}{2}$ is equivalent to

$\sqrt{2} \sin 4x - 1 = 0$ $0 \le 4x < 2\pi$

Solve for sin 4x: $\sqrt{2} \sin 4x = 1$

$\sin 4x = \dfrac{1}{\sqrt{2}}$

Solve over $0 \le 4x < 2\pi$:   Sketch a graph of

$y = \sin u$ and $y = \dfrac{1}{\sqrt{2}}$ in the same coordinate

system. We see that the solutions are in the
first and second quadrants.

$4x = \dfrac{\pi}{4}, \dfrac{3\pi}{4}$

Solutions: $x = \dfrac{\pi}{16}, \dfrac{3\pi}{16}$

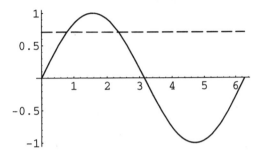

23. $\tan \dfrac{\theta}{2} + \sqrt{3} = 0$ $-180° < \theta < 180°$ is equivalent to

$\tan \dfrac{\theta}{2} + \sqrt{3} = 0$ $-90° < \dfrac{\theta}{2} < 90°$

Solve for $\tan \dfrac{\theta}{2}$: $\tan \dfrac{\theta}{2} = -\sqrt{3}$

Solve over $-90° < \dfrac{\theta}{2} < 90°$:   Sketch a graph

of $y = \tan u$ and $y = -\sqrt{3}$ in the same
coordinate system. We see that the solution
is in the fourth quadrant.

$\dfrac{\theta}{2} = -60°$

Solution: $\theta = -120°$

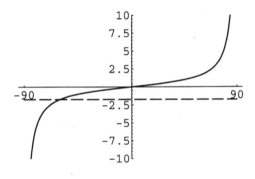

24. $\cos^{-1} x = 25$ is equivalent to $\cos 25 = x$. Since the calculator is in degree mode, calculate
$\cos 25° = 0.906308$.

25. $\cos(\cos^{-1} 0.315) = 0.315$ from the cosine-inverse cosine identity.

26. $\tan^{-1} [\tan(-1.5)] = -1.5$ from the tangent-inverse tangent identity.

27.   Let $y = \tan^{-1}\left(-\dfrac{3}{4}\right)$, then $\tan y = -\dfrac{3}{4}, -\dfrac{\pi}{2} < y < \dfrac{\pi}{2}$.

Draw the reference triangle associated with y, then $\sin y = \sin\left[\tan^{-1}\left(-\dfrac{3}{4}\right)\right]$ can be determined
directly from the triangle.

$a^2 + b^2 = c^2$ $c = \sqrt{4^2 + (-3)^2} = 5$

$\sin\left[\tan^{-1}\left(-\dfrac{3}{4}\right)\right] = \sin y = \dfrac{-3}{5} = -\dfrac{3}{5}$

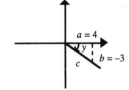

28. Let $y = \arccos\left(-\frac{2}{3}\right)$, then $\cos y = -\frac{2}{3}$, $0 \le y \le \pi$.

    Sketch the reference triangle associated with y, then
    $$\cot y = \cot\left[\cos^{-1}\left(-\frac{2}{3}\right)\right]$$
    can be determined directly from the triangle.

    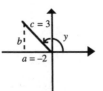

    $a^2 + b^2 = c^2 \qquad b = \sqrt{3^2 - (-2)^2} = \sqrt{5}$

    $\cot\left[\arccos\left(-\frac{2}{3}\right)\right] = \cot y = \frac{-2}{\sqrt{5}} = -\frac{2}{\sqrt{5}}$

29. Let $y = \cot^{-1}\left(-\frac{1}{3}\right)$, then $\cot y = -\frac{1}{3}$, $0 < y < \pi$.

    Draw the reference triangle associated with y, then
    $$\csc y = \csc\left[\cot^{-1}\left(-\frac{1}{3}\right)\right]$$ can be determined directly from the
    triangle.

    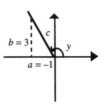

    $a^2 + b^2 = c^2 \qquad c = \sqrt{(-1)^2 + 3^2} = \sqrt{10}$

    $\csc\left[\cot^{-1}\left(-\frac{1}{3}\right)\right] = \csc y = \frac{\sqrt{10}}{3}$

30. Let $y = \text{arccsc } 5$, then $\csc y = 5$, $-\frac{\pi}{2} \le y \le \frac{\pi}{2}$.

    Draw the reference triangle associated with y, then
    $\cos y = \cos(\text{arccsc } 5)$ can be determined directly from the triangle.

    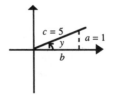

    $a^2 + b^2 = c^2 \qquad c = \sqrt{5^2 + 1^2} = \sqrt{24} = 2\sqrt{6}$

    $\cos(\text{arccsc } 5) = \cos y = \frac{2\sqrt{6}}{5}$

31. Calculator in radian mode: $\sin^{-1}(\cos 22.37) = -1.192$

32. Calculator in radian mode: $\sin^{-1}(\tan 1.345) = $ error
    $\sin^{-1}(\tan 1.345) = \sin^{-1}(4.353)$ is not defined.
    4.353 is not in the domain of the inverse sine function.

33. Calculator in radian mode: $\sin[\tan^{-1}(-14.00)] = -0.9975$

34. Calculator in radian mode: $\csc[\cos^{-1}(-0.4081)] = \dfrac{1}{\sin[\cos^{-1}(-0.4081)]} = 1.095$

35. Calculator in radian mode: $\cos(\cot^{-1} 6.823) = \cos\left(\tan^{-1}\dfrac{1}{6.823}\right) = 0.9894$

36.   Calculator in radian mode: $\sec[\text{arccsc}(-25.52)] = \dfrac{1}{\cos[\text{arccsc}(-25.52)]} = \dfrac{1}{\cos\left[\text{arcsin}\left(\dfrac{1}{-25.52}\right)\right]} = 1.001$

37.   (a) is in degree mode and illustrates a sine-inverse sine identity, since 2 (meaning 2°) is in the domain for this identity, $-90° \le \theta \le 90°$. (b) is in radian mode and does not illustrate this identity, since 2 (meaning 2 radians) is not in the domain for the identity, $-\dfrac{\pi}{2} \le x \le \dfrac{\pi}{2}$.

38. $\sin^2 \theta = -\cos 2\theta \qquad 0° \le \theta \le 360°$

   Solve for sin θ and/or cos θ: $\sin^2 \theta = -(1 - 2\sin^2 \theta) = -1 + 2\sin^2 \theta$ Use double-angle identity
$$0 = -1 \; \sin^2 \theta$$
$$\sin^2 \theta = 1$$
$$\sin \theta = \pm 1$$

   $\sin \theta = 1 \qquad\qquad \sin \theta = -1$
   $\theta = 90° \qquad\qquad \theta = 270°$

   Solutions: $\theta = 90°, 270°$

39. $\sin 2x = \dfrac{1}{2} \qquad 0 \le x < \pi$ is equivalent to

   $\sin 2x = \dfrac{1}{2} \qquad 0 \le 2x < 2\pi$

   Solve over $0 \le 2x < 2\pi$: Sketch a graph of

   $y = \sin u$ and $y = \dfrac{1}{2}$ in the same coordinate

   system. We see that the solutions are in the first and second quadrants.
$$2x = \dfrac{\pi}{6}, \dfrac{5\pi}{6}$$

   Solutions: $x = \dfrac{\pi}{12}, \dfrac{5\pi}{12}$

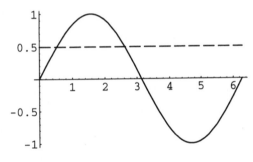

40. $2\cos x + 2 = -\sin^2 x \qquad -\pi \le x < \pi$

   Solve for sin x and/or cos x: $2\cos x + 2 = -(1 - \cos^2 x) = -1 + \cos^2 x$ Use Pythagorean identity
$$0 = \cos^2 x - 2\cos x - 3$$
$$0 = (\cos x - 3)(\cos x + 1)$$

   $\cos x - 3 = 0 \qquad\qquad \cos x + 1 = 0$
   $\cos x = 3 \qquad\qquad \cos x = -1$
   No solution $\qquad\qquad x = -\pi$

   Solution: $x = -\pi$

209

41. $2 \sin^2 \theta - \sin \theta = 0$ All θ

Solve over $0° \leq \theta < 360°$: $\sin \theta (2 \sin \theta - 1) = 0$

$$\sin \theta = 0 \qquad\qquad 2 \sin \theta - 1 = 0$$
$$\theta = 0°, 180° \qquad\qquad \sin \theta = \frac{1}{2}$$
$$\theta = 30°, 150°$$

Thus, the solutions over one period are $0°, 180°, 30°, 150°$.

Write an expression for all solutions: Because the sine function is periodic with period $360°$, all solutions are given by

$$\theta = 0° + k360°, \ 180° + k360°, \ 30° + k360°, \ 150° + k\,360°, \ k \text{ any integer.}$$

42. $\sin 2x = \sqrt{3} \sin x$ All real x

Solve over $0 \leq x < 2\pi$: $2 \sin x \cos x = \sqrt{3} \sin x$ Use double-angle identity
$$2 \sin x \cos x - \sqrt{3} \sin x = 0$$
$$\sin x (2 \cos x - \sqrt{3}) = 0$$

$$\sin x = 0 \qquad\qquad 2 \cos x - \sqrt{3} = 0$$
$$x = 0, \pi \qquad\qquad \cos x = \frac{\sqrt{3}}{2}$$
$$x = \frac{\pi}{6}, \frac{11\pi}{6}$$

Thus, the solutions over one period are $0, \pi, \dfrac{\pi}{6}, \dfrac{11\pi}{6}$.

Write an expression for all solutions: Because the sine and cosine functions are periodic with period 2π, all solutions are given by

$$x = 0 + 2k\pi, \ \pi + 2k\pi, \ \frac{\pi}{6} + 2k\pi, \ \frac{11\pi}{6} + 2k\pi, \ k \text{ any integer.}$$

(The last set of solutions is completely equivalent to the text answer $-\dfrac{\pi}{6} + 2k\pi$. Why?)

43. $2 \sin^2 \theta + 5 \cos \theta + 1 = 0$ $0° \leq \theta < 360°$

Solve for $\sin \theta$ and/or $\cos \theta$: $2(1 - \cos^2 \theta) + 5 \cos \theta + 1 = 0$ Use Pythagorean identity
$$2 - 2 \cos^2 \theta + 5 \cos \theta + 1 = 0$$
$$-2 \cos^2 \theta + 5 \cos \theta + 3 = 0$$
$$2 \cos^2 \theta - 5 \cos \theta - 3 = 0$$
$$(2 \cos \theta + 1)(\cos \theta - 3) = 0$$

$$2 \cos \theta + 1 = 0 \qquad\qquad \cos \theta - 3 = 0$$
$$\cos \theta = -\frac{1}{2} \qquad\qquad \cos \theta = 3$$
$$\theta = 120°, 240° \qquad\qquad \text{No solution}$$

Solutions: $\theta = 120°, 240°$

44. $3 \sin 2x = -2 \cos^2 2x$ $0 \le x \le \pi$ is equivalent to
       $3 \sin 2x = -2 \cos^2 2x$ $0 \le 2x \le 2\pi$
       Solve for sin 2x and/or cos 2x: $3 \sin 2x = -2(1 - \sin^2 2x) = -2 + 2 \sin^2 2x$ Use Pythagorean
                                                                                                        identity

$$0 = 2 \sin^2 2x - 3 \sin 2x - 2$$

$$0 = (2 \sin 2x + 1)(\sin 2x - 2)$$

$2 \sin 2x + 1 = 0$ $\sin 2x - 2 = 0$

$\sin 2x = -\dfrac{1}{2}$ $0 \le 2x \le 2\pi$ $\sin 2x = 2$

$2x = \dfrac{7\pi}{6}, \dfrac{11\pi}{6}$ No solution

$x = \dfrac{7\pi}{12}, \dfrac{11\pi}{12}$

Solutions: $x = \dfrac{7\pi}{12}, \dfrac{11\pi}{12}$

45. $\sin x = 0.7088$
       Solve over $0 \le x < 2\pi$: Sketch a graph of
       $y = \sin x$ and $y = 0.7088$ in the same
       coordinate system. We see that the solutions
       are in the first and second quadrants.

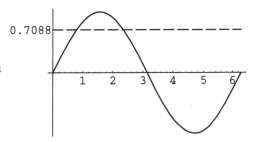

       $x = \sin^{-1} 0.7088 = 0.7878$
       $x = \pi - 0.7878 = 2.354$

Write an expression for all solutions: Because the sine function is periodic with period 2π, all
solutions are given by
       $x = 0.7878 + 2k\pi, \ x = 2.354 + 2k\pi, \ k$ any integer.

46. $\tan x = -4.318$

       Solve over $-\dfrac{\pi}{2} < x < \dfrac{\pi}{2}$: Sketch a graph of

       $y = \tan x$ and $y = -4.318$ in the same
       coordinate system. We see that the only
       solution is in the fourth quadrant.

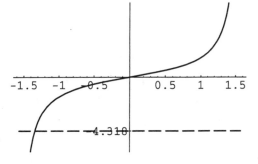

       $x = \tan^{-1} (-4.318) = -1.343$

Write an expression for all solutions:
Because the tangent function is periodic with
period π, all solutions are given by

       $x = -1.343 + k\pi, \ k$ any integer.

47. $\sin^2 x + 2 = 4 \sin x$

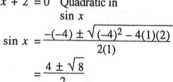

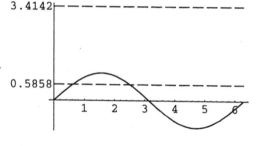

Solve over $0 \le x < 2\pi$:

$\sin^2 x - 4 \sin x + 2 = 0$ Quadratic in
$\sin x$

$\sin x = \dfrac{-(-4) \pm \sqrt{(-4)^2 - 4(1)(2)}}{2(1)}$ 0.5858

$= \dfrac{4 \pm \sqrt{8}}{2}$

$\sin x = 3.4142$ or $\sin x = 0.5858$

Sketch a graph of $y = \sin x$, $y = 3.4142$, and $y = 0.5858$ in the same coordinate system.

$\sin x = 3.4142$ No solution (3.4142 is not in the domain of the sine function.)

$\sin x = 0.5858$ From the graph, we see that the solutions are in the first and second quadrants.

$x = \sin^{-1}(0.5858) = 0.6259$

$x = \pi - 0.6259 = 2.516$

Write an expression for all solutions: Because the sine function is periodic with period 2π, all solutions are given by

$x = 0.6259 + 2k\pi,\ x = 2.516 + 2k\pi,\ k$ any integer.

48. $\tan^2 x = 2 \tan x + 1$

Solve over $-\dfrac{\pi}{2} < x < \dfrac{\pi}{2}$:

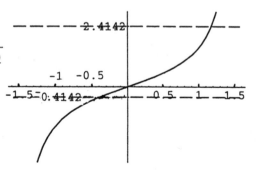

$\tan^2 x - 2 \tan x - 1 = 0$ Quadratic in $\tan x$

$\tan x = \dfrac{-(-2) \pm \sqrt{(-2)^2 - 4(1)(-1)}}{2(1)}$

$= \dfrac{2 \pm \sqrt{8}}{2}$

$\tan x = 2.4142$ or $\tan x = -0.4142$

Sketch a graph of $y = \tan x$, $y = 2.4142$, and $y = -0.4142$ in the same coordinate system.

$\tan x = 2.4142 \qquad x = \tan^{-1} 2.4142 = 1.178$
$\tan x = -0.4142 \qquad x = \tan^{-1}(-0.4142) = -0.3927$

Because the tangent function is periodic with period π, all solutions are given by
$x = 1.178 + k\pi,\ x = -0.3927 + k\pi,\ k$ any integer.

49. Graph y1 = sin x and y2 = 0.25 in the viewing window in a graphing utility and find the point(s) of intersection using an automatic intersection routine. The intersection points are found to be 0.253 and 2.889.

Check: sin(0.253) = 0.250
        sin(2.889) = 0.250

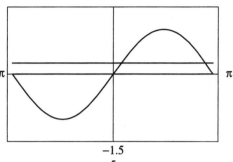

50. Graph y1 = cot x = $\dfrac{1}{\tan x}$ and y2 = –4 in the same viewing window in a graphing utility and find the point(s) of intersection using an automatic intersection routine. The intersection points are found to be –0.245 and 2.897.

Check: $\cot(-0.245) = \dfrac{1}{\tan(-0.245)} = -4.000$

        $\cot(2.897) = \dfrac{1}{\tan 2.897} = -4.007$

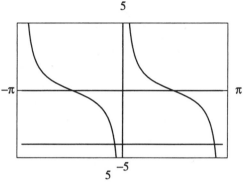

51. Graph y1 = sec x = $\dfrac{1}{\cos x}$ and y2 = 2 in the same viewing window in a graphing utility and find the point(s) of intersection using an automatic intersection routine. The intersection points are found to be ±1.047.

Check: $\sec(\pm 1.047) = \dfrac{1}{\cos(\pm 1.047)} = 1.999$

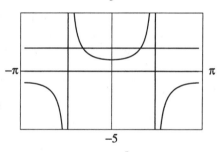

52. Graph y1 = cos x and y2 = x^2 in the same viewing window in a graphing utility and find the point(s) of intersection using an automatic intersection routine. The intersection points are found to be ±0.824.

Check: cos(±0.824) = 0.679
        $(\pm 0.824)^2$ = 0.679

Since |cos x| ≤ 1, while $x^2 > 1$ for real x not shown, there can be no other solutions.

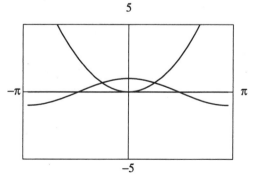

53. Graph y1 = sin x and y2 = $\sqrt{x}$ in the same viewing window in a graphing utility and find the points of intersection using an automatic intersection routine. The single point of intersection appears to be 0.

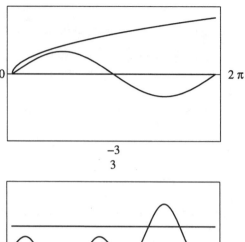

Check: sin 0 = 0 = $\sqrt{0}$.

Since |sin x| ≤ 1, while $\sqrt{x}$ > 1 for real x not shown, there can be no other solutions.

54. Graph y1 = 2 sin x cos 2x and y2 = 1 in the same viewing window in a graphing utility and find the points of intersection using an automatic intersection routine. The intersection points are found to be 4.227 and 5.197.

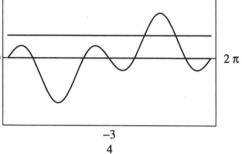

Check: 2 sin 4.227 cos 2(4.227) = 0.999
2 sin 5.197 cos 2(5.197) = 1.002

55. Graph y1 = sin $\frac{x}{2}$ + 3 sin x and y2 = 2 in the same viewing window in a graphing utility and find the points of intersection using an automatic intersection routine. The intersection points are found to be 0.604, 2.797, 7.246, and 8.203.

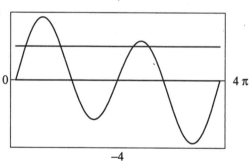

Check: sin $\frac{0.604}{2}$ + 3 sin 0.604 = 2.001

sin $\frac{2.797}{2}$ + 3 sin 2.797 = 1.999

sin $\frac{7.246}{2}$ + 3 sin 7.246 = 1.999

sin $\frac{8.203}{2}$ + 3 sin 8.203 = 2.000

56.   Graph y1 = sin x + 2 sin 2x + 3 sin 3x and y2 = 3
in the same viewing window in a graphing utility
and find the points of intersection using an automatic
intersection routine. The intersection points are
found to be 0.228 and 1.008.

Check:
sin 0.228 + 2 sin 2(0.228) + 3 sin 3(0.228)= 3.002
sin 1.008 + 2 sin 2(1.008) + 3 sin 3(1.008)= 3.003

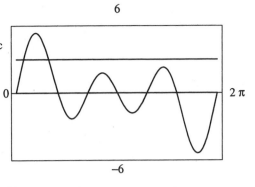

57.   The domain for y1 is the domain for $\sin^{-1} x$, $-1 \le x \le 1$.

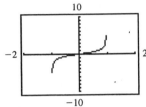
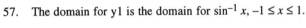

58.   (A) $\sin^{-1} x$ has domain $-1 \le x \le 1$, therefore, $\sin^{-1}\left(\dfrac{x-2}{2}\right)$ has domain given by

$$-1 \le \frac{x-2}{2} \le 1$$
$$-2 \le x - 2 \le 2$$
$$0 \le x \le 4$$

(B) The graph appears only for the domain values $0 \le x \le 4$.

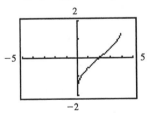

59.   No, $\tan^{-1} 23.255$ represents only one number and one solution. The equation has infinitely many
solutions, which are given by $x = \tan^{-1} 23.255 + k\pi$, k any integer.

60. $\cos x = 1 - \sin x$ $0 \le x < 2\pi$

$$\cos^2 x = (1 - \sin x)^2 \qquad \text{Squaring both sides}$$
$$\cos^2 x = 1 - 2 \sin x + \sin^2 x$$
$$1 - \sin^2 x = 1 - 2 \sin x + \sin^2 x \qquad \text{Pythagorean identity}$$
$$0 = 2 \sin^2 x - 2 \sin x$$
$$0 = 2 \sin x (\sin x - 1)$$

215

$$2 \sin x = 0 \qquad \sin x - 1 = 0$$
$$\sin x = 0 \qquad \sin x = 1$$
$$x = 0, \pi \qquad x = \frac{\pi}{2}$$

In squaring both sides, we may have introduced extraneous solutions; hence, it is necessary to check solutions of these equations in the original equation.

$x = 0$	$x = \pi$	$x = \frac{\pi}{2}$
$\cos x = 1 - \sin x$	$\cos x = 1 - \sin x$	$\cos x = 1 - \sin x$
$\cos 0 = 1 - \sin 0$	$\cos \pi = 1 - \sin \pi$	$\cos \frac{\pi}{2} = 1 - \sin \frac{\pi}{2}$
$1 = 1 - 0$	$-1 \neq 1 - 0$	$0 = 1 - 1$
A solution	Not a solution	A solution

Solutions: $x = 0, \frac{\pi}{2}$

61. $\cos^2 2x = \cos 2x + \sin^2 2x \qquad 0 \leq x < \pi$ is equivalent to

$\cos^2 2x = \cos 2x + \sin^2 2x \qquad 0 \leq 2x < 2\pi$

$\cos^2 2x = \cos 2x + 1 - \cos^2 2x \qquad$ Pythagorean identity

$2 \cos^2 2x - \cos 2x - 1 = 0$

$(2 \cos 2x + 1)(\cos 2x - 1) = 0$

$2 \cos 2x + 1 = 0 \qquad 0 \leq 2x < 2\pi \qquad \cos 2x - 1 = 0 \qquad 0 \leq 2x < 2\pi$

$\cos 2x = -\frac{1}{2} \qquad\qquad\qquad\qquad \cos 2x = 1$

$2x = \frac{2\pi}{3}, \frac{4\pi}{3} \qquad\qquad\qquad\qquad 2x = 0$

$x = \frac{\pi}{3}, \frac{2\pi}{3} \qquad\qquad\qquad\qquad x = 0$

Solutions: $x = 0, \frac{\pi}{3}, \frac{2\pi}{3}$

62. $2 + 2 \sin x = 1 + 2 \cos^2 x \qquad 0 \leq x \leq 2\pi$

Solve for sin x and/or cos x: $\quad 2 + 2 \sin x = 1 + 2(1 - \sin^2 x) \qquad$ Pythagorean identity

$2 + 2 \sin x = 1 + 2 - 2 \sin^2 x$

$2 \sin^2 x + 2 \sin x - 1 = 0 \qquad\qquad$ Quadratic in $\sin x$

$$\sin x = \frac{-2 \pm \sqrt{(2)^2 - 4(2)(-1)}}{2(2)} = \frac{-2 \pm \sqrt{12}}{4}$$

$\sin x = -1.3660 \qquad \sin x = 0.3660$

Solve over $0 \leq x \leq 2\pi$: Sketch a graph of $y = \sin x$, $y = -1.3660$, and $y = 0.3660$ in the same coordinate system.

$\sin x = -1.3660$ No solution (-1.3660 is not in the domain of the sine function.)

$\sin x = 0.3660$ From the graph, we see that the solutions are in the first and second quadrants.

$x = \sin^{-1}(0.3660) = 0.375$

$x = \pi - 0.375 = 2.77$

63.   Let $y = \tan^{-1}\left(-\dfrac{3}{4}\right)$. Then, $\tan y = -\dfrac{3}{4}$. We are asked to evaluate $\sin(2y)$, which is $2 \sin y \cos y$ from the double-angle identity. Sketch a reference triangle associated with y; then,

$\cos y = \cos\left[\tan^{-1}\left(-\dfrac{3}{4}\right)\right]$ and $\sin y = \sin\left[\tan^{-1}\left(-\dfrac{3}{4}\right)\right]$ can be

determined directly from the triangle. Note: $-\dfrac{\pi}{2} < y < \dfrac{\pi}{2}$.

$c = \sqrt{4^2 + (-3)^2} = 5$ $\cos y = \dfrac{4}{5}$ $\sin y = \dfrac{-3}{5} = -\dfrac{3}{5}$

Thus, $\sin\left[2 \tan^{-1}\left(-\dfrac{3}{4}\right)\right] = 2 \sin\left[\tan^{-1}\left(-\dfrac{3}{4}\right)\right] \cos\left[\tan^{-1}\left(-\dfrac{3}{4}\right)\right] = 2\left(-\dfrac{3}{5}\right)\left(\dfrac{4}{5}\right) = -\dfrac{24}{25}$

64.   Let $u = \sin^{-1}\left(\dfrac{3}{5}\right)$ and $v = \cos^{-1}\left(\dfrac{4}{5}\right)$. Then we are asked to evaluate $\sin(u + v)$, which is

$\sin u \cos v + \cos u \sin v$ from the sum identity. We know $\sin u = \sin\left[\sin^{-1}\left(\dfrac{3}{5}\right)\right] = \dfrac{3}{5}$ and

$\cos v = \cos\left[\cos^{-1}\left(\dfrac{4}{5}\right)\right] = \dfrac{4}{5}$ from the function-inverse function identities. It remains to find

$\cos u$ and $\sin v$. Note: $-\dfrac{\pi}{2} \leq u \leq \dfrac{\pi}{2}$ and $0 \leq v \leq \pi$.

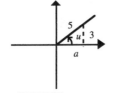

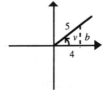

$a = \sqrt{5^2 - 3^2} = 4$ $\cos u = \dfrac{4}{5}$ $b = \sqrt{5^2 - 4^2} = 3$ $\sin v = \dfrac{3}{5}$ $(u = v)$

Then, $\sin\left[\sin^{-1}\left(\dfrac{3}{5}\right) + \cos^{-1}\left(\dfrac{4}{5}\right)\right] = \sin(u + v) = \sin u \cos v + \cos u \sin v$

$$= \left(\dfrac{3}{5}\right)\left(\dfrac{4}{5}\right) + \left(\dfrac{4}{5}\right)\left(\dfrac{3}{5}\right) = \dfrac{12}{25} + \dfrac{12}{25} = \dfrac{24}{25}$$

65. Let $y = \sin^{-1} x$ $-\dfrac{\pi}{2} \le y \le \dfrac{\pi}{2}$ or, equivalently, $\sin y = x$ $-\dfrac{\pi}{2} \le y \le \dfrac{\pi}{2}$
     Geometrically,

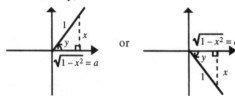

or

In either case, $a = \sqrt{1 - x^2}$. Thus, $\tan(\sin^{-1} x) = \tan y = \dfrac{b}{a} = \dfrac{x}{\sqrt{1 - x^2}}$.

66. Let $y = \tan^{-1} x$ $-\dfrac{\pi}{2} < y < \dfrac{\pi}{2}$ or, equivalently, $x = \tan y$ $-\dfrac{\pi}{2} < y < \dfrac{\pi}{2}$
     Geometrically,

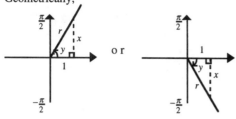

o r

In either case, $r = \sqrt{x^2 + 1}$. Thus, $\cos(\tan^{-1} x) = \cos y = \dfrac{a}{r} = \dfrac{x}{\sqrt{x^2 + 1}}$.

67. (A)

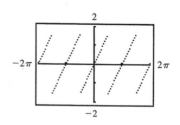

(B) The domain for $\tan x$ is the set of all real numbers,
     except $x = \dfrac{\pi}{2} + k\pi$, k an integer. The range is R, which
     is the domain for $\tan^{-1} x$. Thus, $y = \tan^{-1} (\tan x)$ has a
     graph for all real x, except $x = \dfrac{\pi}{2} + k\pi$, k an integer.
     But $\tan^{-1} (\tan x) = x$ only on the restricted domain of
     $\tan x$, $-\dfrac{\pi}{2} < x < \dfrac{\pi}{2}$.

68. We are to solve $0.05 = 0.08 \sin 880\pi t$.

$$\sin 880\pi t = \dfrac{0.05}{0.08}$$

$$880\pi t = \sin^{-1} \dfrac{0.05}{0.08} \text{ will yield the smallest positive } t$$

$$t = \dfrac{1}{880\pi} \sin^{-1} \dfrac{0.05}{0.08} \quad \text{(calculator in radian mode)}$$

$$= 0.00024 \text{ sec}$$

69.   We are to solve $20 = 30 \sin 120\pi t$.

$$\sin 120\pi t = \frac{20}{30}$$

$\qquad 120\pi t = \sin^{-1}\dfrac{20}{30}$ will yield the smallest positive t

$$t = \frac{1}{120\pi}\sin^{-1}\frac{20}{30} \qquad \text{(calculator in radian mode)}$$

$$= 0.001936 \text{ sec}$$

70.   (A) From the figure in the text we can see that $\tan\dfrac{\theta}{2} = \dfrac{500}{x}$.

Thus, $\dfrac{\theta}{2} = \arctan\dfrac{500}{x}$ $\theta = 2\arctan\dfrac{500}{x}$

(B) We use the above formula with $x = 1200$.

$$\theta = 2\arctan\frac{500}{1200} = 45.2°.$$

71.   (A) θ will increase rapidly from $0°$ to some maximum value, then decrease slowly after that.

(B) We redraw the text figure in a side view. We note:

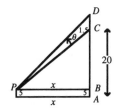

$$\tan BPC = \frac{BC}{x} = \frac{20-5}{x} = \frac{15}{x}$$

$$\tan BPD = \frac{BD}{x} = \frac{15+1.5}{x} = \frac{16.5}{x}$$

$$\theta = BPD - BPC$$

Hence, $\tan\theta = \tan(BPD - BPC)$

$$= \frac{\tan BPD - \tan BPC}{1 + \tan BPD \tan BPC} \text{ by the subtraction identity for tangent.}$$

$$\tan\theta = \frac{\dfrac{16.5}{x} - \dfrac{15}{x}}{1 + \dfrac{16.5}{x}\cdot\dfrac{15}{x}} = \frac{16.5x - 15x}{x^2 + (16.5)(15)} = \frac{1.5x}{x^2 + 247.5}$$

$$\theta = \tan^{-1}\frac{1.5x}{x^2 + 247.5}$$

(C)

x ft	0	5	10	15	20	25	30
$\theta°$	0.00	1.58	2.47	2.73	2.65	2.46	2.25

The maximum value of θ from the table is $2.73°$ when x is 15 feet.

(D)

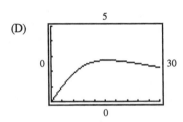

As x increases from 0 ft to 30 ft, θ increases rapidly at first to a maximum of about 2.73° at 15 feet then decreases more slowly.

(E)

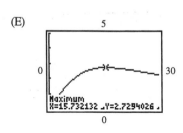

From the automatic maximum routine, the maximum value of θ is 2.73° when $x = 15.73$ feet.

(F)   We are to solve $2.5 \tan^{-1} \dfrac{1.5x}{x^2 + 247.5}$. We graph $y1 = 2.5$ and $y2 = \tan^{-1} \dfrac{1.5x}{x^2 + 247.5}$ in the same viewing window.

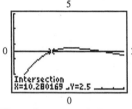   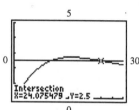

From the automatic intersection routine, $\theta = 2.5°$ when $x = 10.28$ feet or when $x = 24.08$ feet.

72.   Label the figure in the text as shown:

(A)   The area of segment $PABCP$ = Area of sector $OABC$ + Area of triangle OAC. Area of sector $OABC$ =

$$\frac{1}{2}r^2(2\pi - \theta)$$

Since $\sin \dfrac{\theta}{2} = \dfrac{AP}{OA} = \dfrac{d}{2} \div r = \dfrac{d}{2r}, \dfrac{\theta}{2} = \sin^{-1} \dfrac{d}{2r},$

$\theta = 2 \sin^{-1} \dfrac{d}{2r}$. Hence,

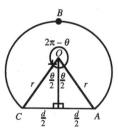

area of sector $OABC = \dfrac{1}{2}r^2 \left(2\pi - 2 \sin^{-1} \dfrac{d}{2r} \right)$

$= \pi r^2 - r^2 \sin^{-1} \dfrac{d}{2r}$

Area of triangle $OAC = \dfrac{1}{2}$ (base)(altitude) $= \dfrac{1}{2}(AC)(OP)$. $AC = d$. From the Pythagorean theorem applied to triangle OAP, $OP^2 + AP^2 = OA^2$, hence

$$OP^2 = OA^2 - AP^2 = r^2 - \left(\frac{d}{2}\right)^2. \quad OP = \sqrt{r^2 - \frac{d^2}{4}} = \sqrt{\frac{4r^2 - d^2}{4}} = \frac{1}{2}\sqrt{4r^2 - d^2}$$

Hence, area of triangle $AOC = \dfrac{1}{2} \cdot d \cdot \dfrac{1}{2}\sqrt{4r^2 - d^2} = \dfrac{d}{4}\sqrt{4r^2 - d^2}$.

Thus, area of segment $PABCP$ = area of sector $OABC$ + Area of triangle OAC

$= \pi r^2 - r^2 \sin^{-1} \dfrac{d}{2r} + \dfrac{d}{4}\sqrt{4r^2 - d^2}$.

(B)

d ft	8	10	12	14	16	18	20
A ft^2	704	701	697	690	682	670	654

(C) We are to solve

$$675 = \pi 15^2 - 15^2 \sin^{-1} \dfrac{d}{2 \cdot 15} + \dfrac{d}{4}\sqrt{4 \cdot 15^2 - d^2}$$

or

$$675 = 225\pi - 225 \sin^{-1}\dfrac{d}{30} + \dfrac{d}{4}\sqrt{900 - d^2}$$

We graph y1 = 675 and y2 = $225\pi - 225 \sin^{-1}\dfrac{d}{30} + \dfrac{d}{4}\sqrt{900 - d^2}$ in the same viewing window.

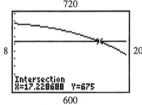

From the automatic intersection routine, $d = 17.2$ ft for $A = 675$ ft^2.

73. Analyzing triangle OAB, we see the following:

$$OA = r - h$$
$$OA^2 + AB^2 = r^2$$
$$(r - h)^2 + 5^2 = r^2$$
$$r^2 - 2hr + h^2 + 25 = r^2$$
$$h^2 + 25 = 2hr$$
$$r = \dfrac{h^2 + 25}{2h}$$

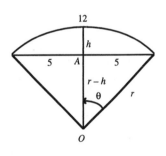

$\sin \theta = \dfrac{5}{r}$ $\theta = \sin^{-1}\dfrac{5}{r}$. Using the formula $s = r\theta$, we can write $12 = 2\theta \cdot r$; $\theta = \dfrac{6}{r}$.

Hence, $\dfrac{6}{r} = \sin^{-1}\dfrac{5}{r}$.

$6 \div \dfrac{h^2 + 25}{2h} = \sin^{-1}\left(5 \div \dfrac{h^2 + 25}{2h}\right)$. Thus, $\dfrac{12h}{h^2 + 25} = \sin^{-1}\dfrac{10h}{h^2 + 25}$ is the equation that must be satisfied by h.

We graph $y1 = \dfrac{12h}{h^2 + 25}$ and $y2 = \sin^{-1}\dfrac{10h}{h^2 + 25}$. From the figure, we see that y1 and y2 intersect once for nonzero h on the interval from 0 to 4. Using zoom and trace procedures or the automatic intersection routine, the intersection is found to be $h = 2.82$ ft.

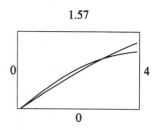

74. (A)

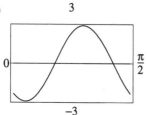

(B) We graph $y1 = -1.8 \sin 4t - 2.4 \cos 4t$ and $y2 = 2$. From the figure, we see that y1 and y2 intersect twice on the indicated interval.

Using zoom and trace procedures or the automatic intersection routine, the intersections are found to be $t = 0.74$ sec and $t = 1.16$ sec.

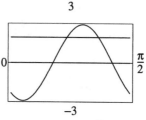

(C) We graph $y1 = -1.8 \sin 4t - 2.4 \cos 4t$ and $y2 = -2$. From the figure, we see that y1 and y2 intersect twice on the indicated interval.

Using zoom and trace procedures or the automatic intersection routine, the intersections are found to be $t = 0.37$ sec and $t = 1.52$ sec.

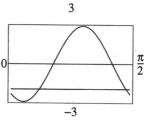

CUMULATIVE REVIEW EXERCISE CHAPTERS 1—5

1. $\theta_d = \dfrac{180°}{\pi\,\text{rad}}\,\theta_r$. Thus, if $\theta_r = 4.21$, $\theta_d = \dfrac{180}{\pi}(4.21) = 241.22°$

2. $\theta_r = \dfrac{\pi\,\text{rad}}{180°}\,\theta_d$. Thus, if $\theta_d = 505°42'$, or $\left(505 + \dfrac{42}{60}\right)°$,

$$\theta_r = \dfrac{\pi}{180}\left(505 + \dfrac{42}{60}\right) = 8.83 \text{ rad}$$

3. An angle of radian measure 0.5 is the central angle of a circle subtended by an arc with measure half that of the radius of the circle.

4. Let $c = 7.6$ m, $b = 4.5$ m

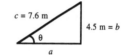

Solve for θ: We will use the sine: $\sin\theta = \dfrac{b}{c} = \dfrac{4.5 \text{ m}}{7.6 \text{ m}} = 0.5921$

$$\theta = \sin^{-1}(0.5921) = 36°$$

Solve for the complementary angle: $90° - \theta = 90° - 36° = 54°$

Solve for a: We choose the cosine to find a. $\cos\theta = \dfrac{a}{c}$

$$a = c\cos\theta = (7.6 \text{ m})(\cos 36°) = 6.1 \text{ m}$$

5. $P(a, b) = (-5, -12)$

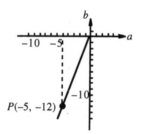

$$r = \sqrt{a^2 + b^2} = \sqrt{(-5)^2 + (-12)^2} = 13$$

$$\cos\theta = \frac{a}{r} = \frac{-5}{13} = -\frac{5}{13}$$

$$\cot\theta = \frac{a}{b} = \frac{-5}{-12} = \frac{5}{12}$$

6. (A) Degree mode: $\cos 67°45' = \cos 67.75°$ Convert to decimal degrees, if necessary
$$= 0.3786$$

(B) Degree mode: $\csc 176.2° = \dfrac{1}{\sin 176.2°} = 15.09$

(C) Radian mode: $\cot 2.05 = \dfrac{1}{\tan 2.05} = -0.5196$

Cumulative Review Exercise Chapters 1—5

7. (A)

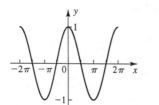

(B)

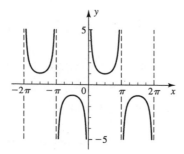

(C)

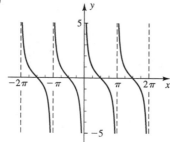

8. No, because $\sin \theta = \dfrac{1}{\csc \theta}$ and either both are positive or both are negative.

9. $\tan x \csc x = \dfrac{\sin x}{\cos x} \dfrac{1}{\sin x}$ Quotient and reciprocal identities

$= \dfrac{1}{\cos x}$ Algebra

$= \sec x$ Reciprocal identity

10. $\csc \theta - \sin \theta = \dfrac{1}{\sin \theta} - \sin \theta$ Reciprocal identity

$= \dfrac{1}{\sin \theta} - \dfrac{\sin^2 \theta}{\sin \theta}$ Algebra

$= \dfrac{1 - \sin^2 \theta}{\sin \theta}$ Algebra

$= \dfrac{\cos^2 \theta}{\sin \theta}$ Pythagorean identity

$= \cos \theta \cdot \dfrac{\cos \theta}{\sin \theta}$ Algebra

$= \cos \theta \cot \theta$ Quotient identity

11. $(\sin^2 u)(\tan^2 u + 1)$ $= \sin^2 u \sec^2 u$ Pythagorean identity

$$= \sin^2 u \, \frac{1}{\cos^2 u}$$ Reciprocal identity

$$= \frac{\sin^2 u}{\cos^2 u}$$ Algebra

$$= \tan^2 u$$ Quotient identity

$$= \sec^2 u - 1$$ Pythagorean identity

12. $\dfrac{\sin^2 \alpha - \cos^2 \alpha}{\sin \alpha \cos \alpha} = \dfrac{\dfrac{\sin^2 \alpha}{\sin \alpha \cos \alpha} - \dfrac{\cos^2 \alpha}{\sin \alpha \cos \alpha}}{\dfrac{\sin \alpha \cos \alpha}{\sin \alpha \cos \alpha}}$ Algebra

$$= \frac{\dfrac{\sin \alpha}{\cos \alpha} - \dfrac{\cos \alpha}{\sin \alpha}}{\dfrac{\sin \alpha \cos \alpha}{\cos \alpha \sin \alpha}}$$ Algebra

$$= \frac{\tan \alpha - \cot \alpha}{\tan \alpha \cot \alpha}$$ Quotient identities

13. $\cos x = \sin x$ is not an identity, because $\sin x$ and $\cos x$ are not equal for other values of x for which both sides are defined, for example, they are not equal for $x = 0$ or $x = \dfrac{\pi}{2}$.

14. Locate the 45° reference triangle, determine (a, b) and r, then evaluate.

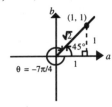

$$\cos\left(-\frac{7\pi}{4}\right) = \frac{1}{\sqrt{2}}$$

15. Locate the 30°–60° reference triangle, determine (a, b) and r, then evaluate.

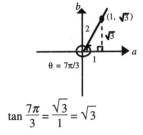

$$\tan \frac{7\pi}{3} = \frac{\sqrt{3}}{1} = \sqrt{3}$$

16. $(a, b) = (0, -1)$ $r = 1$

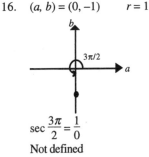

$$\sec \frac{3\pi}{2} = \frac{1}{0}$$
Not defined

17. $y = \arctan 0$ is equivalent to $\tan y = 0$. No reference triangle can be drawn, but the only y between $-\dfrac{\pi}{2}$ and $\dfrac{\pi}{2}$ which has tangent equal to 0 is $y = 0$. Thus, $\arctan 0 = 0$.

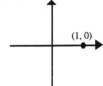

18. $y = \cos^{-1}\left(-\dfrac{\sqrt{3}}{2}\right)$ is equivalent to $\cos y = -\dfrac{\sqrt{3}}{2}$. What y between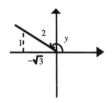

    0 and π has cosine equal to $-\dfrac{\sqrt{3}}{2}$? y must be associated with a

    reference triangle in the second quadrant.
    Reference triangle is a special 30°–60° triangle.

    $$y = \frac{5\pi}{6} \qquad \cos^{-1}\left(-\frac{\sqrt{3}}{2}\right) = \frac{5\pi}{6}$$

19. 3 is not in the domain of the inverse sine function. $-1 \le 3 \le 1$ is false. arcsin 3 is not defined.

20. $y = \text{arccot}\,(-\sqrt{3})$ is equivalent to $\cot y = -\sqrt{3}$ and $0 < y < \pi$.
    What number between 0 and π has cotangent equal to $-\sqrt{3}$?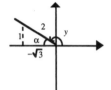
    y must be positive and in the second quadrant.

    $$\cot y = -\sqrt{3} = \frac{-\sqrt{3}}{1} \qquad \alpha = \frac{\pi}{6} \quad y = \frac{5\pi}{6}$$

    Thus, $\cot^{-1}(-\sqrt{3}) = \dfrac{5\pi}{6}$

21. Calculator in radian mode: $\sin^{-1}(0.0505) = 0.0505$

22. Calculator in radian mode: $\cos^{-1}(-0.7228) = 2.379$

23. Calculator in radian mode: $\arctan(-9) = -1.460$

24. Calculator in radian mode: $\text{arccot}\,3 = \arctan\dfrac{1}{3} = 0.3218$

25. Calculator in radian mode: $\sec^{-1} 2.6 = \cos^{-1}\dfrac{1}{2.6} = 1.176$

26. $\tan^{-1} x = -1.000000$ is equivalent to $\tan(-1.000000) = x$. Since the calculator is in radian mode, calculate
    $x = \tan(-1.000000) = -1.557408$.

27. $2\sin\theta - 1 = 0 \qquad 0° \le \theta < 360°$

    Solve for $\sin\theta$: $\quad 2\sin\theta = 1$

    $$\sin\theta = \frac{1}{2}$$

    Solve over $0° \le \theta < 360°$: Sketch a
    graph of $y = \sin\theta$ and $y = \dfrac{1}{2}$ in the same
    coordinate system. We see that the
    solutions are in the first and second
    quadrants.
    $\qquad \theta = 30°, 150°$

    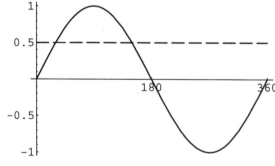

28. $3 \tan x + \sqrt{3} = 0 \quad -\frac{\pi}{2} < x < \frac{\pi}{2}$

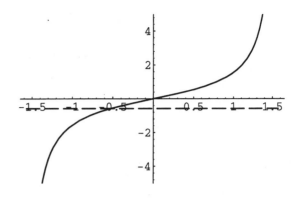

Solve for tan x: $\quad 3 \tan x = -\sqrt{3}$

$$\tan x = -\frac{\sqrt{3}}{3}$$

Solve over $-\frac{\pi}{2} < x < \frac{\pi}{2}$: Sketch a

graph of $y = \tan x$ and $y = -\frac{\sqrt{3}}{3}$ in the

same coordinate system. We see that the single solution is in the fourth quadrant.

$$x = -\frac{\pi}{6}$$

29. $2 \cos x + 2 = 0 \quad -\pi \leq x < \pi$

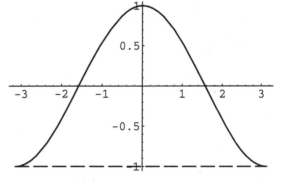

Solve for cos x: $\quad 2 \cos x = -2$

$\cos x = -1$

Solve over $-\pi \leq x < \pi$: Sketch a graph of $y = \cos x$ and $y = -1$ in the same coordinate system. We see that the single solution is $x = -\pi$.

30. $\sin x \cos y = \frac{1}{2} [\sin(x + y) + \sin(x - y)]$. Let $x = 7u$ and $y = 3u$

$\sin 7u \cos 3u = \frac{1}{2} [\sin(7u + 3u) + \sin(7u - 3u)] = \frac{1}{2} (\sin 10u + \sin 4u)$

31. $\cos x - \cos y = -2 \sin \frac{x + y}{2} \sin \frac{x - y}{2}$. Let $x = 5w$ and $y = w$

$\cos 5w - \cos w = -2 \sin \frac{5w + w}{2} \sin \frac{5w - w}{2} = -2 \sin 3w \sin 2w$

32. Yes, because, using the formula $\theta_d = \frac{360°}{2\pi} \theta_r$, if θ_r is halved, then θ_d will also be halved.

33. $92.462° = 92°(0.462 \times 60)' = 92°27.72' = 92°27'(0.72 \times 60)" = 92°27'43"$

34. Since $\frac{s}{C} = \frac{\theta°}{360°}$ then $\frac{12 \text{ in}}{30 \text{ in}} = \frac{\theta}{360°}$ $\quad \theta = \frac{12}{30} \cdot 360° = 144°$

35. Since the two right triangles are similar, we can write:

$$\frac{4}{x} = \frac{x}{x+3}$$

$$x(x+3)\frac{4}{x} = x(x+3)\frac{x}{x+3}$$

$$(x+3)4 = x^2$$

$$4x + 12 = x^2$$

$$0 = x^2 - 4x - 12$$

$$0 = (x-6)(x+2)$$

$$x - 6 = 0 \quad \text{or} \quad x + 2 = 0$$
$$x = 6 \qquad\qquad x = -2 \qquad \text{We discard the negative answer.}$$

Since $\tan\theta = \frac{4}{x}$, we can write $\tan\theta = \frac{4}{6}$ $\qquad \theta = \tan^{-1}\frac{4}{6} = 33.7°$

36. $\theta_r = \frac{\pi\,\text{rad}}{180°}\,\theta_d$ $\qquad$ Thus, if $\theta_d = 72°$, $\theta_r = \frac{\pi}{180}\cdot 72 = \frac{2\pi}{5}$

37. We sketch a reference triangle and label what we know. Since $\tan\theta = \frac{b}{a} = \frac{1}{2} = \frac{-1}{-2}$ and $\sin\theta < 0$, the terminal side of θ must lie in quadrant III. Hence, $b = -1$ and $a = -2$.
Use the Pythagorean theorem to find r: $\quad (-2)^2 + (-1)^2 = r^2$
$$r^2 = 5$$
$$r = \sqrt{5}\ (r \text{ is never negative})$$

We can now find the other five functions.

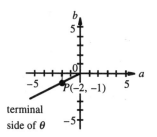

$$\sin\theta = \frac{b}{r} = \frac{-1}{\sqrt{5}} \qquad \cot\theta = \frac{a}{b} = \frac{2}{1} = 2$$

$$\cos\theta = \frac{a}{r} = -\frac{2}{\sqrt{5}} \qquad \sec\theta = \frac{r}{a} = \frac{\sqrt{5}}{-2} = -\frac{\sqrt{5}}{2}$$

$$\csc\theta = \frac{r}{b} = \frac{\sqrt{5}}{-1} = -\sqrt{5}$$

terminal
side of θ

38. $y = 2 - 2\sin\frac{x}{2}$. Amplitude $= |-2| = 2$. Period $= 2\pi \div \frac{1}{2}$

$= 4\pi$. This graph is the graph of $y = -2\sin\frac{x}{2}$ moved up 2
units. We start by drawing a horizontal broken line 2 units
above the x axis, then graph $y = -2\sin\frac{x}{2}$
(an upside down sine curve with amplitude 2 and period
4π) relative to the broken line and the original y axis.

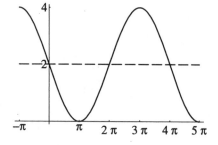

39. Amplitude $= |A| = |3| = 3$. Phase Shift and Period: Solve

$$Bx + C = 0 \quad \text{and} \quad Bx + C = 2\pi$$
$$2x - \pi = 0 \qquad\qquad 2x - \pi = 2\pi$$
$$2x = \pi \qquad\qquad 2x = \pi + 2\pi$$
$$x = \frac{\pi}{2} \qquad\qquad x = \frac{\pi}{2} + \pi$$
$$\text{Phase Shift} = \frac{\pi}{2} \qquad \text{Period} = \pi$$

Graph one cycle over the interval from $\frac{\pi}{2}$ to $\frac{\pi}{2} + \pi = \frac{3\pi}{2}$. Then extend the graph from $-\pi$ to 2π.

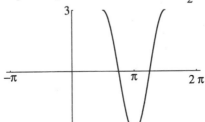 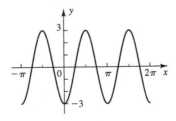

40. We find the period and phase shift by solving $\pi x - \frac{\pi}{4} = 0$ and $\pi x - \frac{\pi}{4} = \pi$

$$\pi x = \frac{\pi}{4} \qquad\qquad \pi x = \frac{\pi}{4} + \pi$$
$$x = \frac{1}{4} \qquad\qquad x = \frac{1}{4} + 1$$
$$\text{Phase Shift} = \frac{1}{4} \qquad \text{Period} = 1$$

We then sketch one period of the graph starting at $x = \frac{1}{4}$ (the phase shift) and ending at $x = \frac{1}{4} + 1 = \frac{5}{4}$ (the phase shift plus one period). Note that a vertical asymptote is at $x = \frac{3}{4}$. We then extend the graph from 0 to 3.

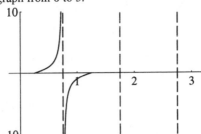 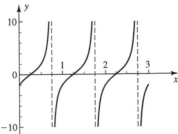

41. Amplitude $= \frac{1}{2} (y_{max} - y_{min}) = \frac{1}{2} (4 - (-2)) = 3.$ $k = \frac{1}{2} (y_{max} + y_{min}) = \frac{1}{2} (4 + (-2)) = 1.$

Period $= \frac{2\pi}{B} = 2.$ Thus, $B = \frac{2\pi}{2} = \pi.$ The form of the graph is that of the basic sine curve.

Thus, $y = |A| \sin Bx + k = 3 \cos \pi x + 1.$

42. Amplitude $= \frac{1}{2}(y_{max} - y_{min}) = \frac{1}{2}(4 - (-3)) = 2.$ $k = \frac{1}{2}(y_{max} + y_{min}) = \frac{1}{2}(1 + (-3)) = -1.$

Period $= \frac{2\pi}{B} = 4.$ Thus, $B = \frac{2\pi}{4} = \frac{\pi}{2}.$ The form of the graph is that of the basic sine curve.

Thus, $y = |A| \sin Bx + k = 2 \sin \frac{\pi}{2}x - 1.$

43. Let $a = 19.4$ cm and $b = 41.7$ cm.

Solve for c: We will use the Pythagorean theorem
$$c^2 = a^2 + b^2$$
$$c = \sqrt{a^2 + b^2} = \sqrt{(19.4)^2 + (41.7)^2} = 46.0 \text{ cm}$$

41.7 cm
θ
19.4 cm

Solve for θ: We will use the tangent $\tan \theta = \frac{b}{a} = \frac{41.7 \text{ cm}}{19.4 \text{ cm}} = 2.1495$
$$\theta = \tan^{-1}(2.1495) = 65°0'$$

Solve for the complementary angle: $90° - \theta = 90° - 65°0' = 25°0'$

44. $\dfrac{\cos x}{1 + \sin x} + \tan x = \dfrac{\cos x}{1 + \sin x} + \dfrac{\sin x}{\cos x}$ \qquad Quotient identity

$$= \frac{\cos x \cos x}{\cos x(1 + \sin x)} + \frac{\sin x(1 + \sin x)}{\cos x(1 + \sin x)} \qquad \text{Algebra}$$

$$= \frac{\cos^2 x + \sin x + \sin^2 x}{\cos x(1 + \sin x)} \qquad \text{Algebra}$$

$$= \frac{\cos^2 x + \sin^2 x + \sin x}{\cos x(1 + \sin x)} \qquad \text{Algebra}$$

$$= \frac{1 + \sin x}{\cos x(1 + \sin x)} \qquad \text{Pythagorean identity}$$

$$= \frac{1}{\cos x} \qquad \text{Algebra}$$

$$= \sec x \qquad \text{Reciprocal identity}$$

45. In this problem, it is more straightforward to start with the right-hand side of the identity to be verified. The student can confirm that the steps would be valid if reversed.

$$(\sec \theta - \tan \theta)(\sec \theta + 1) = \left(\frac{1}{\cos \theta} - \frac{\sin \theta}{\cos \theta}\right)\left(\frac{1}{\cos \theta} + 1\right) \qquad \text{Quotient and Reciprocal identities}$$

$$= \left(\frac{1 - \sin \theta}{\cos \theta}\right)\left(\frac{1}{\cos \theta} + \frac{\cos \theta}{\cos \theta}\right) \qquad \text{Algebra}$$

$$= \frac{1 - \sin \theta}{\cos \theta} \cdot \frac{1 + \cos \theta}{\cos \theta} \qquad \text{Algebra}$$

$$= \frac{(1 - \sin \theta)(1 + \cos \theta)}{\cos^2 \theta} \qquad \text{Algebra}$$

$$= \frac{(1 - \sin\theta)(1 + \cos\theta)}{1 - \sin^2\theta} \qquad \text{Pythagorean identity}$$

$$= \frac{(1 - \sin\theta)(1 + \cos\theta)}{(1 - \sin\theta)(1 + \sin\theta)} \qquad \text{Algebra}$$

$$= \frac{1 + \cos\theta}{1 + \sin\theta} \qquad \text{Algebra}$$

46. $\cot \dfrac{u}{2} = 1 \div \tan \dfrac{u}{2}$ Reciprocal identity

$$= 1 \div \frac{\sin u}{1 + \cos u} \qquad \text{Half-angle identity for tangent}$$

$$= 1 \cdot \frac{1 + \cos u}{\sin u} \qquad \text{Algebra}$$

$$= \frac{1 + \cos u}{\sin u} \qquad \text{Algebra}$$

$$= \frac{1}{\sin u} + \frac{\cos u}{\sin u} \qquad \text{Algebra}$$

$$= \csc u + \cot u \qquad \text{Quotient and Reciprocal identities}$$

47. $\dfrac{2}{1 + \sec 2\theta} = \dfrac{2}{1 + \dfrac{1}{\cos 2\theta}}$ Reciprocal identity

$$= \frac{2 \cos 2\theta}{2 \cos 2\theta + 1} \qquad \text{Algebra}$$

$$= \frac{2(2 \cos^2\theta - 1)}{2 \cos^2\theta - 1 + 1} \qquad \text{Double-angle identity}$$

$$= \frac{4 \cos^2\theta - 2}{2 \cos^2\theta} \qquad \text{Algebra}$$

$$= \frac{4 \cos^2\theta}{2 \cos^2\theta} - \frac{2}{2 \cos^2\theta} \qquad \text{Algebra}$$

$$= 2 - \frac{1}{\cos^2\theta} \qquad \text{Algebra}$$

$$= 2 - \sec^2\theta \qquad \text{Reciprocal identity}$$

$$= 2 - (\tan^2\theta + 1) \qquad \text{Pythagorean identity}$$

$$= 1 - \tan^2\theta \qquad \text{Algebra}$$

48. $\dfrac{\sin x - \sin y}{\cos x + \cos y}$

$= \dfrac{(\sin x - \sin y) \sin(x-y)}{(\cos x + \cos y) \sin(x-y)}$ Algebra

$= \dfrac{(\sin x - \sin y)(\sin x \cos y - \cos x \sin y)}{(\cos x + \cos y) \sin(x-y)}$ Difference identity for sine

$= \dfrac{\sin^2 x \cos y - \sin x \cos x \sin y - \sin x \sin y \cos y + \cos x \sin^2 y}{(\cos x + \cos y) \sin(x-y)}$ Algebra

$= \dfrac{(1 - \cos^2 x)\cos y - \sin x \cos x \sin y - \sin x \sin y \cos y + \cos x (1 - \cos^2 y)}{(\cos x + \cos y) \sin(x-y)}$ Pythagorean identity

$= \dfrac{\cos y - \cos^2 x \cos y - \sin x \cos x \sin y - \sin x \sin y \cos y + \cos x - \cos x \cos^2 y}{(\cos x + \cos y) \sin(x-y)}$ Algebra

$= \dfrac{(\cos x - \cos^2 x \cos y - \sin x \cos x \sin y) + (\cos y - \sin x \sin y \cos y - \cos x \cos^2 y)}{(\cos x + \cos y) \sin(x-y)}$ Algebra

$= \dfrac{\cos x (1 - \cos x \cos y - \sin x \sin y) + \cos y (1 - \sin x \sin y - \cos x \cos y)}{(\cos x + \cos y) \sin(x-y)}$ Algebra

$= \dfrac{(\cos x + \cos y)(1 - \cos x \cos y - \sin x \sin y)}{(\cos x + \cos y) \sin(x-y)}$ Algebra

$= \dfrac{1 - (\cos x \cos y + \sin x \sin y)}{\sin(x-y)}$ Algebra

$= \dfrac{1 - \cos(x-y)}{\sin(x-y)}$ Difference identity for cosine

$= \tan\dfrac{x-y}{2}$ Half-angle identity

49. (A) From the identities for negatives,

$$\sin(-x) = -\sin x$$

Hence

$$\sin(-x) = -0.4969$$

(B) From the Pythagorean identity, $\cos^2 x + \sin^2 x = 1$, thus

$$\cos^2 x = 1 - \sin^2 x$$

Hence

$$(\cos x)^2 = 1 - 0.2469 = 0.7531$$

50. $\sin 3x \cos x - \cos 3x \sin x = \sin(3x - x)$ Difference identity for sine

                                       $= \sin 2x$ Algebra

Graph $y1 = \sin 3x \cos x - \cos 3x \sin x$ and $y2 = \sin 2x$.

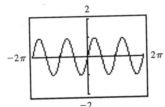

51. First draw a reference triangle in the third quadrant and find $\cos x$.

$r = \sqrt{(-15)^2 + (-8)^2} = 17$. $\cos x = \dfrac{-15}{17}$. We can now find $\sin\dfrac{x}{2}$

from the half-angle identity, after determining its sign, as follows:

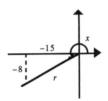

If $\pi < x < \dfrac{3\pi}{2}$, then $\dfrac{\pi}{2} < \dfrac{x}{2} < \dfrac{3\pi}{4}$.

Thus, $\dfrac{x}{2}$ is in the second quadrant, where sine is positive.

Using half-angle identities, we obtain

$$\sin\frac{x}{2} = \sqrt{\frac{1 - \cos x}{2}} = \sqrt{\frac{1 - (-15/17)}{2}} = \sqrt{\frac{32/17}{2}} = \sqrt{\frac{16}{17}} \text{ or } \frac{4}{\sqrt{17}}$$

To find $\cos 2x$, we use a double-angle identity.

$$\cos 2x = 2\cos^2 x - 1 = 2\left(-\frac{15}{17}\right)^2 - 1 = \frac{2}{1}\cdot\frac{225}{289} - 1 = \frac{450 - 289}{289} = \frac{161}{289}$$

52. $[-2\pi, \pi) \cup [\pi, 2\pi]$

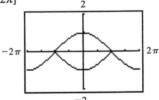

53. (A) Graph both sides of the equation in the same viewing window.

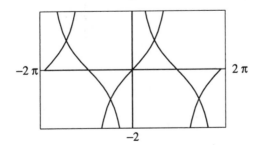

$\dfrac{\sin x}{1 + \cos x} = \dfrac{1 + \cos x}{\sin x}$ is not an identity, since the graphs do not match.

Try $x = \dfrac{\pi}{4}$

Left side: $\dfrac{\sin\left(\dfrac{\pi}{4}\right)}{1 + \cos\left(\dfrac{\pi}{4}\right)} = \dfrac{\dfrac{1}{\sqrt{2}}}{1 + \dfrac{1}{\sqrt{2}}} = \dfrac{1}{\sqrt{2} + 1} = \sqrt{2} - 1$

Right side: $\dfrac{1 + \cos\left(\dfrac{\pi}{4}\right)}{\sin\dfrac{\pi}{4}} = \dfrac{1 + \dfrac{1}{\sqrt{2}}}{\dfrac{1}{\sqrt{2}}} = \sqrt{2} + 1$

(B) Graph both sides of the equation in the same viewing window.

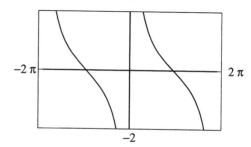

-2π 2π

$\dfrac{\sin x}{1 - \cos x} = \dfrac{1 + \cos x}{\sin x}$ appears to be an identity, which we verify.

$\dfrac{\sin x}{1 - \cos x} = \dfrac{\sin x(1 + \cos x)}{(1 - \cos x)(1 + \cos x)}$ Algebra

$= \dfrac{\sin x(1 + \cos x)}{1 - \cos^2 x}$ Algebra

$= \dfrac{\sin x(1 + \cos x)}{\sin^2 x}$ Pythagorean identity

$= \dfrac{1 + \cos x}{\sin x}$ Algebra

54. (A) Let $y = \cos^{-1} 0.4$, then $\cos y = 0.4 = \dfrac{2}{5}$, $0 \le y \le \pi$. Draw the
reference triangle associated with y. Then,
$\sin y = \sin(\cos^{-1} 0.4)$ can be determined directly from the
triangle.

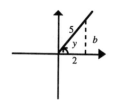

$b = \sqrt{5^2 - 2^2} = \sqrt{21}$

$\sin(\cos^{-1} 0.4) = \sin y = \dfrac{b}{r} = \dfrac{\sqrt{21}}{5}$

(B) Let $y = \arctan(-\sqrt{5})$, then $\tan y = -\sqrt{5}$, $-\frac{\pi}{2} < y < \frac{\pi}{2}$. Draw the reference triangle associated with y. Then, $\sec y = \sec[\arctan(-\sqrt{5})]$ can be determined directly from the triangle.

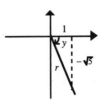

$$r = \sqrt{1^2 + (-\sqrt{5})^2} = \sqrt{6}$$

$$\sec[\arctan(-\sqrt{5})] = \sec y = \frac{r}{a} = \frac{\sqrt{6}}{1} = \sqrt{6}$$

(C) Let $y = \sin^{-1}\frac{1}{3}$, then $\sin y = \frac{1}{3}$, $-\frac{\pi}{2} \le y \le \frac{\pi}{2}$. Then, $\csc y = \csc\left(\sin^{-1}\frac{1}{3}\right) = \frac{1}{\sin y} = \frac{1}{\frac{1}{3}} = 3$

(D) Let $y = \sec^{-1} 4$, then $\sec y = 4$, $0 \le y \le \pi$. Draw the reference triangle associated with y. Then, $\tan y = \tan(\sec^{-1} 4)$ can be determined directly from the triangle.

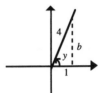

$$b = \sqrt{4^2 - 1^2} = \sqrt{15}$$

$$\tan(\sec^{-1} 4) = \tan y = \frac{b}{a} = \frac{\sqrt{15}}{1} = \sqrt{15}$$

55. $\theta = \tan^{-1}\sqrt{3}$ is equivalent to $\tan\theta = \sqrt{3}$ $\quad -90° < \theta < 90°$. Thus, $\theta = 60°$.

56. Calculator in degree mode: $\theta = \sin^{-1} 0.8989 = 64.01°$

57. $2 + 3\sin x = \cos 2x$ $\qquad 0 \le x \le 2\pi$

    Solve for sin x and/or cos x: $\qquad 2 + 3\sin x = 1 - 2\sin^2 x$ $\qquad$ Use double-angle identity

$$2\sin^2 x + 3\sin x + 1 = 0$$

$$(2\sin x + 1)(\sin x + 1) = 0$$

$$2\sin x + 1 = 0 \qquad\qquad \sin x + 1 = 0$$

$$\sin x = -\frac{1}{2} \qquad\qquad \sin x = -1$$

$$x = \frac{7\pi}{6}, \frac{11\pi}{6} \qquad\qquad x = \frac{3\pi}{2}$$

Solutions: $x = \dfrac{7\pi}{6}, \dfrac{3\pi}{2}, \dfrac{11\pi}{6}$

(The student should sketch graphs as appropriate to confirm these values.)

58. First solve for θ over one period, $0 \le \theta < 360°$. Then add integer multiplies of $360°$ to find all solutions.

$\sin 2\theta = 2\cos\theta \qquad 0 \le \theta \le 360°$

$$2\sin\theta\cos\theta = 2\cos\theta \qquad\qquad \text{Use double-angle identity}$$

$$2\sin\theta\cos\theta - 2\cos\theta = 0$$

$$2\cos\theta(\sin\theta - 1) = 0$$

$$2 \cos \theta = 0 \qquad \sin \theta - 1 = 0$$
$$\cos \theta = 0 \qquad \sin \theta = 1$$
$$\theta = 90°, 270° \qquad \theta = 90°$$

Thus, all solutions over one period, $0° \le \theta < 360°$, are $x = 90°, 270°$. Thus, the solutions, if θ is allowed to range over all degree values, can be written as
$$\theta = 90° + k360°, \ 270° + k360°, \ k \text{ any integer}$$

More concisely, since the solutions are 90°, 270°, 450°, and so on, we can write
$$\theta = 90° + k180° \qquad k \text{ any integer}$$

(The student should sketch graphs as appropriate to confirm these values.)

59. First solve for x over one period of $\tan x$, $0 \le x < \pi$. Then add integer multiples of π to find all solutions.

$$4 \tan^2 x - 3 \sec^2 x = 0$$
$$4 \tan^2 x - 3(\tan^2 x + 1) = 0 \qquad \text{Use Pythagorean identity}$$
$$4 \tan^2 x - 3 \tan^2 x - 3 = 0$$
$$\tan^2 x - 3 = 0$$
$$\tan^2 x = 3$$
$$\tan x = \pm\sqrt{3}$$
$$x = \frac{\pi}{3}, \frac{2\pi}{3}$$

Thus, all solutions over one period, $0 \le x < \pi$, are $x = \frac{\pi}{3}, \frac{2\pi}{3}$.

Because the tangent function is periodic with period π, all solutions are given by
$$x = \frac{\pi}{3} + k\pi, \quad x = \frac{2\pi}{3} + k\pi, \ k \text{ any integer}$$

(The student should sketch graphs as appropriate to confirm these values.)

60. $\sin x = -0.5678$

Solve over $0 \le x < 2\pi$: Sketch a graph of $y = \sin x$ and $y = -0.5678$ in the same coordinate system. We see that the solutions are in the third and fourth quadrants.
$$x = 2\pi + \sin^{-1}(-0.5678) = 5.679$$
$$x = \pi + \sin^{-1}(0.5678) = 3.745$$

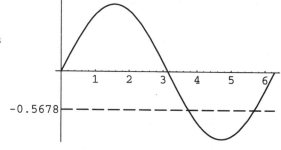

Write an expression for all solutions: Because the sine function is periodic with period 2π, all solutions are given by
$$x = 3.745 + 2k\pi, \quad x = 5.679 + 2k\pi,$$
k any integer.

61. $\sec x = 2.345$

    Solve for cos x: $\dfrac{1}{\cos x} = 2.345$

    $\cos x = \dfrac{1}{2.345}$

    $\cos x = 0.4264$

    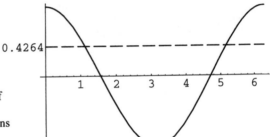

    Solve over $0 \le x < 2\pi$: Sketch a graph of
    $y = \cos x$ and $y = 0.4264$ in the same
    coordinate system. We see that the solutions
    are in the first and fourth quadrants.
    $x = \cos^{-1} 0.4264 = 1.130$
    $x = 2\pi - 1.130 = 5.153$

    Write an expression for all solutions: Because the cosine function is periodic with period 2π, all
    solutions are given by:
    $x = 1.130 + 2k\pi, \quad x = 5.153 + 2k\pi, \quad k$ any integer.

62. First solve for x over one period, $0 \le x < 2\pi$. Then add integer multiplies of 2π to find all
    solutions.

    $2 \cos 2x = 7 \cos x$

    $\qquad 2(2 \cos^2 x - 1) = 7 \cos x \qquad\qquad$ Use double-angle identity

    $\qquad\quad 4 \cos^2 x - 2 = 7 \cos x$

    $\quad 4 \cos^2 x - 7 \cos x - 2 = 0$

    $(4 \cos x + 1)(\cos x - 2) = 0$

    $\quad 4 \cos x + 1 = 0 \qquad 0 \le x < 2\pi \qquad \cos x - 2 = 0$

    $\qquad\quad 4 \cos x = -1 \qquad\qquad\qquad\qquad \cos x = 2$

    $\qquad\qquad \cos x = -\dfrac{1}{4} \qquad\qquad\qquad$ No solution

    $\qquad\qquad x = \cos^{-1}\left(-\dfrac{1}{4}\right)$ or $\pi + \cos^{-1}\left(\dfrac{1}{4}\right)$

    $\qquad\qquad x = 1.823$ or 4.460

    Thus, all solutions over one period, $0 \le x < 2\pi$, are $x = 1.823, 4.460$.

    Because the cosine function is periodic with period 2π, all solutions are given by $x = 1.823 + 2k\pi$,
    $x = 4.460 + 2k\pi$, k any integer.

63. (a) does not illustrate a cosine-inverse cosine identity, because -3 is not in $0 \le x \le \pi$, the restricted
    domain of the cosine function.

    (b) does illustrate a cosine-inverse cosine identity, because 2.51 is in $0 \le x \le \pi$, the restricted
    domain of the cosine function.

64. Graph y1 = tan x and y2 = 3 in the same viewing window in a graphing utility and find the points of intersection using an automatic intersection routine. The intersection points are found to be −1.893 and 1.249.

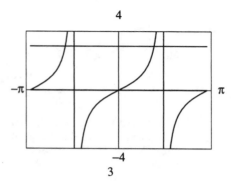

Check: $\tan(-1.893) = 2.995$
$\tan(1.249) = 3.000$

65. Graph y1 = cos x and y2 = $\sqrt{x}$ in the same viewing window in a graphing utility and find the points of intersection using an automatic intersection routine. The intersection point is found to be 0.642.

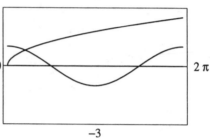

Check: $\cos(0.642) = 0.801$
$\sqrt{0.642} = 0.801$

Since $|\cos x| \le 1$, while $\sqrt{x} > 1$ for real x not shown, there can be no other solutions.

66. Graph y1 = $\cos\dfrac{x}{2} - 2\sin x$ and y2 = 1 in the same viewing window in a graphing utility and find the points of intersection using an automatic intersection routine. The intersection points are found to be 0, 3.895, 5.121, 9.834, 12.566.

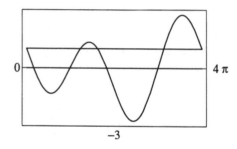

Check: $\cos\dfrac{0}{2} - 2\sin 0 = 1 - 0 = 1$

$\cos\dfrac{3.895}{2} - 2\sin 3.895 = 1.000$

(The remaining checking is left to the student.)

67. The coordinates are (cos 28.703, sin 28.703) = (−0.9095, −0.4157). The point P is in the third quadrant, since both coordinates are negative.

68. Since (x, y) is on a unit circle with $(x, y) = (0.5313, 0.8472) = (\cos s, \sin s)$, we can solve $\cos s = 0.5313$ or $\sin s = 0.8472$. Then $s = \cos^{-1}(0.5313) = 1.0107$ or $s = \sin^{-1}(0.8472) = 1.0107$

69. Draw a reference triangle and label what we know. Since $\cos\theta = a = \dfrac{a}{1}$, we can write $r = 1$. Use the Pythagorean theorem to find b.

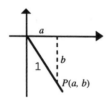

$a^2 + b^2 = 1$
$b^2 = 1 - a^2$
$b = -\sqrt{1 - a^2}$

238

b is negative since $P(a, b)$ is in quadrant IV. We can now find the other five functions.

$$\sin \theta = \frac{b}{r} = \frac{-\sqrt{1 - a^2}}{1} = -\sqrt{1 - a^2}$$

$$\cot \theta = \frac{a}{b} = \frac{a}{-\sqrt{1 - a^2}} = -\frac{a}{\sqrt{1 - a^2}}$$

$$\tan \theta = \frac{b}{a} = \frac{-\sqrt{1 - a^2}}{a}$$

$$\sec \theta = \frac{r}{a} = \frac{1}{a}$$

$$\csc \theta = \frac{r}{b} = \frac{1}{-\sqrt{1 - a^2}} = -\frac{1}{\sqrt{1 - a^2}}$$

70. $\quad \tan 3x = \tan(x + 2x)$ Algebra

$$= \frac{\tan x + \tan 2x}{1 - \tan x \tan 2x} \qquad\qquad \text{Sum identity for tangent}$$

$$= \frac{\tan x + \dfrac{2 \tan x}{1 - \tan^2 x}}{1 - \tan x \cdot \dfrac{2 \tan x}{1 - \tan^2 x}} \qquad\qquad \text{Double-angle identity}$$

$$= \frac{(1 - \tan^2 x) \tan x + 2 \tan x}{(1 - \tan^2 x) - \tan x \cdot 2 \tan x} \qquad\qquad \text{Algebra}$$

$$= \frac{\tan x - \tan^3 x + 2 \tan x}{1 - \tan^2 x - 2 \tan^2 x} \qquad\qquad \text{Algebra}$$

$$= \frac{3 \tan x - \tan^3 x}{1 - 3 \tan^2 x} \qquad\qquad \text{Algebra}$$

$$= \tan x \frac{3 - \tan^2 x}{1 - 3 \tan^2 x} \qquad\qquad \text{Algebra}$$

$$= \tan x \frac{3 - \dfrac{\sin^2 x}{\cos^2 x}}{1 - 3 \dfrac{\sin^2 x}{\cos^2 x}} \qquad\qquad \text{Quotient identity}$$

$$= \tan x \frac{3 \cos^2 x - \sin^2 x}{\cos^2 x - 3 \sin^2 x} \qquad\qquad \text{Algebra}$$

$$= \tan x \frac{3 \left(\dfrac{2 \cos^2 x - 1 + 1}{2} \right) - \left(\dfrac{1 - (1 - 2 \sin^2 x)}{2} \right)}{\dfrac{2 \cos^2 x - 1 + 1}{2} - 3 \left(\dfrac{1 - (1 - 2 \sin^2 x)}{2} \right)} \qquad\qquad \text{Algebra}$$

$$= \tan x \frac{3 \left(\dfrac{\cos 2x + 1}{2} \right) - \dfrac{1 - \cos 2x}{2}}{\dfrac{\cos 2x + 1}{2} - 3 \left(\dfrac{1 - \cos 2x}{2} \right)} \qquad\qquad \text{Double-angle identities}$$

$$= \tan x \ \frac{\frac{3}{2}\cos 2x + \frac{3}{2} - \frac{1}{2} + \frac{1}{2}\cos 2x}{\frac{1}{2}\cos 2x + \frac{1}{2} - \frac{3}{2} + \frac{3}{2}\cos 2x} \qquad \text{Algebra}$$

$$= \tan x \ \frac{2\cos 2x + 1}{2\cos 2x - 1} \qquad \text{Algebra}$$

71. Let $y = \sin^{-1}\frac{1}{3}$. Then we are asked to evaluate $\cos(2y)$ which is $1 - 2\sin^2 y$ from the double-angle

identity. Since $\sin y = \sin\left(\sin^{-1}\frac{1}{3}\right) = \frac{1}{3}$ from the sine-inverse sine identity, we can write

$$\cos\left(2\sin^{-1}\frac{1}{3}\right) = \cos 2y = 1 - 2\sin^2 y = 1 - 2\left(\frac{1}{3}\right)^2 = 1 - 2\cdot\frac{1}{9} = 1 - \frac{2}{9} = \frac{7}{9}$$

72. Let $u = \cos^{-1} x$ and $v = \tan^{-1} x$ \quad $-1 \le x \le 1$ \qquad or, equivalently

$x = \cos u \quad 0 \le u \le \pi$ \quad and \quad $x = \tan v \quad -\frac{\pi}{4} \le v \le \frac{\pi}{4}$

Then, $\sin(\cos^{-1} x - \tan^{-1} x) = \sin(u - v) = \sin u \cos v - \cos u \sin v$

For u, geometrically, we have

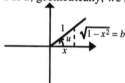 \qquad or \qquad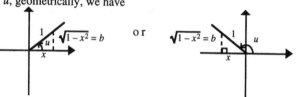

In either case, $b = \sqrt{1 - x^2}$ \quad $\sin u = \frac{b}{r} = \sqrt{1 - x^2}$ \quad $\cos u = x$

For v, geometrically, we have

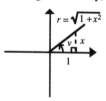 \qquad or \qquad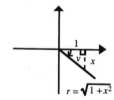

In either case, $r = \sqrt{1 + x^2}$; \quad $\sin v = \frac{x}{\sqrt{1 + x^2}}$; \qquad $\cos v = \frac{1}{\sqrt{1 + x^2}}$

Thus, $\sin(\cos^{-1} x - \tan^{-1} x) = \sin(u - v) = \sin u \cos v - \cos u \sin v$

$$= \sqrt{1 - x^2} \cdot \frac{1}{\sqrt{1 + x^2}} - x \cdot \frac{x}{\sqrt{1 + x^2}} = \frac{\sqrt{1 - x^2} - x^2}{\sqrt{1 + x^2}}$$

73. First solve for x over one period, $0 \leq x < 2\pi$. Then add integer multiples of 2π to find all solutions.

$$\sin x = 1 + \cos x$$
$$\pm\sqrt{1 - \cos^2 x} = 1 + \cos x$$
$$1 - \cos^2 x = (1 + \cos x)^2 \qquad \text{Squaring both sides}$$
$$1 - \cos^2 x = 1 + 2\cos x + \cos^2 x$$
$$0 = 2\cos x + 2\cos^2 x$$
$$0 = 2\cos x\,(1 + \cos x)$$

$2\cos x = 0$	$1 + \cos x = 0$
$\cos x = 0$	$\cos x = -1$
$x = \dfrac{\pi}{2}, \dfrac{3\pi}{2}$	$x = \pi$

In squaring both sides, we may have introduced extraneous solutions; hence, it is necessary to check solutions of these equations in the original equation.

Check:
$$x = \frac{\pi}{2} \qquad\qquad x = \frac{3\pi}{2} \qquad\qquad x = \pi$$

$$\sin x = 1 + \cos x \qquad \sin x = 1 + \cos x \qquad \sin \pi = 1 + \cos x$$
$$\sin\frac{\pi}{2} = 1 + \cos\frac{\pi}{2} \qquad \sin\frac{3\pi}{2} = 1 + \cos\frac{3\pi}{2} \qquad \sin \pi = 1 + \cos \pi$$
$$1 = 1 + 0 \qquad\qquad -1 \neq 1 + 0 \qquad\qquad 0 = 1 + (-1)$$

A solution $\qquad\qquad$ Not a solution $\qquad\qquad$ A solution

Thus, all solutions over one period, $0 \leq x < 2\pi$, are $x = \dfrac{\pi}{2}, \pi$.

Thus, the solutions, if x is allowed to range over all real numbers are

$$x = \begin{cases} \dfrac{\pi}{2} + 2k\pi \\ \pi + 2k\pi \end{cases} \quad k \text{ any integer}$$

74. $\sin x = \cos^2 x \qquad 0 \leq x \leq 2\pi$

Solve for $\sin x$ and/or $\cos x$: $\quad \sin x = 1 - \sin^2 x \qquad$ Use Pythagorean identity

$$\sin^2 x + \sin x - 1 = 0 \qquad\qquad \text{Quadratic in } \sin x$$
$$\sin x = \frac{-1 \pm \sqrt{1^2 - 4(1)(-1)}}{2 \cdot 1} = \frac{-1 \pm \sqrt{5}}{2}$$

$$\sin x = -1.6180 \qquad \sin x = 0.6180$$

Sketch a graph of $y = \sin x$, $y = -1.6180$, and $y = 0.6180$ in the same coordinate system.

$\sin x = -1.6180$ No solution (-1.618 is not in the domain of the sine function)

$\sin x = 0.6180$ From the graph, we see that the solutions are in the first and second quadrants.

$x = \sin^{-1}(0.6180) = 0.6662$

$x = \pi - 0.6662 = 2.475$

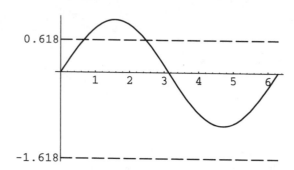

75. The graph of $f(x)$ is shown in the figure.

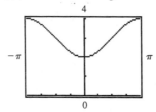

The graph appears to be an upside down cosine curve with period 2π, amplitude $= \frac{1}{2}(y_{max} - y_{min}) = \frac{1}{2}(4-2) = 1$, displaced upward by $k = 3$ units. It appears that $g(x) = 3 - \cos x$ would be an appropriate choice.

We verify $f(x) = g(x)$ as follows:

$$f(x) = \frac{\sin^2 x}{1 - \cos x} + \frac{2 \tan^2 x \cos^2 x}{1 + \cos x}$$

$$= \frac{1 - \cos^2 x}{1 - \cos x} + \frac{2 \dfrac{\sin^2 x}{\cos^2 x} \cos^2 x}{1 + \cos x} \qquad \text{Pythagorean and quotient identities}$$

$$= \frac{1 - \cos^2 x}{1 - \cos x} + \frac{2 \sin^2 x}{1 + \cos x} \qquad \text{Algebra}$$

$$= \frac{1 - \cos^2 x}{1 - \cos x} + \frac{2(1 - \cos^2 x)}{1 + \cos x} \qquad \text{Pythagorean identity}$$

$$= \frac{(1 - \cos x)(1 + \cos x)}{1 - \cos x} + \frac{2(1 - \cos x)(1 + \cos x)}{1 + \cos x} \qquad \text{Algebra}$$

$$= 1 + \cos x + 2(1 - \cos x) \qquad \text{Algebra}$$

$$= 1 + \cos x + 2 - 2 \cos x \qquad \text{Algebra}$$

$$= 3 - \cos x = g(x) \qquad \text{Algebra}$$

76. The graph of $f(x)$ is shown in the figure.

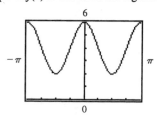

The graph appears to be a basic cosine curve with period π, amplitude $= \dfrac{1}{2}(y_{max} - y_{min})$ $= \dfrac{1}{2}(6 - 2) = 2$, displaced upward by $k = 4$ units. It appears that $g(x) = 4 + 2\cos 2x$ would be an appropriate choice.

We verify $f(x) = g(x)$ as follows:

$$
\begin{aligned}
f(x) &= 2\sin^2 x + 6\cos^2 x & \\
&= 2(1 - \cos^2 x) + 6\cos^2 x & \text{Pythagorean identity}\\
&= 2 - 2\cos^2 x + 6\cos^2 x & \text{Algebra}\\
&= 2 + 4\cos^2 x & \text{Algebra}\\
&= 2 + 2(2\cos^2 x - 1) + 2 & \text{Algebra}\\
&= 4 + 2(2\cos^2 x - 1) & \text{Algebra}\\
&= 4 + 2\cos 2x = g(x) & \text{Double-angle identity}
\end{aligned}
$$

77. The graph of $f(x)$ is shown in the figure.

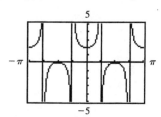

The graph appears to have vertical asymptotes $x = -\dfrac{3\pi}{4}, -\dfrac{\pi}{4}, \dfrac{\pi}{4},$ and $\dfrac{3\pi}{4}$ and period π. It appears to have high and low points with y coordinates 0 and 2, respectively.

It appears that $g(x) = 1 + \sec 2x$ would be an appropriate choice.

We verify $f(x) = g(x)$ as follows:

$$
\begin{aligned}
f(x) &= \dfrac{2 - 2\sin^2 x}{2\cos^2 x - 1} & \\
&= \dfrac{1 + 1 - 2\sin^2 x}{2\cos^2 x - 1} & \text{Algebra}\\
&= \dfrac{1 + \cos 2x}{\cos 2x} & \text{Double-angle identity}\\
&= \dfrac{1}{\cos 2x} + \dfrac{\cos 2x}{\cos 2x} & \text{Algebra}\\
&= \sec 2x + 1 = g(x) & \text{Quotient identity, algebra}
\end{aligned}
$$

78. The graph of $f(x)$ is shown in the figure.

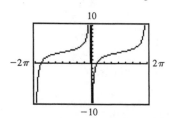

The graph appears to have vertical asymptotes $x = -2\pi$, $x = 0$, and $x = 2\pi$, and period 2π. Its x intercepts are difficult to determine, but since there appears to be symmetry with respect to the points where the curve crosses the line $y = 3$, it appears to be in upside down cotangent curve displaced upward by $k = 3$ units.

It appears that $g(x) = 3 - \cot \dfrac{x}{2}$ would be an appropriate choice.

We verify $f(x) = g(x)$ as follows:

$$f(x) = \frac{3 \cos x + \sin x - 3}{\cos x - 1}$$

$$= \frac{3(\cos x - 1) + \sin x}{\cos x - 1} \qquad\qquad \text{Algebra}$$

$$= \frac{3(\cos x - 1)}{\cos x - 1} - \frac{\sin x}{1 - \cos x} \qquad\qquad \text{Algebra}$$

$$= 3 - \frac{\sin x}{1 - \cos x} \qquad\qquad \text{Algebra}$$

$$= 3 - 1 + \frac{1 - \cos x}{\sin x} \qquad\qquad \text{Algebra}$$

$$= 3 - 1 + \tan \frac{x}{2} \qquad\qquad \text{Half-angle identity}$$

$$= 3 - \cot \frac{x}{2} = g(x) \qquad\qquad \text{Reciprocal identity}$$

79. Graph $y1 = 2 \sin \dfrac{\pi x}{2} - 3 \cos \dfrac{\pi x}{2}$ and $y2 = 2x - 1$ in the same viewing window in a graphing utility. Finding the points of intersection using an automatic intersection routine, we see that the graph of $y1$ is not below the graph of $y2$ for the interval $[0.694, 2.000]$.

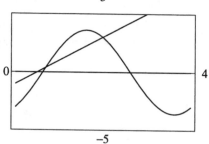

80. Labeling the diagram, we note: In triangle ACD, $\cot 32° = \dfrac{1000 + x}{h}$

    In triangle BCD, $\cot 48° = \dfrac{x}{h}$

We solve the system of equations

$$\cot 32° = \frac{1000 + x}{h} \qquad \cot 48° = \frac{x}{h}$$

by clearing of fractions, then eliminating x.

$$h \cot 32° = 1000 + x \qquad h \cot 48° = x$$

$$h \cot 32° = 1000 + h \cot 48°$$

$$h \cot 32° - h \cot 48° = 1000$$

$$h(\cot 32° - \cot 48°) = 1000$$

$$h = \frac{1000}{\cot 32° - \cot 48°} = 1429 \text{ meters}$$

81. $\sqrt{u^2 - a^2} = \sqrt{(a \csc x)^2 - a^2}$ Using the given substitution

    $\qquad\quad = \sqrt{a^2 \csc^2 x - a^2}$ Algebra

    $\qquad\quad = \sqrt{a^2(\csc^2 x - 1)}$ Algebra

    $\qquad\quad = \sqrt{a^2 \cot^2 x}$ Pythagorean identity

    $\qquad\quad = |a|\,|\cot x|$ Algebra

    $\qquad\quad = a \cot x$ Since $a > 0$ and x is in quadrant I $\left(\text{given } 0 < x < \dfrac{\pi}{2}\right)$, thus, $\cot x > 0$.

82. We note that $\tan \theta = \dfrac{2}{x}$ and $\tan 2\theta = \dfrac{2 + 4}{x} = \dfrac{6}{x}$ (see figure).

    Then, $\qquad \tan 2\theta = \dfrac{2 \tan \theta}{1 - \tan^2 \theta}$

$$\frac{6}{x} = \frac{2\left(\dfrac{2}{x}\right)}{1 - \left(\dfrac{2}{x}\right)^2} = \frac{\dfrac{4}{x}}{1 - \left(\dfrac{2}{x}\right)^2} = \frac{x^2 \cdot \dfrac{4}{x}}{x^2 \cdot 1 - x^2 \cdot \dfrac{4}{x^2}} = \frac{4x}{x^2 - 4}$$

$$x(x^2 - 4) \cdot \frac{6}{x} = x(x^2 - 4) \cdot \frac{4x}{x^2 - 4}$$

$$6(x^2 - 4) = x \cdot 4x$$

$$6x^2 - 24 = 4x^2$$

$$2x^2 - 24 = 0$$

$$x^2 = 12$$

$$x = \sqrt{12} \quad \text{(we discard the negative solution)}$$

$$= 2\sqrt{3}$$

Since $\tan \theta = \dfrac{2}{x} = \dfrac{2}{2\sqrt{3}} = \dfrac{1}{\sqrt{3}}$, $\theta = 30°$

83. We are to find the smallest positive solution to $0 = -7.2 \sin 5t - 9.6 \cos 5t$

$$7.2 \sin 5t = -9.6 \cos 5t$$

$$\frac{7.2 \sin 5t}{7.2 \cos 5t} = \frac{-9.6 \cos 5t}{7.2 \cos 5t}$$

$$\tan 5t = -\frac{9.6}{7.2}$$

$$5t = \tan^{-1}\left(-\frac{9.6}{7.2}\right) + \pi$$

$$t = \frac{1}{5}\left[\tan^{-1}\left(-\frac{9.6}{7.2} + \pi\right)\right] = 0.443 \text{ sec}$$

84. We graph $y1 = -7.2 \sin 5t - 9.6 \cos 5t$ and $y2 = 5$. From the figure, we see that $y1$ and $y2$ intersect twice on the indicated interval. From the automatic intersection routine, the intersections are found to be $t = 0.529$ sec and $t = 0.985$ sec.

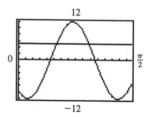

85. Following the hint, label the text diagram as shown at the right.

Then $\tan \theta = \dfrac{y_B}{x_B} = \dfrac{3}{5} \qquad \tan(\alpha + \theta) = \dfrac{y_A}{x_A} = \dfrac{5}{2}$

$\tan \theta = \tan(\alpha + \theta - \alpha) = \tan[(\alpha + \theta) - \alpha]$

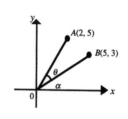

$$= \frac{\tan(\alpha + \theta) - \tan \alpha}{1 + \tan(\alpha + \theta) \tan \alpha}$$

$$= \frac{\dfrac{5}{2} - \dfrac{3}{5}}{1 + \dfrac{5}{2} \cdot \dfrac{3}{5}} = \frac{10 \cdot \dfrac{5}{2} - 10 \cdot \dfrac{3}{5}}{10 \cdot 1 + 10 \cdot \dfrac{5}{2} \cdot \dfrac{3}{5}}$$

$$= \frac{25 - 6}{10 + 15} = \frac{19}{25}$$

Therefore, $\theta = \tan^{-1} \dfrac{19}{25} = 0.650$

86. $I = 60 \sin\left(90\pi t - \dfrac{\pi}{2}\right)$

(A) We compute amplitude, period, frequency, and phase shift as follows:
Amplitude $= |A| = |60| = 60$.
Phase Shift and Period: Solve $Bx + C = 0$ and $Bx + C = 2\pi$

$$90\pi t - \frac{\pi}{2} = 0 \qquad\qquad 90\pi t - \frac{\pi}{2} = 2\pi$$

$$90\pi t = \frac{\pi}{2} \qquad\qquad 90\pi t = \frac{\pi}{2} + 2\pi$$

$$t = \frac{1}{180} \qquad\qquad t = \frac{1}{180} + \frac{1}{45}$$

Phase Shift   Period

$\text{Frequency} = \dfrac{1}{\text{Period}} = \dfrac{1}{1/45} = 45 \text{ Hz.}$

(B)   (C) $t = 0.0078$ sec

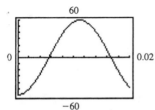

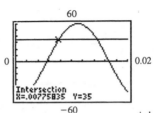

87. (A) From the figure, it should be clear that $\cot\theta = \dfrac{d}{200}$.

Thus, $d = 200 \cot\theta$.

(B) Period $= \pi$
One period of the graph would therefore extend from 0 to π, with vertical asymptotes at $t = 0$ and $t = \pi$. We sketch half of one period, since the required interval is from 0 to $\dfrac{\pi}{2}$ only.
Ordinates can be determined from a calculator, thus:

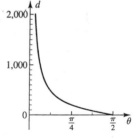

θ	$\dfrac{\pi}{20}$	$\dfrac{\pi}{10}$	$\dfrac{3\pi}{20}$	$\dfrac{\pi}{5}$	$\dfrac{\pi}{4}$	$\dfrac{3\pi}{10}$	$\dfrac{7\pi}{20}$	$\dfrac{2\pi}{5}$	$\dfrac{9\pi}{20}$	$\dfrac{\pi}{2}$
$200\cos\theta$	1263	616	393	275	200	145	101	65	32	0

88. (A) Labeling the diagram as shown, we note
In triangle ABC, $\tan\alpha = \dfrac{100}{x}$

In triangle ABD, $\tan(\theta + \alpha) = \dfrac{200}{x}$

Hence,

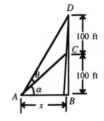

247

$$\tan\theta = \tan(\theta + \alpha - \alpha) = \frac{\tan(\theta + \alpha) - \tan\alpha}{1 + \tan(\theta + \alpha)\tan\alpha} = \frac{\dfrac{200}{x} - \dfrac{100}{x}}{1 + \dfrac{200}{x}\dfrac{100}{x}}$$

$$= \frac{x^2 \cdot \dfrac{200}{x} - x^2 \cdot \dfrac{100}{x}}{x^2 + x^2 \cdot \dfrac{200}{x} \cdot \dfrac{100}{x}} = \frac{200x - 100x}{x^2 + 20,000} = \frac{100x}{x^2 + 20,000}$$

$$\theta = \arctan\frac{100x}{x^2 + 20,000}$$

(B) We are given $x = 50$ feet. Thus, $\theta = \arctan\dfrac{100 \cdot 50}{50^2 + 20,000} = 12.5°$

89. We graph $y1 = \tan^{-1}\dfrac{100x}{x^2 + 20,000}$ and $y2 = 15$ on the interval from 0 to 400. We see that the curves intersect twice on the interval.

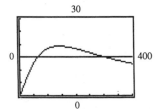

From the automatic intersection routine, the solutions are found to be $x = 64.9$ ft and $x = 308.3$ ft.

90. Since the sprockets are connected by the bicycle chain, the distance (arc length) that the rear sprocket turns is equal to the distance that the pedal sprocket turns. Let R_1 = radius of rear sprocket and R_2 = radius of pedal sprocket.

$$s = R_1\theta_1 \qquad s = R_2\theta_2 \qquad R_1\theta_1 = R_2\theta_2 \qquad \theta_1 = \frac{R_2}{R_1}\theta_2$$

Note that the angle through which the rear wheel turns is equal to the angle thorugh which the rear sprocket turns.

Thus, $\theta_1 = \dfrac{11.0}{4.0}\,18\pi = 49.5\,\pi$ rad

91. We use the formula $\theta_1 = \dfrac{R_2}{R_1}\theta_2$ from the previous problem, and note:

$\omega_1 = \dfrac{\theta_1}{t}$ angular velocity of rear wheel and sprocket

$\omega_2 = \dfrac{\theta_2}{t}$ angular velocity of pedal sprocket.

Hence, $\dfrac{\theta_1}{t} = \dfrac{R_2}{R_1}\dfrac{\theta_2}{t}$, $\omega_1 = \dfrac{R_2}{R_1}\omega_2 = \dfrac{11.0}{4.0}\,60.0$ rpm = 165 rpm or $165 \cdot 2\pi$ rad/min

Then, the linear velocity of the rear wheel (i.e., that of the bicycle), v, is given by

$$v = R_{\text{wheel}}\omega_1 = \frac{70.0\text{ cm}}{2} \cdot 165 \cdot 2\pi \text{ rad/min} = 11,550\pi \text{ cm/min} \approx 36,300 \text{ cm/min}$$

92. (A) For small θ (near $0°$), L is extremely large. As θ increases from $0°$, L decreases to some minimum value, then increases again beyond all bounds as θ approaches $90°$.

    (B) Label the text figure as shown. Note that ABC and CDE are right triangles, and that

    $AE = AC + CE = a + b$.

    Then, $\csc \theta = \dfrac{a}{50}$ from triangle ABC, so

    $a = 50 \csc \theta$, $\sec \theta = \dfrac{b}{25}$ from triangle CDE,

    so $b = 25 \sec \theta$. Thus, $AE = 50 \csc \theta + 25 \sec \theta$.

    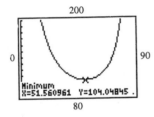

    (C)

$\theta°$	35	40	45	50	55	60	65
L ft	117.7	110.4	106.1	104.1	104.6	107.7	114.3

    (D) According to the table, Min $L = 104.1$ ft for $\theta = 50°$. The length of the longest log that will go around the corner is the minimum length L.

    (E) According to the graph, Min $L = 104.0$ ft for $\theta = 51.6°$.

    200

    0   90

    Minimum
    X=51.560961 Y=104.04845

    80

93. (A)

x (months)	1, 13	2, 14	3, 15	4, 16	5, 17	6, 18	7, 19	8, 20	9, 21
$y \left(\dfrac{\text{daylight}}{\text{duration}}\right)$	6.52	9.17	11.83	14.60	17.55	19.27	18.45	15.85	12.95

x (months)	10, 22	11, 23	12, 24
$y \left(\dfrac{\text{daylight}}{\text{duration}}\right)$	10.15	7.32	5.60

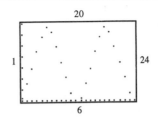

(B) From the table, Max $x = 19.27$ and Min $y = 5.60$. Then,

$$A = \frac{\text{Max } y - \text{Min } y}{2} = \frac{19.27 - 5.60}{2} = 6.835$$

$$B = \frac{2\pi}{\text{Period}} = \frac{2\pi}{12} = \frac{\pi}{6}$$

$$k = \text{Min } y + A = 5.60 + 6.835 = 12.435$$

From the plot in (A) or the table, we estimate the smallest value of x for which $y = k = 12.435$ to be approximately 3.1. Then, this is the phase-shift for the graph. Substitute $B = \frac{\pi}{6}$ and $x = 3.1$ into the phase-shift equation $x = -\frac{C}{B}$, $3.1 = \frac{-C}{\frac{\pi}{6}}$, $C = \frac{-3.1\pi}{6} \approx -1.6$. Thus, the

equation required is $y = 12.435 + 6.835 \sin\left(\frac{\pi x}{6} - 1.6\right)$.

(C)

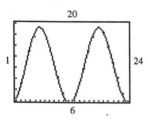

Chapter 6   Additional Triangle Topics; Vectors

EXERCISE 6.1   Law of Sines

1. *Solve for α:* $\alpha + \beta + \gamma = 180°$
   $\alpha = 180° - (36° + 43°) = 101°$

   Solve for b: $\dfrac{\sin \alpha}{a} = \dfrac{\sin \beta}{b}$

   $b = \sin \beta \dfrac{a}{\sin \alpha} = (\sin 43°) \dfrac{92 \text{ cm}}{\sin 101°}$
   $= 64 \text{ cm}$

   Solve for c: $\dfrac{\sin \alpha}{a} = \dfrac{\sin \gamma}{c}$

   $c = \sin \gamma \dfrac{a}{\sin \alpha} = (\sin 36°) \dfrac{92 \text{ cm}}{\sin 101°}$
   $= 55 \text{ cm}$

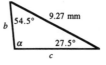

3. *Solve for α:* $\alpha + \beta + \gamma = 180°$
   $\alpha = 180° - (27.5° + 54.5°) = 98.0°$

   Solve for b: $\dfrac{\sin \alpha}{a} = \dfrac{\sin \beta}{b}$

   $b = \sin \beta \dfrac{a}{\sin \alpha} = (\sin 27.5°) \dfrac{9.27 \text{ mm}}{\sin 98.0°}$
   $= 4.32 \text{ mm}$

   Solve for c: $\dfrac{\sin \alpha}{a} = \dfrac{\sin \gamma}{c}$

   $c = \sin \gamma \dfrac{a}{\sin \alpha} = (\sin 54.5°) \dfrac{9.27 \text{ mm}}{\sin 98.0°}$
   $= 7.62 \text{ mm}$

5. *Solve for γ:* $\alpha + \beta + \gamma = 180°$
   $\gamma = 180° - (122.7° + 34.4°) = 22.9°$

   Solve for a: $\dfrac{\sin \alpha}{a} = \dfrac{\sin \beta}{b}$

   $a = \sin \alpha \dfrac{b}{\sin \beta} = (\sin 122.7°) \dfrac{18.3 \text{ cm}}{\sin 34.4°}$
   $= 27.3 \text{ cm}$

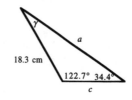

   Solve for c: $\dfrac{\sin \beta}{b} = \dfrac{\sin \gamma}{c}$

   $c = \sin \gamma \dfrac{b}{\sin \beta} = (\sin 22.9°) \dfrac{18.3 \text{ cm}}{\sin 34.4°}$
   $= 12.6 \text{ cm}$

7.   *Solve for α:* $\alpha + \beta + \gamma = 180°$
$\alpha = 180° - (100°0' + 12°40') = 67°20'$

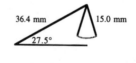

Solve for a: $\dfrac{\sin \alpha}{a} = \dfrac{\sin \beta}{b}$

$a = \sin \alpha \dfrac{b}{\sin \beta} = (\sin 67°20') \dfrac{13.1 \text{ km}}{\sin 12°40'}$

$= 55.1 \text{ km}$

Solve for c: $\dfrac{\sin \beta}{b} = \dfrac{\sin \gamma}{c}$

$c = \sin \gamma \dfrac{b}{\sin \beta} = (\sin 100°0') \dfrac{13.1 \text{ km}}{\sin 12°40'} = 58.8 \text{ km}$

9.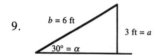

$h = b \sin \alpha = 6 \sin 30° = 3 = a.$
This is the case where α is acute and $a = h$.
1 triangle can be constructed.

11.   This is the case where α is acute and $a \ge b$.
1 triangle can be constructed.

13.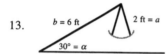

$h = b \sin \alpha = 6 \sin 30° = 3 > 2 = a.$
This is the case where α is acute and
$0 < a < h$.
0 triangles can be constructed.

15. $h = b \sin \alpha = 6 \sin 30° = 3 < 5 = a$
This is the case where α is acute and $h < a < b$.
2 triangles can be constructed.

17.   *Solve for β:* $\dfrac{\sin \beta}{b} = \dfrac{\sin \alpha}{a}$

$\sin \beta = \dfrac{b \sin \alpha}{a} = \dfrac{(36.4 \text{ mm})(\sin 25.5°)}{15.0 \text{ mm}}$

$= 1.045$

Since $\sin \beta = 1.045$ has no solution, no triangle exists with the given measurements. No solution.

19.   *Solve for β:* $\dfrac{\sin \beta}{b} = \dfrac{\sin \alpha}{a}$

$\sin \beta = \dfrac{b \sin \alpha}{a} = \dfrac{(18.3 \text{ m})(\sin 135°20')}{(14.6 \text{ m})} = 0.8811$

$\beta = \sin^{-1}(0.8811) = 61°50'$

Since there is not enough room in a triangle for an angle of 135°20' and an angle of 61°50' (their sum is more than 180°), no triangle exists with the given measurements. No solution.

21. β is acute $b > a$ Only one triangle is possible.

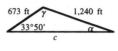

Solve for α: $\dfrac{\sin \alpha}{a} = \dfrac{\sin \beta}{b}$

$$\sin \alpha = \frac{a \sin \beta}{b} = \frac{(673 \text{ ft})(\sin 33°50')}{1,240 \text{ ft}}$$

$$= 0.3022$$

$$\alpha = \sin^{-1} 0.3022 = 17°40'$$

(There is another solution of $\sin \alpha = 0.3022$ that deserves brief consideration:

$\alpha' = 180° - \sin^{-1} 0.3022 = 162°20'$. However, there is not enough room in a triangle for an angle of $162°20'$ and an angle of $33°50'$, since their sum is greater than $180°$.)

Solve for γ: $\alpha + \beta + \gamma = 180°$
$$\gamma = 180° - (17°40' + 33°50') = 128°30'$$

Solve for c: $\dfrac{\sin \beta}{b} = \dfrac{\sin \gamma}{c}$

$$c = \frac{b \sin \gamma}{\sin \beta} = \frac{(1,240 \text{ ft})(\sin 128°30')}{\sin 33°50'} = 1,740 \text{ ft}$$

23.   *Solve for α:* $\dfrac{\sin \alpha}{a} = \dfrac{\sin \beta}{b}$

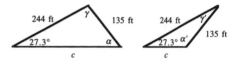

$$\sin \alpha = \frac{a \sin \beta}{b}$$

$$= \frac{(244 \text{ ft})(\sin 27.3°)}{135 \text{ ft}} = 0.829$$

Angle α can be either acute or obtuse.

$\alpha = \sin^{-1} 0.829 = 56.0°$ $\alpha' = 180° - \sin^{-1} 0.829 = 124°$

Solve for γ and γ':

$\gamma = 180° - (\alpha + \beta)$ $\gamma' = 180° - (\alpha' + \beta)$
$\ \ = 180° - (56.0° + 27.3°) = 96.7°$ $\ \ = 180° - (124° + 27.3°) = 28.7°$

Solve for c and c':

$\dfrac{\sin \alpha}{a} = \dfrac{\sin \gamma}{c}$ $\dfrac{\sin \alpha'}{a} = \dfrac{\sin \gamma'}{c'}$

$c = \dfrac{a \sin \gamma}{\sin \alpha} = \dfrac{(244 \text{ ft})(\sin 96.7°)}{\sin 56.0°} = 292 \text{ ft}$ $c' = \dfrac{a \sin \gamma'}{\sin \alpha'} = \dfrac{(244 \text{ ft})(\sin 28.7°)}{\sin 124°} = 141 \text{ ft}$

25.   Using the given and the calculated data, we have

$$(a - b) \cos \frac{\gamma}{2} = c \sin \frac{\alpha - \beta}{2}$$

$$(92 - 64) \cos \frac{36°}{2} = 55 \sin \frac{101° - 43°}{2}$$

$$26.6 = 26.6$$

27.   Using the law of sines in the form $\dfrac{a}{\sin \alpha} = \dfrac{b}{\sin \beta}$, we have $b = \dfrac{a \sin \beta}{\sin \alpha}$, which we use as follows:

$$\frac{a-b}{a+b} = \frac{a - \dfrac{a \sin \beta}{\sin \alpha}}{a + \dfrac{a \sin \beta}{\sin \alpha}} \qquad \text{Law of sines as above}$$

$$= \frac{a \sin \alpha - a \sin \beta}{a \sin \alpha + a \sin \beta} \qquad \text{Algebra}$$

$$= \frac{a(\sin \alpha - \sin \beta)}{a(\sin \alpha + \sin \beta)} \qquad \text{Algebra}$$

$$= \frac{\sin \alpha - \sin \beta}{\sin \alpha + \sin \beta} \qquad \text{Algebra}$$

$$= \frac{2 \cos \dfrac{\alpha + \beta}{2} \sin \dfrac{\alpha - \beta}{2}}{2 \sin \dfrac{\alpha + \beta}{2} \cos \dfrac{\alpha - \beta}{2}} \qquad \text{Sum-product identities}$$

$$= \frac{\dfrac{\sin \dfrac{\alpha - \beta}{2}}{\cos \dfrac{\alpha - \beta}{2}}}{\dfrac{\sin \dfrac{\alpha + \beta}{2}}{\cos \dfrac{\alpha + \beta}{2}}} \qquad \text{Algebra}$$

$$= \frac{\tan \dfrac{\alpha - \beta}{2}}{\tan \dfrac{\alpha + \beta}{2}} \qquad \text{Quotient identities}$$

29.   From the diagram, we can see that k = altitude of any possible triangle.
Thus, $\sin \beta = \dfrac{k}{a}$, $k = a \sin \beta$.

$k = (66.8 \text{ yd}) \sin 46.8° = 48.7 \text{ yd}$
If $0 < b < k$, there is no solution: (1) in the diagram.

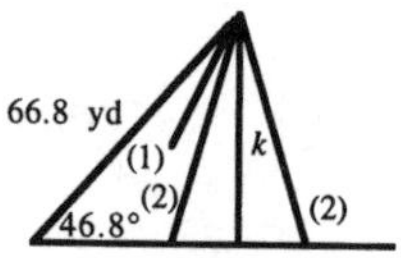

If $b = k$, there is one solution.

If $k < b < a$, there are two solutions: (2) in the diagram.

66.8 yd
46.8°

31. $\angle BAC + \angle ABC + \angle ACB = 180°$
$\angle ACB = 180° - (\angle BAC + \angle ABC) = 180° - (118.1° + 58.1°) = 3.8°$

Now apply the law of sines to find AC

$$\frac{\sin ABC}{AC} = \frac{\sin ACB}{AB}$$

$$AC = \frac{AB \sin ABC}{\sin ACB} = \frac{1.00 \sin 58.1°}{\sin 3.8°} = 12.8 \text{ mi}$$

33. First draw a figure and label known information.

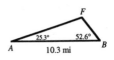

$\angle ABF + \angle BAF + \angle BFA = 180°$

$\angle BFA = 180° - (\angle ABF + \angle BAF)$

$= 180° - (52.6° + 25.3°) = 102.1°$

Now apply the law of sines to find AF and BF:

$$\frac{\sin ABF}{AF} = \frac{\sin BFA}{AB} \qquad\qquad \frac{\sin BAF}{BF} = \frac{\sin BFA}{AB}$$

$$AF = \frac{AB \sin ABF}{\sin BFA} \qquad\qquad BF = \frac{AB \sin BAF}{\sin BFA}$$

$$= \frac{(10.3 \text{ mi})(\sin 52.6°)}{\sin 102.1°} = 8.37 \text{ mi} \qquad\qquad = \frac{(10.3 \text{ mi})(\sin 25.3°)}{\sin 102.1°} = 4.50 \text{ mi}$$

The fire is 8.37 mi from A, 4.50 mi from B.

35. Label known information in the figure.

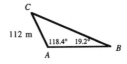

$\angle ABC + \angle CAB + \angle ACB = 180°$

$\angle ACB = 180° - (\angle ABC + \angle CAB)$

$= 180° - (19.2° + 118.4°) = 42.4°$

Now apply the law of sines to find AB.

$$\frac{\sin ACB}{AB} = \frac{\sin ABC}{AC}$$

$$AB = \frac{AC \sin ACB}{\sin ABC} = \frac{(112 \text{ m}) \sin 42.4°}{\sin 19.2°} = 230 \text{ m}$$

37. In the figure, note: Triangle RB_2B_3 is a right triangle, hence, $\dfrac{d}{x} = \sin 58°$.

Triangle RB_1B_2 is not a right triangle; but, from the law of sines,

$$\frac{\sin \alpha}{B_1B_2} = \frac{\sin \beta}{x}.$$

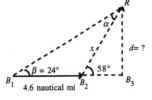

Hence, $x = \dfrac{B_1B_2 \sin \beta}{\sin \alpha}$

$d = x \sin 58° = \dfrac{B_1B_2 \sin \beta \sin 58°}{\sin \alpha}$

Given $B_1B_2 = 4.6$ nautical miles, $\beta = 24°$, we can find α since the exterior angle of a triangle has measure equal to the sum of the two nonadjacent interior angles.

Hence, $\alpha + \beta = 58°$, $\alpha = 58° - \beta = 58° - 24° = 34°$.

Hence, $d = \dfrac{(4.6 \text{ naut. mi})(\sin 24°)(\sin 58°)}{\sin 34°} = 2.8$ nautical miles

39. In the figure, note: Triangle ADC is a right triangle, hence $\angle ACB = 90° - 42° = 48°$. Triangle ABC is not a right triangle; but, from the law of sines,

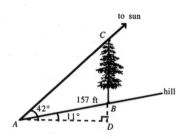

$$\frac{\sin CAB}{BC} = \frac{\sin ACB}{AB}$$

$$\angle CAB = \angle CAD - \angle BAD = 42° - 11° = 31°$$

$$BC = \frac{AB \sin CAB}{\sin ACB} = \frac{(157 \text{ ft}) \sin 31°}{\sin 48°} = 109 \text{ ft}$$

41. Labeling the figure as shown, we have, from the law of sines,

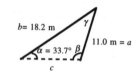

$$\frac{\sin \beta}{b} = \frac{\sin \alpha}{a}$$

$$\sin \beta = \frac{b \sin \alpha}{a} = \frac{(18.2 \text{ m})(\sin 33.7°)}{11.0 \text{ m}} = 0.9180$$

$$\beta = 180° - \sin^{-1}(0.9180) \quad (\text{since } \beta \text{ is obtuse})$$
$$= 113.4°$$

Since $\gamma = 180° - (\alpha + \beta) = 180° - (33.7° + 113.4°) = 32.9°$, we have, from the law of sines,

$$\frac{\sin \gamma}{c} = \frac{\sin \alpha}{a}$$

$$c = \frac{a \sin \gamma}{\sin \alpha} = \frac{(11.0 \text{ m})(\sin 32.9°)}{\sin 33.7°} = 10.8 \text{ m}$$

43. Labeling the diagram as shown, we note: traingle OAB is isosceles, hence $\alpha = \beta$. Thus,

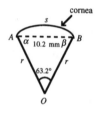

$$\alpha + \alpha + 63.2° = 180°$$
$$2\alpha = 116.8°$$
$$\alpha = 58.4°$$

Now apply the law of sines to find r.

$$\frac{\sin \alpha}{r} = \frac{\sin AOB}{AB}$$

$$r = \frac{AB \sin \alpha}{\sin AOB} = \frac{(10.2 \text{ mm})(\sin 58.4°)}{\sin 63.2°} = 9.73 \text{ mm}$$

To find s, we use the formula $s = \frac{\pi}{180°} \theta r$ from Chapter 2, thus

$$s = \frac{\pi}{180°} (63.2°)(9.73 \text{ mm}) = 10.7 \text{ mm}$$

45. Following the hint, we find all angles for triangle ACS first.
$$\angle SAC = \theta + 90° = 24.9° + 90° = 114.9°$$

For $\angle ACS$, we use the formula $s = \frac{\pi}{180°} \theta R$ from Chapter 2, with $s = 504$ miles, $\theta = \angle ACS$, $R = 3{,}964$ miles. Then,

$$\angle ACS = \frac{180°s}{\pi R} = \frac{180°(504 \text{ mi})}{\pi(3{,}964 \text{ mi})} = 7.28°$$

Since $\angle SAC + \angle ACS + \angle ASC = 180°$, we have $\angle ASC = 180° - (\angle ACS + \angle SAC)$
$$= 180° - (114.9° + 7.28°) = 57.8°$$

Now apply the law of sines to find side CS:

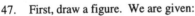

$$\frac{\sin SAC}{CS} = \frac{\sin ASC}{AC}$$

$$CS = \frac{AC \sin SAC}{\sin ASC} = \frac{(3{,}964 \text{ mi}) \sin 114.9°}{\sin 57.8°} = 4{,}248 \text{ miles}$$

Hence, the height of the satellite above $B = BS = CS - BC$.
$BS = 4{,}248 \text{ mi} - 3{,}964 \text{ mi} = 284 \text{ miles}$

47.   First, draw a figure. We are given:

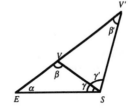

$$SE = \frac{1}{2}(\text{diameter of earth's orbit}) = \frac{1}{2}(2.99 \times 10^8 \text{ km})$$
$$= 1.495 \times 10^8 \text{ km}$$
$$SV = SV' = \frac{1}{2}(\text{diameter of Venus' orbit})$$
$$= \frac{1}{2}(2.17 \times 10^8 \text{ km}) = 1.085 \times 10^8 \text{ km}$$
$$\alpha = \angle SEV = 18°40'$$

There are two possible triangles; hence, two possible values of the required distance. Call them EV and EV'. The law of sines gives two possible values for angle VSE or $V'SE$; we denote them by γ and γ' in the figure. We start by calculating angle EVS, or β, and angle $EV'S$, or β', from the law of sines.

$$\frac{\sin \beta}{SE} = \frac{\sin \alpha}{SV} \qquad \sin \beta = \frac{SE \sin \alpha}{SV} = \frac{(1.495 \times 10^8 \text{ km})(\sin 18°40')}{1.085 \times 10^8 \text{ km}} = 0.4410$$

Hence, the two possibilities are:

β obtuse $\qquad\qquad\qquad\qquad\qquad \beta'$ acute
$\beta = 180° - \sin^{-1} 0.4410 = 153°50' \qquad \beta' = \sin^{-1} 0.4410 = 26.2° = 26°10'$

Hence, the two possibilities for angle VSE or γ, become:

$\gamma = 180° - (\beta + \alpha) \qquad\qquad\qquad \gamma' = 180° - (\beta' + \alpha')$
$= 180° - (153°50' + 18°40') = 7°30' \qquad = 180° - (26°10' + 18°40') = 135°10'$

Applying the law of sines again to calculate EV and EV' from these two values of γ, we have

$$\frac{\sin \gamma}{EV} = \frac{\sin \alpha}{SV} \qquad\qquad\qquad \frac{\sin \gamma'}{EV'} = \frac{\sin \alpha'}{SV'}$$
$$EV = \frac{SV \sin \gamma}{\sin \alpha} \qquad\qquad\qquad EV' = \frac{SV' \sin \gamma'}{\sin \alpha'}$$
$$= \frac{(1.085 \times 10^8 \text{ km})(\sin 7°30')}{\sin 18°40'} \qquad = \frac{(1.085 \times 10^8 \text{ km})(\sin 135°10')}{\sin 18°40'}$$
$$= 4.42 \times 10^7 \text{ km} \qquad\qquad\qquad = 2.39 \times 10^8 \text{ km}$$

49.   In the diagram, we are to calculate c based on the given information. There are two possible triangles, but the requirement that the distance c be as long as possible leads to the choice of α acute.

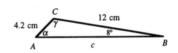

Solve for α: $\dfrac{\sin \alpha}{a} = \dfrac{\sin \beta}{b}$

$$\sin \alpha = \frac{a \sin \beta}{b} = \frac{(12 \text{ cm})(\sin 8°)}{4.2 \text{ cm}} = 0.397...$$

$$\alpha = \sin^{-1} 0.397 = 23°$$

Solve for γ: $\alpha + \beta + \gamma = 180°$ $\gamma = 180° - (8° + 23°) = 149°$

Solve for c: $\dfrac{\sin \beta}{b} = \dfrac{\sin \gamma}{c}$ $c = \dfrac{b \sin \gamma}{\sin \beta} = \dfrac{(4.2 \text{ cm})(\sin 149°)}{\sin 8°} = 16 \text{ cm}$

51. In the figure, note: ABC is a right triangle; hence, $\dfrac{h}{x} = \tan \gamma$

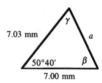

(1) $h = x \tan \gamma$

BCD is not a right triangle, but, from the law of sines,

$$\frac{\sin \delta}{d} = \frac{\sin \alpha}{x}$$

(2) $x = \dfrac{d \sin \alpha}{\sin \delta}$

Since $\alpha + \beta + \delta = 180°$, $\delta = 180° - (\alpha + \beta)$

$$\sin \delta = \sin[180° - (\alpha + \beta)] = \sin 180° \cos(\alpha + \beta) - \cos 180° \sin(\alpha + \beta)$$

$$= 0 \cos(\alpha + \beta) - (-1) \sin(\alpha + \beta) = \sin(\alpha + \beta)$$

Hence, $\dfrac{1}{\sin \delta} = \dfrac{1}{\sin(\alpha + \beta)} = \csc(\alpha + \beta)$. Thus,

$$h = x \tan \gamma = \frac{d \sin \alpha}{\sin \delta} \tan \gamma = d \sin \alpha \frac{1}{\sin \delta} \tan \gamma = d \sin \alpha \csc(\alpha + \beta) \tan \gamma$$

EXERCISE 6.2   Law of Cosines

1. A triangle can have at most one obtuse angle. Since β is acute, then, if the triangle has an obtuse angle it must be the angle opposite the longer of the two sides, a and c. Thus, γ, the angle opposite the shorter of the two sides, c, must be acute.

3. *Solve for* a: We use the law of cosines.

$$a^2 = b^2 + c^2 - 2bc \cos \alpha$$
$$= (7.03)^2 + (7.00)^2 - 2(7.03)(7.00) \cos 50°40'$$
$$= 36.039253...$$
$$a = 6.00 \text{ mm}$$

Since c is the shorter of the remaining sides, γ, the angle opposite c, must be acute.

Solve for γ: We use the law of sines.

$$\frac{\sin \alpha}{a} = \frac{\sin \gamma}{c}$$

$$\sin \gamma = \frac{c \sin \alpha}{a}$$

$$\gamma = \sin^{-1} \frac{c \sin \alpha}{a} = \sin^{-1} \frac{(7.00 \text{ mm})(\sin 50°40')}{6.00 \text{ m}} = 64°30'$$

Solve for β: $\beta = 180° - (\alpha + \gamma) = 180° - (50°40' + 64°30') = 65°10'$

5. *Solve for c:* We use the law of cosines.

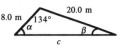

$$c^2 = a^2 + b^2 - 2ab \cos \gamma$$
$$= (20.0)^2 + (8.00)^2 - 2(20.0)(8.00) \cos 134°$$
$$= 686.29068...$$
$$c = 26.2 \text{ m}$$

Since b is the shorter of the remaining sides, β, the angle opposite b, must be acute.

Solve for β: We use the law of sines. $\dfrac{\sin \beta}{b} = \dfrac{\sin \gamma}{c}$

$$\sin \beta = \frac{b \sin \gamma}{c}$$

$$\beta = \sin^{-1} \frac{b \sin \gamma}{c} = \sin^{-1} \frac{(8.00 \text{ m})(\sin 134.0°)}{26.2 \text{ m}} = 12.7°$$

Solve for γ: $\gamma = 180° - (\alpha + \beta) = 180° - (134° + 12.7°) = 33.3°$

7. If the triangle has an obtuse angle, then it must be the angle opposite the longest side; in this case, α.

9. Find the measure of the angle opposite the longest side first, using the law of cosines. In this problem this is angle γ.

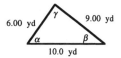

$$c^2 = a^2 + b^2 - 2ab \cos \gamma$$
$$\cos \gamma = \frac{a^2 + b^2 - c^2}{2ab}$$
$$\gamma = \cos^{-1} \frac{a^2 + b^2 - c^2}{2ab}$$
$$= \cos^{-1} \frac{(9.00)^2 + (6.00)^2 - (10.0)^2}{2(9.00)(6.00)}$$
$$= 80.9°$$

Solve for β: Both α and β must be acute, since they are smaller than γ. We choose to solve for β using the law of sines.

$$\frac{\sin \beta}{b} = \frac{\sin \gamma}{c} \qquad\qquad \sin \beta = \frac{b \sin \gamma}{c}$$

$$\beta = \sin^{-1} \frac{b \sin \gamma}{c}$$
$$= \sin^{-1} \frac{6.00 \sin 80.9°}{10.0}$$
$$= 36.3°$$

Solve for α: $\alpha = 180° - (\beta + \gamma) = 180° - (36.3° + 80.9°) = 62.8°$

11. Find the measure of the angle opposite the longest side first, using the law of cosines. In this problem this is angle γ.

$$c^2 = a^2 + b^2 - 2ab \cos \gamma$$
$$\cos \gamma = \frac{a^2 + b^2 - c^2}{2ab}$$
$$\gamma = \cos^{-1} \frac{a^2 + b^2 - c^2}{2ab}$$
$$= \cos^{-1} \frac{(420.0)^2 + (770.0)^2 - (860.0)^2}{2(420.0)(770.0)} = 87°22'$$

259

Solve for β: Both α and β must be acute, since they are smaller than γ. We choose to solve for β using the law of sines.

$$\frac{\sin \beta}{b} = \frac{\sin \gamma}{c} \qquad\qquad \sin \beta = \frac{b \sin \gamma}{c}$$

$$\beta = \sin^{-1}\frac{b \sin \gamma}{c}$$

$$= \sin^{-1}\frac{770.0 \sin 87°22'}{860.0} = 63°26'$$

Solve for α: $\alpha = 180° - (\beta + \gamma) = 180° - (63°26' + 87°22') = 29°12'$

13. We are given two angles and a non-included side (*AAS*).

    Solve for α: $\quad \alpha + \beta + \gamma = 180°$

    $\qquad\qquad\quad \alpha = 180° - (17.3° + 132.4°) = 30.3°$

    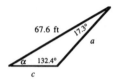

    Now use the law of sines to find the remaining sides.

    Solve for a: $\quad \dfrac{\sin \alpha}{a} = \dfrac{\sin \beta}{b}$

    $\qquad a = \dfrac{b \sin \alpha}{\sin \beta} = \dfrac{(67.6 \text{ ft})(\sin 30.3°)}{\sin 132.4°} = 46.2 \text{ ft}$

    Solve for c: $\quad \dfrac{\sin \beta}{b} = \dfrac{\sin \gamma}{c} \qquad c = \dfrac{b \sin \gamma}{\sin \beta} = \dfrac{(67.6 \text{ ft})(\sin 17.3°)}{\sin 132.4°} = 27.2 \text{ ft}$

15. We are given two sides and the included angle (*SAS*). We use the law of cosines to find the third side, then the law of sines to find a second angle.

    Solve for b: $\quad b^2 = a^2 + c^2 - 2ac \cos \beta$

    $\qquad\qquad\quad = (13.7)^2 + (20.1)^2 - 2(13.7)(20.1) \cos 66.5°$

    $\qquad\qquad\quad = 372.09294...$

    $\qquad\qquad b = 19.3 \text{ m}$

    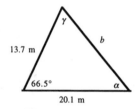

    Since a is the shorter of the remaining sides, α, the angle opposite a, must be acute.

    Solve for α: $\quad \dfrac{\sin \alpha}{a} = \dfrac{\sin \beta}{b}$

    $\qquad \sin \alpha = \dfrac{a \sin \beta}{\sin b} = \dfrac{(13.7 \text{ m})(\sin 66.5°)}{19.3 \text{ m}} = 0.6513$

    $\qquad \alpha = \sin^{-1} 0.6513 = 40.6°$

    Solve for γ: $\quad \alpha + \beta + \gamma = 180°$

    $\qquad\qquad \gamma = 180° - (66.5° + 40.6°) = 72.9°$

17. It is impossible to draw or form a triangle with this data, since angles $\beta + \gamma$ together add up to more than 180°. No solution.

19. We are given three sides (*SSS*). We solve for the largest angle, α (largest because it is opposite the largest side, *a*) using the law of cosines. We then solve for a second angle using the law of sines, because it involves simpler calculations.

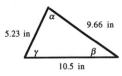

Solve for α: $a^2 = b^2 + c^2 - 2bc \cos \alpha$

$$\cos \alpha = \frac{b^2 + c^2 - a^2}{2bc}$$

$$\alpha = \cos^{-1}\frac{(5.23)^2 + (9.66)^2 - (10.5)^2}{2(5.23)(9.66)}$$

$$= \cos^{-1} 0.1031 = 84.1°$$

Both β and γ must be acute, since they are smaller than α.

Solve for β: $\dfrac{\sin \alpha}{a} = \dfrac{\sin \beta}{b}$

$$\sin \beta = \frac{b \sin \alpha}{a} = \frac{(5.23 \text{ in})(\sin 84.1°)}{10.5 \text{ in}} = 0.4954$$

$$\beta = \sin^{-1} 0.4954 = 29.7°$$

Solve for γ: $\gamma = 180° + (\alpha + \beta) = 180° - (84.1° + 29.7°) = 66.2°$

21. We are given two sides and a non-included angle (*SSA*).

Solve for α: $\dfrac{\sin \alpha}{a} = \dfrac{\sin \gamma}{c}$

$$\sin \alpha = \frac{a \sin \gamma}{c} = \frac{(14.5 \text{ mm})(\sin 80.3°)}{10.0 \text{ mm}} = 1.429$$

Since $\sin \alpha = 1.429$ has no solution, no triangle exists with the given measurements. No solution.

23. We are given two angles and the included side (*ASA*). We use the law of sines.

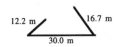

Solve for β: $\alpha + \beta + \gamma = 180°$

$$\beta = 180° - (46.3° + 105.5) = 28.2°$$

Solve for α: $\dfrac{\sin \alpha}{a} = \dfrac{\sin \beta}{b}$ 　　　*Solve for c:* $\dfrac{\sin \beta}{b} = \dfrac{\sin \gamma}{c}$

$$a = \frac{b \sin \alpha}{\sin \beta} \qquad\qquad c = \frac{b \sin \gamma}{\sin \beta}$$

$$= \frac{(643 \text{ m})(\sin 46.3°)}{\sin 28.2°} \qquad = \frac{(643 \text{ m})(\sin 105.5°)}{\sin 28.2°}$$

$$= 984 \text{ m} \qquad\qquad = 1310 \text{ m}$$

25. It is impossible to form a triangle with this data, since the triangle inequality $(a + b > c)$ is not satisfied. No solution.

Chapter 6   Additional Triangle Topics; Vectors

27. We are given two sides and a non-included angle (*SSA*).
     Two triangles are possible. There are two possible
     values for β. We use the law of sines.

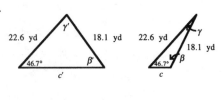

Solve for β: $\dfrac{\sin \alpha}{a} = \dfrac{\sin \beta}{b}$

$\quad\quad\quad\quad\sin \beta = \dfrac{b \sin \alpha}{a}$

$\quad\quad\quad\quad\quad\quad = \dfrac{(22.6 \text{ yd})(\sin 46.7°)}{18.1 \text{ yd}} = 0.9087$

Angle β can be either obtuse or acute.
$\quad\beta = 180° - \sin^{-1} 0.9087 = 114.7°$ $\quad\quad\quad\quad \beta' = \sin^{-1} 0.9087 = 65.3°$

Solve for γ and γ': $\gamma = 180° - (\alpha + \beta)$ $\quad\quad\quad\quad\quad\quad \gamma' = 180° - (\alpha' + \beta')$
$\quad\quad\quad\quad\quad\quad\quad\quad = 180° - (46.7° + 114.7°) = 18.6°$ $\quad\quad = 180° - (46.7° + 65.3°) = 68.0°$

Solve for c and c': $\dfrac{\sin \alpha}{a} = \dfrac{\sin \gamma}{c}$ $\quad\quad\quad\quad\quad\quad\quad \dfrac{\sin \alpha'}{a'} = \dfrac{\sin \gamma'}{c'}$

$\quad\quad\quad\quad\quad\quad c = \dfrac{a \sin \gamma}{\sin \alpha}$ $\quad\quad\quad\quad\quad\quad\quad\quad c' = \dfrac{a' \sin \gamma'}{\sin \alpha'}$

$\quad\quad\quad\quad\quad\quad\quad = \dfrac{(18.1 \text{ yd})(\sin 18.6°)}{\sin 46.7°}$ $\quad\quad\quad\quad\quad\quad = \dfrac{(18.1 \text{ yd})(\sin 68.0°)}{\sin 46.7°}$

$\quad\quad\quad\quad\quad\quad\quad = 7.93 \text{ yd}$ $\quad\quad\quad\quad\quad\quad\quad\quad = 23.1 \text{ yd}$

29. We are given three angles (*AAA*). An infinite number
     of triangles, all similar, can be drawn from the given
     values, but no one triangle is determined. No solution.

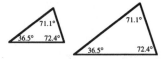

31. The law of cosines states that $b^2 = c^2 + a^2 - 2ac \cos \beta$ for any triangle.
     If $\beta = 90°$, $\cos \beta = 0$; hence, $b^2 = c^2 + a^2 - 2ac \cos (90°) = c^2 + a^2 - 0 = c^2 + a^2$

33. Using the given and the calculated data, we have:
$$(a - b) \cos \frac{\gamma}{2} = c \sin \frac{(\alpha - \beta)}{2}$$
$$(6.00 - 7.03) \cos \frac{64°20'}{2} = 7.00 \sin \frac{(50°40' - 65°0')}{2}$$
$$-0.872 \approx -0.873$$

35. We can write the law of cosines two different ways as follows:
     (1) $a^2 = b^2 + c^2 - 2bc \cos \alpha$
     (2) $b^2 = a^2 + c^2 - 2ac \cos \beta$

     Adding (1) and (2), we have
$$a^2 + b^2 = a^2 + b^2 + 2c^2 - 2bc \cos \alpha - 2ac \cos \beta$$
$$0 = 2c^2 - 2bc \cos \alpha - 2a \cos \beta$$
$$-2c^2 = -2bc \cos \alpha - 2ac \cos \beta$$

     Dividing both sides by $-2c$ (which is never 0), we obtain $c = b \cos \alpha + a \cos \beta$.

37. We are given two sides and an included angle. We use the law of cosines to find side BC.
$$BC^2 = AB^2 + AC^2 - 2(AB)(AC) \cos CAB = 425^2 + 384^2 - 2(425)(384) \cos 98.3°$$
$$= 375,198.864...$$
$$BC = 613 \text{ m}$$

39. In triangle OAB, we are given $OA = OB = 8.26$ cm and
$AB = 13.8$ cm. From the law of cosines, we can
determine the central angle AOB.

$$\cos AOB = \frac{OA^2 + OB^2 - AB^2}{2(OA)(OB)}$$

$$\angle AOB = \cos^{-1} \frac{OA^2 + OB^2 - AB^2}{2(OA)(OB)}$$

$$= \cos^{-1} \frac{(8.26)^2 + (8.26)^2 - (13.8)^2}{2(8.26)(8.26)} = \cos^{-1}(-0.3956) = 113.3°$$

41. First, complete and label the figure. From the given information, we know:
  γ = angle between west and northwest = $45°$
  d_a = (rate of plane A)(time of plane A) = (250 km/hr)(1 hr) = 250 km
  d_b = (rate of plane B)(time of plane B) = (210 km/hr)(1 hr) = 210 km

Hence, from the law of cosines,
$$c^2 = d_a^2 + d_b^2 - 2d_a d_b \cos \gamma = (250)^2 + (210)^2 - 2(250)(210) \cos 45° = 32353.78798...$$
$$c = 180 \text{ km}$$

43. In the figure, we are given $d = 58.3$ cm, $\alpha = 27.8°$. From the law of cosines, we know:

$$d^2 = r^2 + r^2 - 2r \cdot r \cos \alpha$$
$$d^2 = r^2(2 - 2 \cos \alpha)$$
$$r^2 = \frac{d^2}{2 - 2 \cos \alpha}$$
$$r = \frac{d}{\sqrt{2 - 2 \cos \alpha}}$$
Thus, $$r = \frac{58.3 \text{ cm}}{\sqrt{2 - 2 \cos 27.8°}} = 121 \text{ cm}$$

45. From the figure, it should be clear that the sides of the triangle
are:

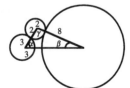

  $a = 2$ cm + 8 cm = 10 cm
  $b = 2$ cm + 3 cm = 5 cm
  $c = 3$ cm + 8 cm = 11 cm

Find the measure of the angle opposite the longest side first,
using the law of cosines. In this problem, this is angle γ.

$$c^2 = a^2 + b^2 - 2ab \cos \gamma \qquad\qquad \cos \gamma = \frac{a^2 + b^2 - c^2}{2ab}$$

$$\gamma = \cos^{-1} \frac{a^2 + b^2 - c^2}{2ab} = \cos^{-1} \frac{10^2 + 5^2 - 11^2}{2 \cdot 10 \cdot 5} = \cos^{-1} .04 = 87°40'$$

Solve for β: Both α and β must be acute, since they are smaller than γ. We choose to solve for β using the law of sines.

$$\frac{\sin \beta}{b} = \frac{\sin \gamma}{c} \qquad \sin \beta = \frac{b \sin \gamma}{c}$$

$$\beta = \sin^{-1}\frac{b \sin \gamma}{c}$$

$$= \sin^{-1}\frac{5 \sin 87°40'}{11}$$

$$= 27°0'$$

Solve for α: $\alpha = 180° - (\beta + \gamma) = 180° - (27°0' + 87°40') = 65°20'$

47.   In triangle ACD, angle $ADC = 8° + 90° = 98°$ (Why?) $DC = 12.0$ ft and $AD = 18.0$ ft. We know two sides and the included angle; hence we can apply the law of cosines to find side AC.

$$AC^2 = AD^2 + DC^2 - 2(AD)(DC)\cos ADC$$
$$= (18.0)^2 + (12.0)^2 - 2(18.0)(12.0)\cos 98° = 528.1227...$$
$$AC = 23.0 \text{ ft}$$

To find side AB, we need to find angle ADB, then apply the law of cosines again, in triangle ADB. Since angle ADB + angle $BDC = 98°$, angle $ADB = 98°$ – angle BDC. By symmetry, angle BDC = angle ACD of triangle ACD, which we can find from the law of sines. Thus,

$$\frac{\sin ACD}{AD} = \frac{\sin ADC}{AC} \qquad \sin ACD = \frac{(AD)(\sin ADC)}{AC}$$

$$\text{angle } ACD = \sin^{-1}\frac{(AD)(\sin ADC)}{AC} = \sin^{-1}\frac{18.0 \sin 98°}{23.0} = 50.9°$$

Hence, angle $ADB = 98° - 50.9° = 47.1°$. Then, applying the law of cosines to triangle ADB, we have ($AC = BD$ by symmetry.)

$$AB^2 = AD^2 + BD^2 - 2(AD)(BD)\cos ADB$$
$$= (18.0)^2 + (23.0)^2 - 2(18.0)(23.0)\cos 47.1° = 290.3765...$$
$$AB = 17.0 \text{ ft}$$

49.   In triangle CST, we are given $TS = 1,034$ miles and $TC = 3,964$ miles.
$$\angle STC = \theta + 90° = 32.4° + 90° = 122.4°$$

We are given two sides and the included angle; hence, we can apply the law of cosines to find side SC.

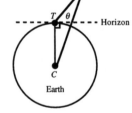

$$SC^2 = TS^2 + TC^2 - 2(TS)(TC)\cos STC$$
$$= (1,034)^2 + (3,964)^2 - 2(1,034)(3,964)\cos 122.4°$$
$$= 21,174,920...$$
$$SC = 4,602 \text{ mi}$$

Hence, the height of the satellite $= SC -$ radius of earth
$$= 4,602 \text{ mi} - 3,964 \text{ mi} = 638 \text{ mi}$$

51.   The three sides of triangle ABC are each in turn the hypotenuse of a right triangle formed with two edges of the solid. Hence, by the Pythagorean theorem.

$$AB^2 = 6.02^2 + 3.0^2 = 45.00 \qquad AB = 6.7 \text{ cm}$$
$$AC^2 = 6.0^2 + 4.0^2 = 52.00 \qquad AC = 7.2 \text{ cm}$$
$$BC^2 = 3.0^2 + 4.0^2 = 25.00 \qquad BC = 5.0 \text{ cm}$$

To find $\angle ABC$, we apply the law of cosines.

$$AC^2 = BC^2 + AB^2 - 2(BC)(AB)\cos ABC \qquad \cos ABC = \frac{AB^2 + BC^2 - AC^2}{2(BC)(AB)}$$

$$\angle ABC = \cos^{-1}\frac{AB^2 + BC^2 - AC^2}{2(BC)(AB)} = \cos^{-1}\frac{6.7^2 + 5.0^2 - 7.2^2}{2(5.0)(6.7)} = 74.4°$$

53. We will first find AD by applying the law of sines to triangle ABD, in which we know all angles and a side. We then will find AC by applying the law of sines to triangle ABC, in which we also know all angles and a side. We can then apply the law of cosines to triangle ADC to find DC. In triangle ABD:

$$\frac{\sin ABD}{AD} = \frac{\sin ADB}{AB} \qquad\qquad AD = \frac{AB \sin ABC}{\sin ADB} = \frac{(120 \text{ m}) \sin 79°}{\sin 29°} = 243.0 \text{ m}$$

In triangle ABC:
$$\frac{\sin ABC}{AC} = \frac{\sin ACB}{AB} \qquad\qquad AC = \frac{AB \sin ABC}{\sin ACB} = \frac{(120 \text{ m}) \sin(79° + 44°)}{\sin 26°} = 229.6 \text{ m}$$

In triangle ACD:
$$CD^2 = AC^2 + AD^2 - 2(AC)(AD)\cos CAD$$
$$= (243.0)^2 + (229.6)^2 - 2(243.0)(229.6)\cos 41° = 27544.55...$$
$$CD = 166 \text{ m}$$

EXERCISE 6.3   Areas of Triangles

1. The base and the height of the triangle are given; hence, we use the formula
$$A = \frac{1}{2}bh = \frac{1}{2}(17.0 \text{ m})(12.0 \text{ m}) = 102 \text{ m}^2$$

3. The given information consists of two sides and the included angle; hence, we use the formula
$$A = \frac{ab}{2}\sin\theta \text{ in the form } A = \frac{1}{2}bc\sin\alpha = \frac{1}{2}(6.0 \text{ cm})(8.0 \text{ cm})\sin 30° = 12 \text{ cm}^2$$

5. The given information consists of three sides; hence, we use Heron's formula. First, find the semiperimeter s:
$$s = \frac{a+b+c}{2} = \frac{4.00 + 6.00 + 8.00}{2} = 9.00 \text{ in.}$$

Then, $s - a = 9.00 - 4.00 = 5.00$
$\qquad\quad s - b = 9.00 - 6.00 = 3.00$
$\qquad\quad s - c = 9.00 - 8.00 = 1.00$

Thus, $A = \sqrt{s(s-a)(s-b)(s-c)} = \sqrt{(9.00)(5.00)(3.00)(1.00)} = \sqrt{135} = 11.6 \text{ in}^2$

7. The given information consists of two sides and the included angle; hence, we use the formula
$$A = \frac{ab}{2}\sin\theta \text{ in the form } A = \frac{1}{2}bc\sin\alpha = \frac{1}{2}(403)(512)\sin 23°20' = 40,900 \text{ ft}^2$$

9. The given information consists of two sides and the included angle; hence, we use the formula
$$A = \frac{ab}{2}\sin\theta \text{ in the form } A = \frac{1}{2}bc\sin\alpha \qquad \alpha \text{ is obtuse; this does not alter the use of the formula}$$
$$= \frac{1}{2}(12.1)(10.2)\sin 132.67° = 45.4 \text{ cm}^2$$

11. The given information consists of three sides; hence, we use Heron's formula. First, find the semiperimeter s:

$$s = \frac{a + b + c}{2} = \frac{12.7 + 20.3 + 24.4}{2} = 28.7 \text{ m}$$

Then, $s - a = 28.7 - 12.7 = 16.0$

$s - b = 28.7 - 20.3 = 8.4$

$s - c = 28.7 - 24.4 = 4.3$

Thus, $A = \sqrt{s(s - a)(s - b)(s - c)} = \sqrt{(28.7)(16.0)(8.4)(4.3)} = 129 \text{ m}^2$

13. All four triangles have two sides a and b given. For triangles with areas A_1 and A_3, the included angle is $180° - \theta$; hence, $A_1 = A_3 = \frac{1}{2} ab \sin(180° - \theta)$.

For triangles with areas A_2 and A_4, the included angle is θ; hence, $A_2 = A_4 = \frac{1}{2} AB \sin \theta$

But, $\sin(180° - \theta) = \sin 180° \cos \theta - \cos 180° \sin \theta = 0 \cos \theta - (-1) \sin \theta = \sin \theta$

Hence, $A_1 = A_3 = \frac{1}{2} ab \sin(180° - \theta) = \frac{1}{2} ab \sin \theta = A_2 = A_4$.

EXERCISE 6.4   Vectors: Geometrically Defined

Figure for Problems 1, 3, and 5:

1. To find $|\mathbf{u} + \mathbf{v}|$: Apply the Pythagorean theorem to triangle ABC.

$|\mathbf{u} + \mathbf{v}|^2 = AB^2 = AC^2 + BC^2 = |\mathbf{u}|^2 + |\mathbf{v}|^2 = 62^2 + 34^2 = 5{,}000$

$|\mathbf{u} + \mathbf{v}| = \sqrt{5{,}000} = 71 \text{ km/hr}$

Solve triangle ABC for θ: $\tan \theta = \dfrac{BC}{AC} = \dfrac{|\mathbf{u}|}{|\mathbf{v}|}$ $\theta = \tan^{-1} \dfrac{|\mathbf{v}|}{|\mathbf{u}|}$ θ is acute

$= \tan^{-1} \dfrac{34}{62} = 29°$

3. To find $|\mathbf{u} + \mathbf{v}|$: Apply the Pythagorean theorem to triangle ABC.

$|\mathbf{u} + \mathbf{v}|^2 = AB^2 = AC^2 + BC^2 = |\mathbf{u}|^2 + |\mathbf{v}|^2 = 48^2 + 31^2 = 3{,}265$

$|\mathbf{u} + \mathbf{v}| = 57 \text{ lb}$

Solve triangle ABC for θ: $\tan \theta = \dfrac{BC}{AC} = \dfrac{|\mathbf{v}|}{|\mathbf{u}|}$ $\theta = \tan^{-1} \dfrac{|\mathbf{v}|}{|\mathbf{u}|}$ θ is acute

$= \tan^{-1} \dfrac{31}{48} = 33°$

5. To find $|\mathbf{u} + \mathbf{v}|$: Apply the Pythagorean theorem to triangle ABC.

$|\mathbf{u} + \mathbf{v}|^2 = AB^2 = AC^2 + BC^2 = |\mathbf{u}|^2 + |\mathbf{v}|^2 = 143^2 + 57.4^2 = 23{,}743.76$

$|\mathbf{u} + \mathbf{v}| = 154 \text{ knots}$

Solve triangle ABC for θ: $\tan \theta = \dfrac{BC}{AC} = \dfrac{|\mathbf{v}|}{|\mathbf{u}|}$ $\qquad \theta = \tan^{-1} \dfrac{|\mathbf{v}|}{|\mathbf{u}|}$ $\qquad \theta$ is acute

$$= \tan^{-1} \frac{57.4}{143} = 21.9°$$

Figure for Problems 7 and 9:

7. $|\mathbf{v}| = 42$ lb $\qquad \theta = 34°$

   Horizontal component $|\mathbf{H}|$: $\cos 34° = \dfrac{|\mathbf{H}|}{42}$ $\qquad\qquad |\mathbf{H}| = 42 \cos 34° = 35$ lb

   Vertical component $|\mathbf{V}|$: $\sin 34° = \dfrac{|\mathbf{V}|}{42}$ $\qquad\qquad |\mathbf{V}| = 42 \sin 34° = 23$ lb.

9. $|\mathbf{v}| = 244$ km/hr $\qquad \theta = 43.2°$

   Horizontal component $|\mathbf{H}|$: $\cos 43.2° = \dfrac{|\mathbf{H}|}{244}$ $\qquad\qquad |\mathbf{H}| = 244 \cos 43.2° = 178$ km/hr

   Vertical component $|\mathbf{V}|$: $\sin 43.2° = \dfrac{|\mathbf{V}|}{244}$ $\qquad\qquad |\mathbf{V}| = 244 \sin 43.2° = 167$ km/hr

11. The magnitude of a geometric vector is the length (a nonnegative quantity) of a directed line segment, hence it can never be negative.

Figure for Problems 13 and 15:

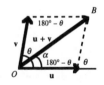

13. $\angle \theta = 44°$. Hence, $\angle OCB = 180° - \theta = 180° - 44° = 136°$. We can find $|\mathbf{u} + \mathbf{v}|$ using the law of cosines:

$$|\mathbf{u} + \mathbf{v}|^2 = |\mathbf{u}|^2 + |\mathbf{v}|^2 - 2|\mathbf{u}||\mathbf{v}| \cos (OCB)$$
$$= 125^2 + 84^2 - 2(125)(84) \cos 136° = 37{,}787.136\ldots$$
$$|\mathbf{u} + \mathbf{v}| = \sqrt{37{,}787.136\ldots} = 190 \text{ lb}$$

To find α, we use the law of sines: $\qquad \dfrac{\sin \alpha}{|\mathbf{v}|} = \dfrac{\sin OCB}{|\mathbf{u} + \mathbf{v}|}$

$$\frac{\sin \alpha}{84} = \frac{\sin 136°}{190}$$
$$\sin \alpha = \frac{84}{190} \sin 136°$$
$$\alpha = \sin^{-1}\left(\frac{84}{190} \sin 136°\right) = 18°$$

15. $\angle\theta = 66.8°$. Hence, $\angle OCB = 180° - 66.8° = 113.2°$. We can find $|\mathbf{u} + \mathbf{v}|$ using the law of cosines:
$$|\mathbf{u} + \mathbf{v}|^2 = |\mathbf{u}|^2 + |\mathbf{v}|^2 - 2|\mathbf{u}||\mathbf{v}|\cos(OCB)$$
$$= 655^2 + (97.3)^2 - 2(655)(97.3)\cos 113.2° = 488,705.31\ldots$$
$$|\mathbf{u} + \mathbf{v}| = \sqrt{488,705.31\ldots} = 699 \text{ km/hr}$$

To find α, we use the law of sines:
$$\frac{\sin\alpha}{|\mathbf{v}|} = \frac{\sin OCB}{|\mathbf{u} + \mathbf{v}|}$$
$$\frac{\sin\alpha}{97.3} = \frac{\sin 113.2°}{699}$$
$$\sin\alpha = \frac{97.3}{699}\sin 113.2°$$
$$\alpha = \sin^{-1}\left(\frac{97.3}{699}\sin 113.2°\right) = 7.4°$$

17. Since the zero vector has an arbitrary direction, it is correct to say that it is parallel to any vector.

19. The actual velocity $\mathbf{v}$ of the boat is the vector sum of the apparent velocity $\mathbf{B}$ of the boat and the velocity $\mathbf{R}$ of the river.
    $$|\mathbf{B}| = 4.0 \text{ km/hr} \qquad |\mathbf{R}| = 3.0 \text{ km/hr}$$

    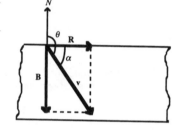

    Using the Pythagorean theorem, we find the magnitude of the resultant vector to be
    $$|\mathbf{v}| = \sqrt{4.0^2 + 3.0^2} = 5.0 \text{ km/hr}$$

    To find θ, we see that
    $$\tan\alpha = \frac{|\mathbf{B}|}{|\mathbf{R}|} = \frac{4.0}{3.0} \qquad \alpha = \tan^{-1}\frac{4.0}{3.0} = 53°$$
    $$\theta = \text{actual heading} = 90° + \alpha = 90° + 53° = 143.$$

21. We require θ such that the actual velocity $\mathbf{R}$ will be the resultant of the apparent velocity $\mathbf{v}$ and the wind velocity $\mathbf{w}$. The heading α will then be $360° - \theta$. From the diagram it should be clear that

    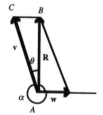

    $$\sin\theta = \frac{|\mathbf{w}|}{|\mathbf{v}|} = \frac{46}{255}$$
    $$\theta = \sin^{-1}\frac{46}{255} = 10° \qquad \alpha = 360° - 10° = 350°$$

    The ground speed for this course will be the magnitude $|\mathbf{R}|$ of the actual velocity. In the right triangle, ABC, we have
    $$\cos\theta = \frac{|\mathbf{R}|}{|\mathbf{v}|} \qquad |\mathbf{R}| = |\mathbf{v}|\cos\theta = 255\cos\left(\sin^{-1}\frac{46}{255}\right)$$
    $$= 255\sqrt{1 - \left(\frac{46}{255}\right)^2} = 250 \text{ mi/hr}$$

23. In triangle ABC, $\beta = 180° - 32° = 148°$. We can find the magnitude of M of the resulting force using the law of cosines.

$$M^2 = |\mathbf{F}_1|^2 + |\mathbf{F}_2|^2 - 2|\mathbf{F}_1||\mathbf{F}_2|\cos\beta = 1{,}500^2 + 1{,}100^2 - 2(1{,}500)(1{,}100)\cos 148°$$
$$= 6{,}258{,}558.7\ldots$$

$$M = \sqrt{6{,}258{,}558.7\ldots} = 2{,}500 \text{ lb}$$

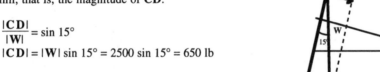

To find α, we use the law of sines.

$$\frac{\sin\alpha}{|\mathbf{F}_2|} = \frac{\sin\beta}{M}$$

$$\frac{\sin\alpha}{1{,}100} = \frac{\sin 148°}{2{,}500}$$

$$\sin\alpha = \frac{1{,}100}{2{,}500}\sin 148° \qquad \alpha = \sin^{-1}\left(\frac{1{,}100}{2{,}500}\sin 148°\right) = 13° \text{ (relative to } \mathbf{F}_1)$$

25. From the analysis in the text, we have $\theta = \tan^{-1}\dfrac{v^2}{gr}$, v in m/sec, r in m, $g = 9.81$ m/sec^2.

Hence, $\theta = \tan^{-1}\dfrac{(29)^2}{(9.81)(138)} = 32°$

27. The force parallel to the hill is the component of $\mathbf{W}$ parallel to the hill, that is, the magnitude of $\mathbf{CD}$.

$$\frac{|\mathbf{CD}|}{|\mathbf{W}|} = \sin 15°$$
$$|\mathbf{CD}| = |\mathbf{W}|\sin 15° = 2500 \sin 15° = 650 \text{ lb}$$

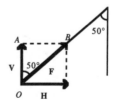

The force perpendicular to the hill is the component of $\mathbf{W}$ perpendicular to the hill, that is, the magnitude of $\mathbf{CH}$.

$$\frac{|\mathbf{CH}|}{|\mathbf{W}|} = \cos 15° \qquad |\mathbf{CH}| = |\mathbf{W}|\cos 15° = 2500 \cos 15° = 2400 \text{ lb}$$

29. From the figure it should be clear that in triangle OAB,

$$\sin 50° = \frac{|\mathbf{H}|}{|\mathbf{F}|} \qquad \cos 50° = \frac{|\mathbf{V}|}{|\mathbf{F}|}$$

Thus

$$|\mathbf{H}| = |\mathbf{F}|\sin 50° = 52 \sin 50° = 40 \text{ lb}$$
$$|\mathbf{V}| = |\mathbf{F}|\cos 50° = 52 \cos 50° = 33 \text{ lb}$$

31. The weights will slide to the left if the horizontal component H_1 pointing left is greater than the horizontal component H_2 pointing right. They will slide to the right if H_2 is greater than H_1.

$$H_1 = |\mathbf{F}_1|\cos 30° = (40\ g)\cos 30° \approx 34\ g$$
$$H_2 = |\mathbf{F}_2|\cos 40° = (30\ g)\cos 40° \approx 23\ g$$

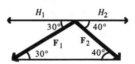

Since H_1 is greater than H_2, they will slide to the left.

EXERCISE 6.5   Vectors: Algebraically Defined

1. The coordinates of $P(x, y)$ are given by
    $$x = x_b - x_a = 5 - 2 = 3$$
    $$y = y_b - y_a = 1 - (-3) = 4$$
    Thus, $P(x, y) = P(3, 4)$

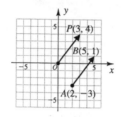

3. The coordinates of $P(x, y)$ are given by
    $$x = x_b - x_a = (-3) - (-1) = -2$$
    $$y = y_b - y_a = (-1) - 3 = -4$$
    Thus, $P(x, y) = P(-2, -4)$

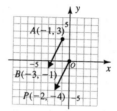

5. The algebraic vector $\langle a, b \rangle$ has coordinates given by
    $$a = x_b - x_a = 3 - (-1) = 4 \qquad b = y_b - y_a = 0 - (-2) = 2$$
    Hence, $\langle a, b \rangle = \langle 4, 2 \rangle$

7. The algebraic vector $\langle a, b \rangle$ has coordinates given by
    $$a = x_b - x_a = 4 - 0 = 4 \qquad b = y_b - y_a = (-2) - 2 = -4$$
    Hence, $\langle a, b \rangle = \langle 4, -4 \rangle$

9. $|\langle a, b \rangle| = \sqrt{a^2 + b^2} = \sqrt{(-3)^2 + 4^2} = 5$

11. $|\langle a, b \rangle| = \sqrt{a^2 + b^2} = \sqrt{(-5)^2 + (-2)^2} = \sqrt{29}$

13. Two geometric vectors are equal if and only if they have the same magnitude and direction.

15. (A) $\mathbf{u} + \mathbf{v} = \langle 1, 4 \rangle + \langle -3, 2 \rangle = \langle -2, 6 \rangle$

    (B) $\mathbf{u} - \mathbf{v} = \langle 1, 4 \rangle - \langle -3, 2 \rangle = \langle 4, 2 \rangle$

    (C) $2\mathbf{u} - 3\mathbf{v} = 2\langle 1, 4 \rangle - 3\langle -3, 2 \rangle = \langle 2, 8 \rangle + \langle 9, -6 \rangle = \langle 11, 2 \rangle$

    (D) $3\mathbf{u} - \mathbf{v} + 2\mathbf{w} = 3\langle 1, 4 \rangle - \langle -3, 2 \rangle + 2\langle 0, 4 \rangle = \langle 3, 12 \rangle + \langle 3, -2 \rangle + \langle 0, 8 \rangle = \langle 6, 18 \rangle$

17. (A) $\mathbf{u} + \mathbf{v} = \langle 2, -3 \rangle + \langle -1, -3 \rangle = \langle 1, -6 \rangle$

    (B) $\mathbf{u} - \mathbf{v} = \langle 2, -3 \rangle - \langle -1, -3 \rangle = \langle 3, 0 \rangle$

    (C) $2\mathbf{u} - 3\mathbf{v} = 2\langle 2, -3 \rangle - 3\langle -1, -3 \rangle = \langle 4, -6 \rangle + \langle 3, 9 \rangle = \langle 7, 3 \rangle$

    (D) $3\mathbf{u} - \mathbf{v} - 2\mathbf{w} = 3\langle 2, -3 \rangle - \langle -1, -3 \rangle + 2\langle -2, 0 \rangle = \langle 6, -9 \rangle + \langle 1, 3 \rangle + \langle -4, 0 \rangle = \langle 3, -6 \rangle$

19. $|\mathbf{v}| = \sqrt{4^2 + (-3)^2} = 5 \qquad \mathbf{u} = \dfrac{1}{|\mathbf{v}|}\mathbf{v} = \dfrac{1}{5}\langle 4, -3 \rangle = \left\langle \dfrac{4}{5}, -\dfrac{3}{5} \right\rangle$

21. $|\mathbf{v}| = \sqrt{2^2 + (-3)^2} = \sqrt{13} \qquad \mathbf{u} = \dfrac{1}{|\mathbf{v}|}\mathbf{v} = \dfrac{1}{\sqrt{13}}\langle 2, -3 \rangle = \left\langle \dfrac{2}{\sqrt{13}}, -\dfrac{3}{\sqrt{13}} \right\rangle$

23. $\mathbf{v} = \langle 3, -2 \rangle = \langle 3, 0 \rangle + \langle 0, -2 \rangle = 3\langle 1, 0 \rangle - 2\langle 0, 1 \rangle = 3\mathbf{i} - 2\mathbf{j}$

25. $\mathbf{v} = \langle 0, 4 \rangle = 4\langle 0, 1 \rangle = 4\mathbf{j}$

27. $\mathbf{v} = \overrightarrow{AB} = \langle 0 - (-2), 2 - (-1) \rangle = \langle 2, 3 \rangle = \langle 2, 0 \rangle + \langle 0, 3 \rangle$

$= 2\langle 1, 0 \rangle + 3\langle 0, 1 \rangle = 2\mathbf{i} + 3\mathbf{j}$

29. $\mathbf{u} - \mathbf{v} = (2\mathbf{i} - 3\mathbf{j}) - (3\mathbf{i} + 4\mathbf{j}) = 2\mathbf{i} - 3\mathbf{j} - 3\mathbf{i} - 4\mathbf{j} = -\mathbf{i} - 7\mathbf{j}$

31. $3\mathbf{u} - 2\mathbf{v} = 3(2\mathbf{i} - 3\mathbf{j}) - 2(3\mathbf{i} + 4\mathbf{j}) = 6\mathbf{i} - 9\mathbf{j} - 6\mathbf{i} - 8\mathbf{j} = -17\mathbf{j}$

33. $\mathbf{u} - 2\mathbf{v} + 2\mathbf{w} = (2\mathbf{i} - 3\mathbf{j}) - 2(3\mathbf{i} + 4\mathbf{j}) + 2(5\mathbf{j}) = 2\mathbf{i} - 3\mathbf{j} - 6\mathbf{i} - 8\mathbf{j} + 10\mathbf{j} = -4\mathbf{i} - \mathbf{j}$

35.   Any one of the force vectors must have the same magnitude as the resultant of the other two force vectors and be oppositely directed to the resultant of the other two.

37. $\mathbf{u} + \mathbf{v} = \langle a, b \rangle + \langle c, d \rangle$

$\quad = \langle a + c, b + d \rangle$ Definition of vector addition

$\quad = \langle c + a, d + b \rangle$ Commutative property for addition of real numbers*

$\quad = \langle c, d \rangle + \langle a, b \rangle$ Definition of vector addition

$\quad = \mathbf{v} + \mathbf{u}$

39. $\mathbf{v} + (-\mathbf{v}) = \langle c, d \rangle + (-\langle c, d \rangle)$

$\quad = \langle c, d \rangle + \langle -c, -d \rangle$ Definition of scalar multiplication

$\quad = \langle c + (-c), d + (-d) \rangle$ Definition of vector addition

$\quad = \langle 0, 0 \rangle$ Additive inverse property for real numbers*

$\quad = \mathbf{0}$

41. $m(\mathbf{u} + \mathbf{v}) = m(\langle a, b \rangle + \langle c, d \rangle)$

$\quad = m\langle a + c, b + d \rangle$ Definition of vector addition

$\quad = \langle m(a + c), m(b + d) \rangle$ Definition of scalar multiplication

$\quad = \langle ma + mc, mb + md \rangle$ Distributive property for real numbers*

$\quad = \langle ma, mb \rangle + \langle mc, md \rangle$ Definition of vector addition

$\quad = m\langle a, b \rangle + m\langle c, d \rangle$ Definition of scalar multiplication

$\quad = m\mathbf{u} + m\mathbf{v}$

43. $1\mathbf{v} = 1\langle a, b \rangle$

$\quad = \langle 1a, 1b \rangle$ Definition of scalar multiplication

$\quad = \langle a, b \rangle$ Multiplicative identity property for real numbers*

$\quad = \mathbf{v}$

---

* The basic properties of the set of real numbers are listed in the text, Appendix A.1.

45.   First, form a force diagram with all force vectors in standard position at the origin.

Let $\mathbf{F}_1$ = the tension in one rope

$\mathbf{F}_2$ = the tension in the other rope

Write each force vector in terms of **i** and **j** unit vectors.

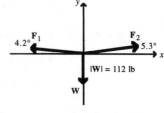

$\mathbf{F}_1 = |\mathbf{F}_1|(-\cos 4.2°)\mathbf{i} + |\mathbf{F}_1|(\sin 4.2°)\mathbf{j}$

$\mathbf{F}_2 = |\mathbf{F}_2|(\cos 5.3°)\mathbf{i} + |\mathbf{F}_2|(\sin 5.3°)\mathbf{j}$

$\mathbf{W} = -112\mathbf{j}$

For the system to be in static equilibrium, we must have $\mathbf{F}_1 + \mathbf{F}_2 + \mathbf{W} = \mathbf{0}$ which becomes, on addition,

$$[-|\mathbf{F}_1|(\cos 4.2°) + |\mathbf{F}_2|(\cos 5.3°)]\mathbf{i} + [|\mathbf{F}_1|(\sin 4.2°) + |\mathbf{F}_2|(\sin 5.3°) - 112]\mathbf{j} = 0\mathbf{i} + 0\mathbf{j}$$

Since two vectors are equal if and only if their corresponding components are equal, we are led to the following system of equations in $|\mathbf{F}_1|$ and $|\mathbf{F}_2|$:

$$-|\mathbf{F}_1| \cos 4.2° + |\mathbf{F}_2| \cos 5.3° = 0$$

$$|\mathbf{F}_1| \sin 4.2° + |\mathbf{F}_1| \sin 5.3° - 112 = 0$$

Solving, $|\mathbf{F}_2| = |\mathbf{F}_1| \dfrac{\cos 4.2°}{\cos 5.3°}$

$$|\mathbf{F}_1| \sin 4.2° + |\mathbf{F}_1| \frac{\cos 4.2°}{\cos 5.3°} \sin 5.3° = 112$$

$$|\mathbf{F}_1| [\sin 4.2° + \cos 4.2° \tan 5.3°] = 112$$

$$|\mathbf{F}_1| = \frac{112}{\sin 4.2° + \cos 4.2° \tan 5.3°} = 676 \text{ lb}$$

$$|\mathbf{F}_2| = 676 \frac{\cos 4.2°}{\cos 5.3°} = 677 \text{ lb}$$

47.   First, form a force diagram with all force vectors in standard position at the origin.

Let $\mathbf{F}_1$ = the force on the horizontal member BC

$\mathbf{F}_2$ = the force on the supporting member AB

$\mathbf{W}$ = the downward force (5,000 lb)

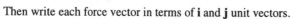

We note: $\cos \theta = \dfrac{5.0}{6.0}$ $\theta = \cos^{-1} \dfrac{5.0}{6.0} = 33.6°$

Then write each force vector in terms of **i** and **j** unit vectors.

$\mathbf{F}_1 = -|\mathbf{F}_1|\,\mathbf{i}$

$\mathbf{F}_2 = |\mathbf{F}_2|(\cos 33.6°)\mathbf{i} + |\mathbf{F}_2|(\sin 33.6°)\mathbf{j}$

$\mathbf{W} = -5,000\mathbf{j}$

For the system to be in static equilibrium, we must have $\mathbf{F}_1 + \mathbf{F}_2 + \mathbf{W} = \mathbf{0}$ which becomes, on addition,

$$[-|\mathbf{F}_1| + |\mathbf{F}_2|(\cos 33.6°)]\mathbf{i} + [|\mathbf{F}_2|(\sin 33.6°) - 5,000]\mathbf{j} = 0\mathbf{i} + 0\mathbf{j}$$

Since two vectors are equal if and only if their corresponding components are equal, we are led to the following system of equations in $|\mathbf{F}_1|$ and $|\mathbf{F}_2|$:

$$-|\mathbf{F}_1| + |\mathbf{F}_2| (\cos 33.6°) = 0$$
$$|\mathbf{F}_1| (\sin 33.6°) - 5,000 = 0$$

Solving, $|\mathbf{F}_2| = \dfrac{5,000}{\sin 33.6°} = 9,050$ lb

$\qquad\qquad |\mathbf{F}_1| = |\mathbf{F}_2| \cos 33.6° = 7,540$ lb

The force in the member AB is directed oppositely to the diagram—a compression of 9,050 lb.

The force in the member BC is also directed oppositely to the diagram—a tension of 7,540 lb.

49. First, form a force diagram with all force vectors in standard position at the origin.

Let $\mathbf{F}_1$ = the force exerted by the weight (10 lb)

$\qquad \mathbf{F}_2$ = the tension on the line

$\qquad \mathbf{W}$ = the weight of the arm (6 lb)

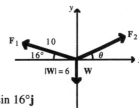

Then write each force vector in terms of $\mathbf{i}$ and $\mathbf{j}$ unit vectors.

$\qquad \mathbf{F}_1 = -|\mathbf{F}_1| \cos 16°\mathbf{i} + |\mathbf{F}_1| \sin 16°\mathbf{j} = -10 \cos 16°\mathbf{i} + 10 \sin 16°\mathbf{j}$

$\qquad \mathbf{F}_2 = |\mathbf{F}_2| \cos \theta \mathbf{i} + |\mathbf{F}_2| \sin \theta \mathbf{j}$

$\qquad \mathbf{W} = -6\mathbf{j}$

For the system to be in static equilibrium, we must have $\mathbf{F}_1 + \mathbf{F}_2 + \mathbf{W} = \mathbf{0}$ which becomes, on addition,

$\qquad [-10 \cos 16° + |\mathbf{F}_2| \cos \theta]\mathbf{i} + [10 \sin 16° + |\mathbf{F}_2| \sin \theta - 6]\mathbf{j} = 0\mathbf{i} + 0\mathbf{j}$

Since two vectors are equal if and only if their corresponding components are equal, we are led to the following system of equations in $|\mathbf{F}_2|$ and θ:

$$-10 \cos 16° + |\mathbf{F}_2| \cos \theta = 0$$
$$10 \sin 16° + |\mathbf{F}_2| \sin \theta - 6 = 0$$

Solving, $|\mathbf{F}_2| \cos \theta = 10 \cos 16°$

$$|\mathbf{F}_2| = \frac{10 \cos 16°}{\cos \theta}$$

$$10 \sin 16° + \frac{10 \cos 16°}{\cos \theta} \sin \theta - 6 = 0$$

$$10 \cos 16° \tan \theta = 6 - 10 \sin 16°$$

$$\tan \theta = \frac{6 - 10 \sin 16°}{10 \cos 16°}$$

$$\theta = \tan^{-1}\frac{6 - 10 \sin 16°}{10 \cos 16°} = 18.6°$$

$$|\mathbf{F}_2| = \frac{10 \cos 16°}{\cos \theta} = 10.1 \text{ lb}$$

EXERCISE 6.6   The Dot Product

1. $\langle 5, 3\rangle \cdot \langle -2, 3\rangle = 5 \cdot (-2) + 3 \cdot 3 = -1$ 3. $(5\mathbf{i} - 3\mathbf{j}) \cdot (-2\mathbf{i} + 3\mathbf{j}) = 5 \cdot (-2) - 3 \cdot 3 = -19$

5. $\langle 2, 8\rangle \cdot \langle 12, -3\rangle = 2 \cdot 12 + 8 \cdot (-3) = 0$ 7. $3\mathbf{i} \cdot 4\mathbf{j} = (3\mathbf{i} + 0\mathbf{j}) \cdot (0\mathbf{i} + 4\mathbf{j}) = 3 \cdot 0 + 0 \cdot 4 = 0$

9. $|\mathbf{u}| = \sqrt{(-3)^2 + 2^2} = \sqrt{13}$ $|\mathbf{v}| = \sqrt{0^2 + 4^2} = 4$

     $\cos\theta = \dfrac{\mathbf{u} \cdot \mathbf{v}}{|\mathbf{u}||\mathbf{v}|} = \dfrac{\langle -3, 2\rangle \cdot \langle 0, 4\rangle}{(\sqrt{13})(4)} = \dfrac{0 + 8}{4\sqrt{13}} = \dfrac{2}{\sqrt{13}}$ $\theta = \cos^{-1}\dfrac{2}{\sqrt{13}} = 56.3°$

11. $|\mathbf{u}| = \sqrt{3^2 + 3^2} = \sqrt{18}$ $|\mathbf{v}| = \sqrt{2^2 + (-5)^2} = \sqrt{29}$

     $\cos\theta = \dfrac{\mathbf{u} \cdot \mathbf{v}}{|\mathbf{u}||\mathbf{v}|} = \dfrac{\langle 3, 3\rangle \cdot \langle 2, -5\rangle}{(\sqrt{18})(\sqrt{29})} = \dfrac{6 + (-15)}{\sqrt{18}\sqrt{29}} = \dfrac{-9}{\sqrt{18}\sqrt{29}}$ $\theta = \cos^{-1}\dfrac{-9}{\sqrt{18}\sqrt{29}} = 113.2°$

13. $|\mathbf{u}| = \sqrt{2^2 + (-3)^2} = \sqrt{13}$ $|\mathbf{v}| = \sqrt{6^2 + 4^2} = \sqrt{52}$

     $\cos\theta = \dfrac{\mathbf{u} \cdot \mathbf{v}}{|\mathbf{u}||\mathbf{v}|} = \dfrac{\langle 2, -3\rangle \cdot \langle 6, 4\rangle}{(\sqrt{13})(\sqrt{52})} = \dfrac{12 - 12}{\sqrt{13}\sqrt{52}} = 0$ $\theta = \cos^{-1} 0 = 90°$

15.   Since $|\mathbf{u}|$ and $|\mathbf{v}|$ are never negative, the sign of the dot product depends only on $\cos\theta$. Since θ is an obtuse angle, $\cos\theta$ and the dot product are negative.

17. $\mathbf{u} \cdot \mathbf{v} = (2\mathbf{i} + \mathbf{j}) \cdot (\mathbf{i} - 2\mathbf{j}) = 2 - 2 = 0$ Thus, $\mathbf{u}$ and $\mathbf{v}$ are orthogonal.

19. $\mathbf{u} \cdot \mathbf{v} = \langle 1, 3\rangle \cdot \langle -3, -1\rangle = -3 - 3 = -6 \neq 0$ Thus, $\mathbf{u}$ and $\mathbf{v}$ are orthogonal.

21. $\text{comp}_{\mathbf{v}}\,\mathbf{u} = \dfrac{\mathbf{u} \cdot \mathbf{v}}{|\mathbf{v}|} = \dfrac{\langle -2, 4\rangle \cdot \langle -3, -1\rangle}{|\langle -3, -1\rangle|} = \dfrac{6 - 4}{\sqrt{(-3)^2 + (-1)^2}} = \dfrac{2}{\sqrt{10}} \approx 0.632$

23. $\text{comp}_{\mathbf{v}}\,\mathbf{u} = \dfrac{\mathbf{u} \cdot \mathbf{v}}{|\mathbf{v}|} = \dfrac{\langle -2, -4\rangle \cdot \langle 6, -3\rangle}{|\langle 6, -3\rangle|} = \dfrac{-12 + 12}{\sqrt{6^2 + (-3)^2}} = 0$

25. $\text{comp}_{\mathbf{v}}\,\mathbf{u} = \dfrac{\mathbf{u} \cdot \mathbf{v}}{|\mathbf{v}|} = \dfrac{(7\mathbf{i} - 2\mathbf{j}) \cdot (\mathbf{i} + \mathbf{j})}{|\mathbf{i} + \mathbf{j}|} = \dfrac{7 - 2}{\sqrt{1^2 + 1^2}} = \dfrac{5}{\sqrt{2}} \approx 3.54$

27. $\mathbf{u} \cdot \mathbf{u} = \langle a, b\rangle \cdot \langle a, b\rangle = a^2 + b^2 = (\sqrt{a^2 + b^2})^2 = |\mathbf{u}|^2$

29. $\mathbf{u} \cdot (\mathbf{v} + \mathbf{w})$ $= \langle a, b\rangle \cdot (\langle c, d\rangle + \langle e, f\rangle)$

                  $= \langle a, b\rangle \cdot \langle c + e, d + f\rangle$ Definition of vector addition

                  $= a(c + e) + b(d + f)$ Definition of dot product

                  $= ac + ae + bd + bf$ Distributive property of real numbers

                  $= (ac + bd) + (ae + bf)$ Commutative and associative properties of real numbers

                  $= \langle a, b\rangle \cdot \langle c, d\rangle + \langle a, b\rangle \cdot \langle e, f\rangle$ Definition of dot product

                  $= \mathbf{u} \cdot \mathbf{v} + \mathbf{u} \cdot \mathbf{w}$

31. $k(\mathbf{u} \cdot \mathbf{v}) = k(\langle a, b \rangle \cdot \langle c, d \rangle)$

$= k(ac + bd)$ Definition of dot product

$= k(ac) + k(bd)$ Distributive property of real numbers

$= (ka)c + (kb)d$ Associative property for multiplication of real numbers

$= \langle ka, kb \rangle \cdot \langle c, d \rangle$ Definition of dot product

$= (k\langle a, b \rangle) \cdot \langle c, d \rangle$ Definition of scalar multiplication

$= (k\mathbf{u}) \cdot \mathbf{v}$

Also, $k(\mathbf{u} \cdot \mathbf{v}) = k(\langle a, b \rangle \cdot \langle c, d \rangle)$

$= k(ac + bd)$ Definition of dot product

$= k(ac) + k(bd)$ Distributive property of real numbers

$= a(kc) + b(kd)$ Commutative and associative properties of real numbers

$= \langle a, b \rangle \cdot \langle kc, kd \rangle$ Definition of dot product

$= \langle a, b \rangle \cdot (k\langle c, d \rangle)$ Definition of scalar multiplication

$= \mathbf{u} \cdot (k\mathbf{v})$

33. $\text{Proj}_{\mathbf{v}}\, \mathbf{u} = \dfrac{\mathbf{u} \cdot \mathbf{v}}{\mathbf{v} \cdot \mathbf{v}}\, \mathbf{v} = \dfrac{\langle 3, 4 \rangle \cdot \langle 4, 0 \rangle}{\langle 4, 0 \rangle \cdot \langle 4, 0 \rangle}\, \langle 4, 0 \rangle = \dfrac{12 + 0}{16 + 0}\, \langle 4, 0 \rangle = \dfrac{3}{4}\, \langle 4, 0 \rangle = \langle 3, 0 \rangle$

35. $\text{Proj}_{\mathbf{v}}\, \mathbf{u} = \dfrac{\mathbf{u} \cdot \mathbf{v}}{\mathbf{v} \cdot \mathbf{v}}\, \mathbf{v} = \dfrac{(-6\mathbf{i} + 3\mathbf{j}) \cdot (-3\mathbf{i} - 2\mathbf{j})}{(-3\mathbf{i} - 2\mathbf{j}) \cdot (-3\mathbf{i} - 2\mathbf{j})}\, (-3\mathbf{i} - 2\mathbf{j}) = \dfrac{18 - 6}{9 + 4}\, (-3\mathbf{i} - 2\mathbf{j}) = \dfrac{12}{13}\, (-3\mathbf{i} - 2\mathbf{j})$

$= -\dfrac{36}{13}\mathbf{i} - \dfrac{24}{13}\mathbf{j}$

37. $W = \left(\begin{array}{l} \text{component of force in} \\ \text{the direction of motion} \end{array} \right) (\text{displacement}) = [(15\ \text{lb})\cos 42°](440\ \text{ft}) = 4{,}900\ \text{ft-lb}$

39. $\mathbf{d} = \langle 8, 1 \rangle$ $W = \mathbf{F} \cdot \mathbf{d} = \langle 10, 5 \rangle \cdot \langle 8, 1 \rangle = 85\ \text{ft-lb}$

41. $\mathbf{d} = \langle -3, 1 \rangle = -3\mathbf{i} + \mathbf{j}$ $W = \mathbf{F} \cdot \mathbf{d} = (-2\mathbf{i} + 3\mathbf{j}) \cdot (-3\mathbf{i} + \mathbf{j}) = 9\ \text{ft-lb}$

43. $\mathbf{d} = \langle 1, 1 \rangle = \mathbf{i} + \mathbf{j}$ $W = \mathbf{F} \cdot \mathbf{d} = (10\mathbf{i} + 10\mathbf{j}) \cdot (\mathbf{i} + \mathbf{j}) = 20\ \text{ft-lb}$

45.   To prove that $\angle ACB$, an angle inscribed in a semicircle, is a right angle, we need only show that $\mathbf{c} - \mathbf{a}$ and $\mathbf{c} + \mathbf{a}$ are orthogonal.

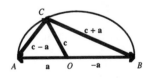

But $(\mathbf{c} - \mathbf{a}) \cdot (\mathbf{c} + \mathbf{a}) = (\mathbf{c} - \mathbf{a}) \cdot \mathbf{c} + (\mathbf{c} - \mathbf{a}) \cdot \mathbf{a}$ Distributive property of dot product

$= \mathbf{c} \cdot \mathbf{c} - \mathbf{a} \cdot \mathbf{c} + \mathbf{c} \cdot \mathbf{a} - \mathbf{a} \cdot \mathbf{a}$ Distributive property of dot product

$= \mathbf{c} \cdot \mathbf{c} - \mathbf{a} \cdot \mathbf{a} - \mathbf{a} \cdot \mathbf{c} + \mathbf{c} \cdot \mathbf{a}$ Commutative and associative properties of real numbers

$= \mathbf{c} \cdot \mathbf{c} - \mathbf{a} \cdot \mathbf{a} - \mathbf{a} \cdot \mathbf{c} + \mathbf{a} \cdot \mathbf{c}$ Commutative property of dot product

$= \mathbf{c} \cdot \mathbf{c} - \mathbf{a} \cdot \mathbf{a}$ Additive inverse and additive identity properties of real numbers

$= |\mathbf{c}||\mathbf{c}| \cos 0° - |\mathbf{a}||\mathbf{a}| \cos 0°$ Definition of dot product

$= (\text{Radius})^2 \cdot 1 - (\text{Radius})^2 \cdot 1$

$= 0$

Therefore, $\mathbf{c} - \mathbf{a}$ and $\mathbf{c} + \mathbf{a}$ are orthogonal, and an arbitrary angle ACB, inscribed in a semicircle, is a right angle.

CHAPTER 6   REVIEW EXERCISE

1. The law of sines needs to have an angle and a side opposite the angle given, which is not the case here.

2. The two forces are oppositely directed, that is, the angle between the two forces is 180°.

3. The forces are acting in the same direction, that is, the angle between the two forces is 0°.

4. (A)

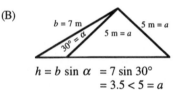

(B)

$h = b \sin \alpha = 8 \sin 30°$
$= 4 = a$

This is the case of SSA where α is acute and $a = h$.
1 triangle can be constructed.

$h = b \sin \alpha = 7 \sin 30°$
$= 3.5 < 5 = a$

This is the case of SSA where α is acute and $h < a < b$.
2 triangles can be constructed.

(C)

$h = b \sin \alpha = 8 \sin 30°$
$= 4 > 3 = a$

This is the case where α is acute and $0 < a < h$.
0 triangles can be constructed.

5. We are given two angles and the included side (*ASA*). We use the law of sines.

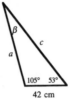

Solve for β: $\alpha + \beta + \gamma = 180°$
$\beta = 180° - (105° + 53°) = 22°$

Solve for β: $\dfrac{\sin \alpha}{a} = \dfrac{\sin \beta}{b}$

$a = \dfrac{b \sin \alpha}{\sin \beta} = \dfrac{(42 \text{ cm})(\sin 53°)}{\sin 22°} = 90 \text{ cm}$

Solve for c: $\dfrac{\sin \beta}{b} = \dfrac{\sin \gamma}{c}$

$c = \dfrac{b \sin \gamma}{\sin \beta} = \dfrac{(42 \text{ cm})(\sin 105°)}{\sin 22°} = 110 \text{ cm}$

6. We are given two angles and a non-included side (*AAS*). We use the law of sines.

Solve for γ: $\alpha + \beta + \gamma = 180°$
$\gamma = 180° - (66° + 32°) = 82°$

Solve for a: $\dfrac{\sin \alpha}{a} = \dfrac{\sin \beta}{b}$

$a = \dfrac{b \sin \alpha}{\sin \beta} = \dfrac{(12 \text{ m})(\sin 66°)}{\sin 32°} = 21 \text{ m}$

Solve for c: $\dfrac{\sin \gamma}{c} = \dfrac{\sin \beta}{b}$

$c = \dfrac{b \sin \gamma}{\sin \beta} = \dfrac{(12 \text{ m})(\sin 82°)}{\sin 32°} = 22 \text{ m}$

7. We are given two sides and the included angle (*SAS*). We use the law of cosines to find the third side, then the law of sines to find a second angle.

Solve for a: $a^2 = b^2 + c^2 - 2bc \cos \alpha$
$= 22^2 + 27^2 - 2(22)(27) \cos 49° = 433.60187\ldots$
$a = \sqrt{433.60187\ldots} = 21 \text{ in}$

Since *b* is the shorter of the remaining sides, β, the angle opposite *b*, must be acute.

Solve for β: We use the law of sines. $\dfrac{\sin \beta}{b} = \dfrac{\sin \alpha}{a}$

$\sin \beta = \dfrac{b \sin \alpha}{a} = \dfrac{(22 \text{ in})(\sin 49°)}{21 \text{ in}} = 0.7974$

$\beta = \sin^{-1} 0.7974 = 53°$

Solve for γ: $\gamma = 180° - (\alpha + \beta) = 180° - (49° + 53°) = 78°$

8. We are given two sides and a non-included angle (*SSA*). α is acute. $a > b$. One triangle is possible. We use the law of sines.

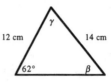

Solve for β: $\dfrac{\sin \alpha}{a} = \dfrac{\sin \beta}{b}$

$\sin \beta = \dfrac{b \sin \alpha}{\sin a} = \dfrac{(12 \text{ cm})(\sin 62°)}{14 \text{ cm}} = 0.7568$

$\beta = \sin^{-1} 0.7568 = 49°$

(There is another solution of $\sin \beta = 0.7568$ that deserves brief consideration:

$\beta' = 180° - \sin^{-1} 0.7568 = 131°$. However, there is not enough room in a triangle for an angle of 62° and an angle of 131°, since their sum is greater than 180°.)

Solve for γ: $\alpha + \beta + \gamma = 180°$ $\gamma = 180° - (62° + 49°) = 69°$

Solve for c: $\dfrac{\sin \beta}{b} = \dfrac{\sin \gamma}{c}$

$c = \dfrac{b \sin \gamma}{\sin \beta} = \dfrac{(12 \text{ cm})(\sin 69°)}{\sin 49°} = 15 \text{ cm}$

9. The given information consists of two sides and the included angle; hence, we use the formula
$A = \dfrac{ab}{2} \sin \theta$ in the form $A = \dfrac{1}{2} bc \sin \alpha = \dfrac{1}{2} (22 \text{ in})(27 \text{ in}) \sin 49° = 224 \text{ in}^2$

10. The given information consists of two sides and a non-included angle. Hence, we use the information computed in Problem 8 in the formula
$A = \dfrac{ab}{2} \sin \theta$ in the form $A = \dfrac{1}{2} ab \sin \gamma = \dfrac{1}{2} (14 \text{ cm})(12 \text{ cm}) \sin 69° = 79 \text{ cm}^2$

11. To find $|\mathbf{u} + \mathbf{v}|$: Apply the Pythagorean theorem to triangle *OCB*.

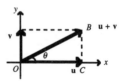

$|\mathbf{u} + \mathbf{v}|^2 = OB^2 = OC^2 + BC^2 = 8.0^2 + 5.0^2 = 89.00$

$|\mathbf{u} + \mathbf{v}| = \sqrt{89.00} = 9.4$

Solve triangle *OCB* for θ: $\tan \theta = \dfrac{BC}{OC} = \dfrac{|\mathbf{v}|}{|\mathbf{u}|}$ $\theta = \tan^{-1} \dfrac{|\mathbf{v}|}{|\mathbf{u}|}$ θ is acute

$\theta = \tan^{-1} \dfrac{5.0}{8.0} = 32°$

12. $|\mathbf{v}| = 12$ $\theta = 35°$

Horizontal component *H*: $\cos 35° = \dfrac{H}{12}$ $H = 12 \cos 35° = 9.8$

Vertical component *V*: $\sin 35° = \dfrac{V}{12}$ $V = 12 \sin 35° = 6.9$

13. The algebraic vector $\langle a, b \rangle$ has coordinates given by
$a = x_b - x_a = (-1) - (-3) = 2$ $b = y_b - y_a = (-3) - 2 = -5$
Hence, $\langle a, b \rangle = \langle 2, -5 \rangle$

14. Magnitude of $\langle -5, 12 \rangle = |\langle -5, 12 \rangle| = \sqrt{a^2 + b^2} = \sqrt{(-5)^2 + 12^2} = 13$

15. $\langle 2, -1 \rangle \cdot \langle -3, 2 \rangle = 2 \cdot (-3) + (-1) \cdot 2 = -8$

16. $(2\mathbf{i} + \mathbf{j}) \cdot (3\mathbf{i} - 2\mathbf{j}) = 2 \cdot 3 + 1 \cdot (-2) = 4$

17. $|\mathbf{u}| = \sqrt{4^2 + 3^2} = 5$ $\qquad\qquad\qquad |\mathbf{v}| = \sqrt{3^2 + 0^2} = 3$

  $\cos \theta = \dfrac{\mathbf{u} \cdot \mathbf{v}}{|\mathbf{u}||\mathbf{v}|} = \dfrac{\langle 4, 3 \rangle \cdot \langle 3, 0 \rangle}{(5)(3)} = \dfrac{12 + 0}{15} = \dfrac{12}{15}$ $\qquad \theta = \cos^{-1} \dfrac{12}{15} = 36.9°$

18. $|\mathbf{u}| = \sqrt{5^2 + 1^2} = \sqrt{26}$ $\qquad\qquad\qquad |\mathbf{v}| = \sqrt{(-2)^2 + 2^2} = \sqrt{8}$

  $\cos \theta = \dfrac{\mathbf{u} \cdot \mathbf{v}}{|\mathbf{u}||\mathbf{v}|} = \dfrac{(5\mathbf{i} + \mathbf{j}) \cdot (-2\mathbf{i} + 2\mathbf{j})}{\sqrt{26}\sqrt{8}} = \dfrac{5(-2) + 1 \cdot 2}{\sqrt{26}\sqrt{8}} = \dfrac{-8}{\sqrt{26}\sqrt{8}}$ $\qquad \theta = \cos^{-1} \dfrac{-8}{\sqrt{26}\sqrt{8}} = 123.7°$

19. No triangle is possible, since $\sin \beta$ cannot exceed one.

20. We are given two sides and the included angle (*SAS*). We use the law of cosines to find the third side, then the law of sines to find a second angle.

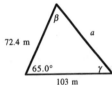

  Solve for a:

  $a^2 = b^2 + c^2 - 2bc \cos \alpha$

  $\quad = (103)^2 + (72.4)^2 - 2(103)(72.4) \cos 65.0° = 9,547.6622...$

  $a = \sqrt{9,547.6622...} = 97.7 \text{ m}$

  Since c is the shorter of the remaining sides, γ, the angle opposite c, must be acute.

  Solve for γ: $\quad \dfrac{\sin \gamma}{c} = \dfrac{\sin \alpha}{a}$

  $\qquad\qquad \sin \gamma = \dfrac{c \sin \alpha}{a} = \dfrac{(72.4 \text{ m})(\sin 65.0°)}{97.7 \text{ m}} = 0.6715$

  $\qquad\qquad\quad \gamma = \sin^{-1} 0.6715 = 42.2°$

  Solve for β: $\alpha + \beta + \gamma = 180°$ $\qquad \beta = 180° - (\alpha + \gamma) = 180° - (65.0° + 42.2°) = 72.8°$

21. We are given two sides and a non-included angle (*SSA*). α is acute. $h = b \sin \alpha = 15.7 \sin 35°20' = 9.08$, $h < a < b$. Two triangles are possible, but β is specified acute. We use the law of sines.

  Solve for β: $\quad \dfrac{\sin \alpha}{a} = \dfrac{\sin \beta}{b}$

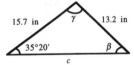

  $\qquad\qquad \sin \beta = \dfrac{b \sin \alpha}{a} = \dfrac{(15.7 \text{ in})(\sin 35°20')}{13.2 \text{ in}} = 0.6879$

  Since β is specified acute, we choose the acute angle solution to this equation,

  $\qquad\qquad \beta = \sin^{-1} 0.6879 = 43°30'$

Solve for γ: $\alpha + \beta + \gamma = 180°$ $\gamma = 180° - (\alpha + \beta) = 180° - (35°20' + 43°30') = 101°10'$

Solve for c: $\dfrac{\sin \gamma}{c} = \dfrac{\sin \alpha}{a}$

$$c = \dfrac{a \sin \gamma}{\sin \alpha} = \dfrac{(13.2 \text{ in})(\sin 101°10')}{\sin 35°20'} = 22.4 \text{ in}$$

22. We are given the same information as in Problem 21, except that β is specified obtuse.

Solve for β: $\dfrac{\sin \alpha}{a} = \dfrac{\sin \beta}{b}$

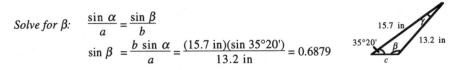

$$\sin \beta = \dfrac{b \sin \alpha}{a} = \dfrac{(15.7 \text{ in})(\sin 35°20')}{13.2 \text{ in}} = 0.6879$$

Since β is specified obtuse, we choose the obtuse angle solution to this equation,

$$\beta = 180° - \sin^{-1} 0.6879 = 136°30'$$

Solve for γ: $\alpha + \beta + \gamma = 180°$ $\gamma = 180° - (\alpha + \beta) = 180° - (35°20' + 136°30') = 8°10'$

Solve for c: $\dfrac{\sin \gamma}{c} = \dfrac{\sin \alpha}{a}$

$$c = \dfrac{a \sin \gamma}{\sin \alpha} = \dfrac{(13.2 \text{ in})(\sin 8°10')}{\sin 35°20'} = 3.24 \text{ in}$$

23. We are given three sides (*SSS*). We solve for the largest angle, γ (largest because it is opposite the largest side, c) using the law of cosines. We then solve for a second angle using the law of sines, because it involves simpler calculations.

Solve for γ: $c^2 = a^2 + b^2 - 2ab \cos \gamma$

$$\cos \gamma = \dfrac{a^2 + b^2 - c^2}{2ab}$$

$$\gamma = \cos^{-1}\dfrac{a^2 + b^2 - c^2}{2ab} = \cos^{-1}\dfrac{43^2 + 48^2 - 53^2}{2 \cdot 43 \cdot 48} = 71°$$

Both β and α must be acute, since they are smaller than γ.

Solve for β: $\dfrac{\sin \beta}{b} = \dfrac{\sin \gamma}{c}$

$$\sin \beta = \dfrac{b \sin \gamma}{c}$$

$$\beta = \sin^{-1}\dfrac{b \sin \gamma}{c}$$

$$= \sin^{-1}\dfrac{(48 \text{ mm}) \sin 71°}{53 \text{ mm}} = 59°$$

Solve for α: $\alpha + \beta + \gamma = 180°$ $\alpha = 180° - (\beta + \gamma) = 180° - (59° + 71°) = 50°$

24. The given information consists of two sides and the included angle; hence, we use the formula
$A = \dfrac{ab}{2} \sin \theta$ in the form $A = \dfrac{1}{2} bc \sin \alpha = \dfrac{1}{2} (103 \text{ m})(72.4 \text{ m}) \sin 65.0° = 3{,}380 \text{ m}^2$

25. The given information consists of three sides; hence, we use Heron's formula. First, find the semiperimeter s:
$$s = \frac{a + b + c}{2} = \frac{43 + 48 + 53}{2} = 72 \text{ mm}$$

Then, $s - a = 72 - 43 = 29$
         $s - b = 72 - 48 = 24$
         $s - c = 72 - 53 = 19$
Thus, $A = \sqrt{s(s - a)(s - b)(s - c)} = \sqrt{72(29)(24)(19)} = 980 \text{ mm}^2$

26. **u** and **v** have the same magnitudes, since $|\mathbf{u}| = |\mathbf{v}|$. However, if they have different directions, then **u** and **v** are not equal, since two geometric vectors are equal if and only if their magnitudes are equal and their directions are equal.

27. $\langle a, b \rangle = \langle c, d \rangle$ if and only if $a = c$ and $b = d$.

28. $\angle DOC = 45.0°$. Hence, $\angle OCB = 180° - 45.0° = 135.0°$.
We can find the magnitude of **u** + **v** using the law of cosines:

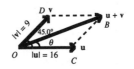

$|\mathbf{u} + \mathbf{v}|^2 = |\mathbf{u}|^2 + |\mathbf{v}|^2 - 2|\mathbf{u}||\mathbf{v}| \cos OCB$
$= 16^2 + 9^2 - 2(16)(9) \cos 135.0° = 540.64675\ldots$
$|\mathbf{u} + \mathbf{v}| = \sqrt{540.64675\ldots} \approx 23.3$

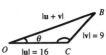

To find θ, we use the law of sines:
$$\frac{\sin \theta}{|\mathbf{v}|} = \frac{\sin OCB}{|\mathbf{u} + \mathbf{v}|}$$
$$\frac{\sin \theta}{9} = \frac{135.0°}{23.3}$$
$$\sin \theta = \frac{9 \sin 135.0°}{23.3}$$
$$\theta = \sin^{-1} \left(\frac{9 \sin 135.0°}{23.3} \right) \approx 15.9°$$

29. (A) $\mathbf{u} + \mathbf{v} = \langle 4, 0 \rangle + \langle -2, -3 \rangle = \langle 2, -3 \rangle$
     (B) $\mathbf{u} - \mathbf{v} = \langle 4, 0 \rangle - \langle -2, -3 \rangle = \langle 6, 3 \rangle$
     (C) $3\mathbf{u} - 2\mathbf{v} = 3\langle 4, 0 \rangle - 2\langle -2, -3 \rangle = \langle 12, 0 \rangle + \langle 4, 6 \rangle = \langle 16, 6 \rangle$
     (D) $2\mathbf{u} - 3\mathbf{v} + \mathbf{w} = 2\langle 4, 0 \rangle - 3\langle -2, -3 \rangle + \langle 1, -1 \rangle = \langle 8, 0 \rangle + \langle 6, 9 \rangle + \langle 1, -1 \rangle = \langle 15, 8 \rangle$

30. (A) $\mathbf{u} + \mathbf{v} = (3\mathbf{i} - \mathbf{j}) + (2\mathbf{i} - 3\mathbf{j}) = 3\mathbf{i} + 2\mathbf{i} - \mathbf{j} - 3\mathbf{j} = 5\mathbf{i} - 4\mathbf{j}$
     (B) $\mathbf{u} - \mathbf{v} = (3\mathbf{i} - \mathbf{j}) - (2\mathbf{i} - 3\mathbf{j}) = 3\mathbf{i} - \mathbf{j} - 2\mathbf{i} + 3\mathbf{j} = \mathbf{i} + 2\mathbf{j}$
     (C) $3\mathbf{u} - 2\mathbf{v} = 3(3\mathbf{i} - \mathbf{j}) - 2(2\mathbf{i} - 3\mathbf{j}) = 9\mathbf{i} - 3\mathbf{j} - 4\mathbf{i} + 6\mathbf{j} = 5\mathbf{i} + 3\mathbf{j}$
     (D) $2\mathbf{u} - 3\mathbf{v} + \mathbf{w} = 2(3\mathbf{i} - \mathbf{j}) - 3(2\mathbf{i} - 3\mathbf{j}) + (-2\mathbf{j}) = 6\mathbf{i} - 2\mathbf{j} - 6\mathbf{i} + 9\mathbf{j} - 2\mathbf{j} = 0\mathbf{i} + 5\mathbf{j} \text{ or } 5\mathbf{j}$

31. $|\mathbf{v}| = \sqrt{(-8)^2 + 15^2} = 17$ $\mathbf{u} = \dfrac{1}{|\mathbf{v}|}\mathbf{v} = \dfrac{1}{17}\langle -8, 15\rangle = \left\langle -\dfrac{8}{17}, \dfrac{15}{17}\right\rangle$

32. (A) $\mathbf{v} = \langle -5, 7\rangle = \langle -5, 0\rangle + \langle 0, 7\rangle = -5\langle 1, 0\rangle + 7\langle 0, 1\rangle = -5\mathbf{i} + 7\mathbf{j}$

    (B) $\mathbf{v} = \langle 0, -3\rangle = -3\langle 0, 1\rangle = -3\mathbf{j}$

    (C) $\mathbf{v} = \overrightarrow{AB} = \langle 0 - 4, (-3) - (-2)\rangle = \langle -4, -1\rangle = \langle -4, 0\rangle + \langle 0, -1\rangle = -4\langle 1, 0\rangle - \langle 0, 1\rangle = -4\mathbf{i} - \mathbf{j}$

33. (A) $\mathbf{u} \cdot \mathbf{v} = \langle -12, 3\rangle \cdot \langle 2, 8\rangle = -24 + 24 = 0$ Thus, $\mathbf{u}$ and $\mathbf{v}$ are orthogonal.

    (B) $\mathbf{u} \cdot \mathbf{v} = (-4\mathbf{i} + \mathbf{j}) \cdot (-\mathbf{i} + 4\mathbf{j}) = 4 + 4 = 8 \neq 0$ Thus, $\mathbf{u}$ and $\mathbf{v}$ are not orthogonal.

34. (A) $\operatorname{comp}_\mathbf{v} |\mathbf{u}| = \dfrac{\mathbf{u} \cdot \mathbf{v}}{|\mathbf{v}|} = \dfrac{\langle 4, 5\rangle \cdot \langle 3, 1\rangle}{|\langle 3, 1\rangle|} = \dfrac{12 + 5}{\sqrt{3^2 + 1^2}} = \dfrac{17}{\sqrt{10}} \approx 5.38$

    (B) $\operatorname{comp}_\mathbf{v} |\mathbf{u}| = \dfrac{\mathbf{u} \cdot \mathbf{v}}{|\mathbf{v}|} = \dfrac{(-\mathbf{i} + 4\mathbf{j}) \cdot (3\mathbf{i} - \mathbf{j})}{|3\mathbf{i} - \mathbf{j}|} = \dfrac{-3 - 4}{\sqrt{3^2 + (-1)^2}} = -\dfrac{7}{\sqrt{10}} \approx -2.21$

35. Any one of the force vectors must have the same magnitude as the resultant of the other two force vectors and be oppositely directed to the resultant of the other two.

36.
$$(a - b)\cos\frac{\gamma}{2} = c\,\sin\frac{\alpha - \beta}{2}$$
$$(8.42 - 11.5)\cos\frac{59.1°}{2} = 10.2\,\sin\frac{45.1° - 75.8°}{2}$$
$$-2.68 \approx -2.70$$
The results agree to two significant digits; the results check.

37. From the diagram, we can see that k = altitude of any possible triangle.
Thus, $\sin\beta = \dfrac{k}{a}$, $k = a\sin\beta$.

$k = (12.7\text{ cm})\sin 52.3° = 10.0\text{ cm}$
If $0 < a < k$, there is no solution: (1) in the diagram.

If $a = k$, there is one solution.

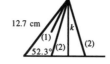

If $k < a < b$, there are two solutions: (2) in the diagram.

38. $\mathbf{u} + \mathbf{v} = \langle a, b\rangle + \langle c, d\rangle$

            $= \langle a + c, b + d\rangle$ Definition of vector addition

            $= \langle c + a, d + b\rangle$ Commutative property for addition of real numbers

            $= \langle c, d\rangle + \langle a, b\rangle$ Definition of vector addition

            $= \mathbf{v} + \mathbf{u}$

39. $\mathbf{u} \cdot \mathbf{v} = \langle a, b\rangle \cdot \langle c, d\rangle$

            $= ac + bd$ Definition of dot product

            $= ca + db$ Commutative property for multiplication of real numbers

            $= \langle c, d\rangle \cdot \langle a, b\rangle$ Definition of dot product

            $= \mathbf{v} \cdot \mathbf{u}$

40. $(mn)\mathbf{v} = (mn)\langle c, d \rangle$

$= \langle (mn)c, (mn)d \rangle$ Definition of scalar multiplication

$= \langle m(nc), m(nd) \rangle$ Associative property for multiplication of real numbers

$= m\langle nc, nd \rangle$ Definition of scalar multiplication

$= m(n\langle c, d \rangle)$ Definition of scalar multiplication

$= m(n\mathbf{v})$

41. $m(\mathbf{u} \cdot \mathbf{v}) = m(\langle a, b \rangle \cdot \langle c, d \rangle)$

$= m(ac + bd)$ Definition of dot product

$= m(ac) + m(bd)$ Distributive property of real numbers

$= (ma)c + (mb)d$ Associative property for multiplication of real numbers

$= \langle ma, mb \rangle \cdot \langle c, d \rangle$ Definition of dot product

$= (m\langle a, b \rangle) \cdot \langle c, d \rangle$ Definition of scalar multiplication

$= (m\mathbf{u}) \cdot \mathbf{v}$

Also, $m(\mathbf{u} \cdot \mathbf{v}) = m(\langle a, b \rangle \cdot \langle c, d \rangle)$

$= m(ac + bd)$ Definition of dot product

$= m(ac) + m(bd)$ Distributive property of real numbers

$= a(mc) + b(md)$ Commutative and associative properties of real numbers

$= \langle a, b \rangle \cdot \langle mc, md \rangle$ Definition of dot product

$= \langle a, b \rangle \cdot (m\langle c, d \rangle)$ Definition of scalar multiplication

$= \mathbf{u} \cdot (m\mathbf{v})$

42. $\mathbf{u} \cdot (\mathbf{v} + \mathbf{w}) = \langle a, b \rangle \cdot (\langle c, d \rangle + \langle e, f \rangle)$

$= \langle a, b \rangle \cdot \langle c + e, \ d + f \rangle$ Definition of vector addition

$= a(c + e) + b(d + f)$ Definition of dot product

$= ac + ae + bd + bf$ Distributive property of real numbers

$= (ac + bd) + (ae + bf)$ Commutative and associative properties of real numbers

$= \langle a, b \rangle \cdot \langle c, d \rangle + \langle a, b \rangle \cdot \langle e, f \rangle$ Definition of dot product

$= \mathbf{u} \cdot \mathbf{v} + \mathbf{u} \cdot \mathbf{w}$

43. $\mathbf{u} \cdot \mathbf{u} = \langle a, b \rangle \cdot \langle a, b \rangle$

$= a^2 + b^2$ Definition of dot product

$= (\sqrt{a^2 + b^2})^2$ Definition of the square root of a real number

$= |\mathbf{u}|^2$ Definition of magnitude

44. Since the diagonals of a parallelogram bisect each other, we see that

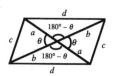

$$a = \frac{1}{2}(16.0 \text{ cm}) = 8.0 \text{ cm} \qquad b = \frac{1}{2}(20.0 \text{ cm}) = 10.0 \text{ cm}$$

$$\theta = 36.4° \qquad 180° - \theta = 143.6°$$

To find c and d, we use the law of cosines.

$$c^2 = a^2 + b^2 - 2ab \cos \theta$$
$$= (8.0)^2 + (10.0)^2 - 2(8.0)(10.0) \cos 36.4$$
$$= 35.216992...$$
$$c = \sqrt{35.216992...} = 5.9 \text{ cm}$$
$$d^2 = a^2 + b^2 - 2ab \cos(180° - \theta) = (8.0)^2 + (10.0)^2 - 2(8.0)(10.0) \cos (143.6°) = 292.78301...$$
$$d = \sqrt{292.78301...} = 17 \text{ cm}$$

45. In triangle ACD, we note $\dfrac{AD}{h} = \cot 31°20'$.

In triangle BCD, we note $\dfrac{BD}{h} = \cot 49°40'$.

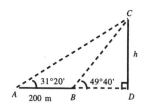

Hence, $AD = h \cot 31°20'$, $BD = h \cot 49°40'$.

Since $AD - BD = AB = 200$ m, we have

200 m = $h \cot 31°20' - h \cot 49°40'$ or

$$h = \frac{200}{\cot 31°20' - \cot 49°40'} = 252 \text{ m}$$

Alternatively, we can note $\angle ABC = 180° - 49°40' = 130°20'$, and use the law of sines to determine BC.

$$\frac{\sin CAB}{BC} = \frac{\sin BCA}{AB}$$

$$\angle BCA = 180° - (\angle CAB + \angle ABC) = 180° - (31°20' + 130°20') = 18°20'$$

$$BC = \frac{AB \sin CAB}{\sin BCA} = \frac{(200 \text{ m})(\sin 31°20')}{\sin 18°20'} = 331 \text{ m}$$

Then, in right triangle BCD, we have

$$\frac{h}{BC} = \sin CBD$$

$$h = BC \sin CBD = (331 \text{ m})(\sin 49°40') = 252 \text{ m}$$

46. We sketch a figure, labeling what we know. In triangle OAB, we know $AB = 34$ cm, $\angle AOB = 85°$, $OA = OB = r$. Thus, from the law of cosines:

$$AB^2 = OA^2 + OB^2 - 2(OA)(OB) \cos AOB$$
$$34^2 = r^2 + r^2 - 2r^2 \cos 85° = 2r^2 - 2r^2 \cos 85° = r^2(2 - 2 \cos 85°)$$

$$r^2 = \frac{34^2}{2 - 2 \cos 85°} \qquad r = \sqrt{\frac{34^2}{2 - 2 \cos 85°}} \quad \text{We discard the negative solution}$$

$$= 25 \text{ cm}$$

47. We will first find AD by applying the law of sines to triangle ABD, in which we know all angles and a side. We will then find AC by applying the law of sines to triangle ABC, in which we also know all angles and a side. We can then apply the law of cosines to triangle ADC to find DC.

In triangle ABD: $\dfrac{\sin ABD}{AD} = \dfrac{\sin ADB}{AB}$

$$AD = \frac{AB \sin ABD}{\sin ADB} = \frac{(450 \text{ ft}) \sin 72°}{\sin 50°} = 558.7 \text{ ft}$$

In triangle ABC: $\dfrac{\sin ABC}{AC} = \dfrac{\sin ACB}{AB}$

$$AC = \frac{AB \sin ABC}{\sin ACB} = \frac{(450 \text{ ft}) \sin (72° + 65°)}{\sin 20°} = 897.3 \text{ ft}$$

In triangle ACD:
$$CD^2 = AC^2 + AD^2 - 2(AC)(AD) \cos CAD$$
$$= (558.7)^2 + (897.3)^2 - 2(558.7)(897.3) \cos 35°$$
$$= 295993.95\ldots$$
$$CD = 540 \text{ ft}$$

48. Area of $ABCD$ = Area of triangle ADC + Area of triangle ABC
$$= \frac{1}{2} (AD)(AC) \sin DAC + \frac{1}{2} (AC)(AB) \sin CAB$$
$$= \frac{1}{2} (558.7)(897.3) \sin 35° + \frac{1}{2} (897.3)(450) \sin 23° = 220{,}000 \text{ ft}^2$$

49. In triangle ABC, we are given two sides and the included angle; hence, we can apply the law of cosines to find side BC.
$$BC^2 = AB^2 + AC^2 - 2(AB)(AC) \cos BAC$$
$$= (5.72)^2 + (6.37)^2 - 2(5.72)(6.37) \cos 38.2°$$
$$= 16.027708\ldots$$
$$BC = 4.00 \text{ km}$$

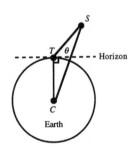

50. In triangle CST, we are given $TS = 1{,}147$ miles and $TC = 3{,}964$ miles.

$$\angle STC = \theta + 90° = 28.6° + 90° = 118.6°.$$

We are given two sides and the included angle; hence, we can apply the law of cosines to find side SC.
$$SC^2 = TS^2 + TC - 2(TS)(TC) \cos STC$$
$$= (1{,}147)^2 + (3{,}964)^2 - 2(1{,}147)(3{,}964) \cos 118.6°$$
$$= 21{,}381{,}849\ldots$$
$$SC = 4{,}624 \text{ mi}$$

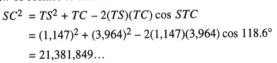

Hence, the height of the satellite = SC – radius of the earth = $4{,}624$ mi – $3{,}964$ mi = 660 mi

51. Following the hint, we find all angles for triangle ACS first.
$$\angle SAC = \theta + 90° = 21.7° + 90° = 111.7°$$

For $\angle ACS$, we use the formula $s = \dfrac{\pi}{180}\,\theta R$ from Chapter 2, with

$$s = 632 \text{ mi} \qquad \theta = \angle ACS \qquad R = 3{,}964 \text{ mi}$$

Then, $\angle ACS = \dfrac{180°s}{\pi R} = \dfrac{180°(632 \text{ mi})}{\pi(3{,}964 \text{ mi})} = 9.13°$. Since $\angle SAC + \angle ACS + \angle ASC = 180°$, we have

$$\angle ASC = 180° - (\angle ACS + \angle SAC) = 180° - (111.7° + 9.13°) = 59.2°$$

Now, apply the law of sines to find side CS:
$$\frac{\sin SAC}{CS} = \frac{\sin ASC}{AC}$$
$$CS = \frac{AC \sin SAC}{\sin ASC} = \frac{(3{,}964 \text{ mi})\sin 111.7°}{\sin 59.2°} = 4{,}289 \text{ mi}$$

Hence, the height of the satellite above $B = BS = CS - BC$.

$$BS = 4{,}289 \text{ mi} - 3{,}964 \text{ mi} = 325 \text{ mi}$$

52. We are given $\varepsilon = 5°$ (wind heading), $\rho = 68°$ (plane heading). We want to find $|R|$, where R is the resultant sum of v, the plane's velocity, and w, the wind velocity. We also want θ, then $\theta + \varepsilon = \theta + 5°$ will be the plane's actual direction relative to north.

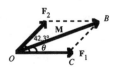

Solve for $|R|$:
In triangle ORA, since $\angle AOP = \rho - \varepsilon = 68° - 5° = 63°$, α must equal

$$180° - \angle AOP = 180° - 63° = 117° \qquad |w| = 55 \text{ km/hr} \qquad AR = |v| = 230 \text{ km/hr}$$

Now, apply the law of cosines.
$$|R|^2 = |w|^2 + |v|^2 - 2|w||v| \cos \alpha = 55^2 + 230^2 - 2(55)(230) \cos 117° = 67{,}410.96\ldots$$
$$|R| = \sqrt{67{,}410.96\ldots} \approx 260 \text{ km/hr}$$

Solve for θ: $\dfrac{\sin \theta}{AR} = \dfrac{\sin \alpha}{|R|}$ $\qquad \dfrac{\sin \theta}{230} = \dfrac{\sin 117°}{260}$

$$\sin \theta = \frac{230}{260}\sin 117° \qquad\qquad \theta = \sin^{-1}\left(\frac{230}{260}\sin 117°\right) \approx 52°$$

The actual direction $= \theta + \varepsilon = 52° + 5° = 57°$.

53. $\angle OCB = 180° - 42.3° = 137.7°$ $\qquad |F_1| = 352 \text{ lb} \qquad |F_2| = 168 \text{ lb}$

We can find the magnitude M of $F_1 + F_2$ using the law of cosines:

$$M^2 = |F_1|^2 + |F_2|^2 - 2|F_1||F_2| \cos OCB$$
$$= 352^2 + 168^2 - 2(352)(168) \cos 137.7° = 239{,}605.65\ldots$$
$$M = \sqrt{239{,}605.65\ldots} = 489 \text{ lb}$$

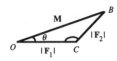

To find θ, we use the law of sines: $\dfrac{\sin \theta}{|F_2|} = \dfrac{\sin OCB}{M}$ $\dfrac{\sin \theta}{168} = \dfrac{\sin 137.7°}{489}$

$\sin \theta = \dfrac{168}{489} \sin 137.7°$ $\theta = \sin^{-1}\left(\dfrac{168}{489} \sin 137.7°\right) \approx 13.4°$ (relative to $\mathbf{F}_1$)

54. $W = Fd = (489 \text{ lb})(22 \text{ ft}) = 10{,}800 \text{ ft-lb}$

55. First, form a force diagram with all force vectors in standard position at the origin.

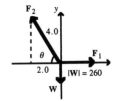

Let $\mathbf{F}_1$ = the force on the horizontal member AB
        $\mathbf{F}_2$ = the force on the supporting member BC
        $\mathbf{W}$ = the downward force (260 lb)

We note: $\cos \theta = \dfrac{2.0}{4.0}$ $\theta = \cos^{-1} \dfrac{2.0}{4.0} = 60°$

Then write each force vector in terms of $\mathbf{i}$ and $\mathbf{j}$ unit vectors:

$\mathbf{F}_1 = |\mathbf{F}_1| \, \mathbf{i}$
$\mathbf{F}_2 = -|\mathbf{F}_2| (\cos 60°)\mathbf{i} + |\mathbf{F}_2| (\sin 60°)\mathbf{j}$
$\mathbf{W} = -260\mathbf{j}$

For the system to be in static equilibrium, we must have $\mathbf{F}_1 + \mathbf{F}_2 + \mathbf{W} = \mathbf{0}$ which becomes, on addition,

$[|\mathbf{F}_1| - |\mathbf{F}_2| (\cos 60°)]\mathbf{i} + [|\mathbf{F}_2| (\sin 60°) - 260]\mathbf{j} = 0\mathbf{i} + 0\mathbf{j}$.

Since two vectors are equal if and only if their corresponding components are equal, we are led to the following system of equations in $|\mathbf{F}_1|$ and $|\mathbf{F}_2|$:

$|\mathbf{F}_1| - |\mathbf{F}_2| (\cos 60°) = 0$
$|\mathbf{F}_2| (\sin 60°) - 260 = 0$

Solving, $|\mathbf{F}_2| = \dfrac{260}{\sin 60°} = 300 \text{ lb}$
            $|\mathbf{F}_1| = |\mathbf{F}_2| \cos 60° = 150 \text{ lb}$

The force in the member AB is directed oppositely to the diagram, a compression of 150 lb.

The force in the member BC is also directed oppositely to the diagram, a tension of 300 lb.

56. (A) We write each force vector in terms of $\mathbf{i}$ and $\mathbf{j}$ unit vectors.

Note: $\alpha = 90° - \theta$.

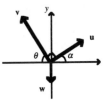

$\mathbf{u} = |\mathbf{u}| \cos (90° - \theta)\mathbf{i} + |\mathbf{u}| \sin (90° - \theta)\mathbf{j}$
    $= |\mathbf{u}| \sin \theta \mathbf{i} + |\mathbf{u}| \cos \theta \mathbf{j}$
$\mathbf{v} = -|\mathbf{v}| \cos \theta \mathbf{i} + |\mathbf{v}| \sin \theta \mathbf{j}$
$\mathbf{w} = -|\mathbf{W}| \mathbf{j}$

For the system to be in static equilibrium, we must have $\mathbf{u} + \mathbf{v} + \mathbf{w} = \mathbf{0}$ which becomes, on addition,

$[|\mathbf{u}| \sin \theta - |\mathbf{v}| \cos \theta]\mathbf{i} + [|\mathbf{u}| \cos \theta + |\mathbf{v}| \sin \theta - |\mathbf{w}|]\mathbf{j} = 0\mathbf{i} + 0\mathbf{j}$

Since two vectors are equal if and only if their corresponding components are equal, we are let to the following system of equations in $|u|$ and $|v|$:

(1) $|u| \sin \theta - |v| \cos \theta = 0$
(2) $|u| \cos \theta + |v| \sin \theta - |w| = 0$

Solving, we have:

(3) $|u| \sin^2 \theta - |v| \sin \theta \cos \theta \qquad\qquad = 0$ Multiplying (1) by $\sin \theta$

(4) $|u| \cos^2 \theta + |v| \sin \theta \cos \theta - |w| \cos \theta = 0$ Multiplying (2) by $\cos \theta$

$\overline{\qquad |u| (\sin^2 \theta + \cos^2 \theta) \qquad\quad - |w| \cos \theta = 0}$ Adding (3) and (4)

Since $\sin^2 \theta + \cos^2 \theta = 1$ by the Pythagorean identity, we have $|u| = |w| \cos \theta$.

Substituting this result into (1), we have
$|w| \cos \theta \sin \theta - |v| \cos \theta = 0$

Hence, $|v| = \dfrac{|w| \cos \theta \sin \theta}{\cos \theta}$ $|v| = |w| \sin \theta$

(B) We are given $|w| = 130$ lb and $\theta = 72°$. Hence,
$\qquad |u| = |w| \cos \theta = (130 \text{ lb}) \cos 72° = 40 \text{ lb}$
and
$\qquad |v| = |w| \sin \theta = (130 \text{ lb}) \sin 72° = 124 \text{ lb}$

(C) We are given $|u| = \dfrac{1}{6} |w|$. Hence
$\qquad \dfrac{1}{6} |w| = |w| \cos \theta \qquad \cos \theta = \dfrac{1}{6} \qquad \theta = \cos^{-1} \dfrac{1}{6} = 80°$

57. $W = \mathbf{F} \cdot \mathbf{d} = \langle -5, 8 \rangle \cdot \langle -8, 2 \rangle = 56$ ft-lb

58. First, form a force diagram with all force vectors in standard position at the origin.

Let $\mathbf{F}_1$ = the force exerted by the weight (20 lb)
         $\mathbf{F}_2$ = the tension on the line
         W = the weight of the leg (12 lb)

Then write each force vector in terms of $\mathbf{i}$ and $\mathbf{j}$ unit vectors.
     $\mathbf{F}_1 = 20 \cos 20° \mathbf{i} + 20 \sin 20° \mathbf{j}$
     $\mathbf{F}_2 = -|\mathbf{F}_2| \cos \theta \mathbf{i} + |\mathbf{F}_2| \sin \theta \mathbf{j}$
     $W = -12\mathbf{j}$

For the system to be in static equilibrium, we must have $\mathbf{F}_1 + \mathbf{F}_2 + W = 0$ which becomes, on addition,
     $[20 \cos 20° - |\mathbf{F}_2| \cos \theta]\mathbf{i} + [20 \sin 20° + |\mathbf{F}_2| \sin \theta - 12]\mathbf{j} = 0$

Since two vectors are equal if and only if their corresponding components are equal, we are led to the following system of equations in $|\mathbf{F}_2|$ and θ:

     $20 \cos 20° - |\mathbf{F}_2| \cos \theta = 0$
     $20 \sin 20° + |\mathbf{F}_2| \sin \theta - 12 = 0$

Solving, $\qquad |\mathbf{F}_2| \cos\theta = 20 \cos 20°$

$$|\mathbf{F}_2| = 20 \frac{\cos 20°}{\cos\theta}$$

$$20 \sin 20° + 20 \frac{\cos 20°}{\cos\theta} \sin\theta - 12 = 0$$

$$20 \cos 20° \tan\theta = 12 - 20 \sin 20°$$

$$\tan\theta = \frac{12 - 20 \sin 20°}{20 \cos 20°}$$

$$\theta = \tan^{-1}\frac{12 - 20 \sin 20°}{20 \cos 20°} = 15.4°$$

$$|\mathbf{F}_2| = 20 \frac{\cos 20°}{\cos\theta} = 19.5 \text{ lb}$$

Chapter 7   Polar Coordinates; Complex Numbers

EXERCISE 7.1   Polar and Rectangular Coordinates

1.

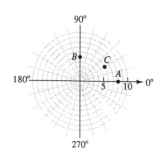

3.

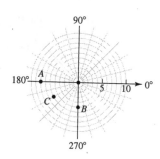

5.

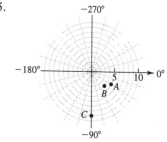

7.

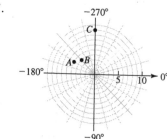

9.

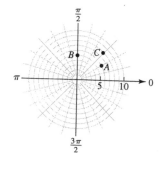

11.

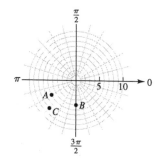

13.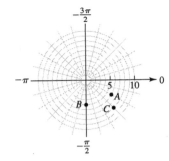

15.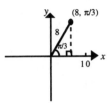

$$x = 8 \cos \frac{\pi}{3} = 8\left(\frac{1}{2}\right) = 4$$

$$y = 8 \sin \frac{\pi}{3} = 8\left(\frac{\sqrt{3}}{2}\right) = 4\sqrt{3}$$

Rectangular coordinates: $(4, 4\sqrt{3})$

17.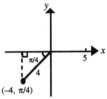

$$x = -4 \cos \frac{\pi}{4} = -4\left(\frac{\sqrt{2}}{2}\right) = -2\sqrt{2}$$

$$y = -4 \sin \frac{\pi}{4} = -4\left(\frac{\sqrt{2}}{2}\right) = -2\sqrt{2}$$

Rectangular coordinates: $(-2\sqrt{2}, -2\sqrt{2})$

19.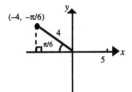

$$x = -4 \cos\left(-\frac{\pi}{6}\right) = -4\left(\frac{\sqrt{3}}{2}\right) = -2\sqrt{3}$$

$$y = -4 \sin\left(-\frac{\pi}{6}\right) = -4\left(-\frac{1}{2}\right) = 2$$

Rectangular coordinates: $(-2\sqrt{3}, 2)$

21. Use $r^2 = x^2 + y^2$ and $\tan \theta = \frac{y}{x}$

$r^2 = (2\sqrt{3})^2 + 2^2 = 16$ $r = 4$ $\tan \theta = \frac{2}{2\sqrt{3}} = \frac{1}{\sqrt{3}}$ $\theta = \frac{\pi}{6}$ since the point is in the first quadrant

Polar coordinates: $\left(4, \frac{\pi}{6}\right)$

23. Use $r^2 = x^2 + y^2$ and $\tan \theta = \frac{y}{x}$

$r^2 = (-4)^2 + (-4\sqrt{3})^2 = 64$ $r = 8$ $\tan \theta = \frac{-4\sqrt{3}}{-4} = \sqrt{3}$ $\theta = -\frac{2\pi}{3}$ since the point is in the third quadrant

Polar coordinates: $\left(8, -\frac{2\pi}{3}\right)$

25. Use $r^2 = x^2 + y^2$ and $\tan \theta = \frac{y}{x}$

$r^2 = 0^2 + (-7)^2 = 49$ $r = 7$ $\tan \theta = \frac{-7}{0}$ is undefined $\theta = -\frac{\pi}{2}$ since the point is on the negative y axis

Polar coordinates: $\left(7, -\frac{\pi}{2}\right)$

27. See figure at the right. The point with coordinates
     (6, –30°) can equally well be described as (–6, –210°)
     or (–6, 150°) or (6, 330°). Thus, (–6, –210°):
     The polar axis is rotated 210° clockwise (negative
     direction) and the point is located 6 units from the
     pole along the negative polar axis. (–6, 150°):
     The polar axis is rotated 150° counterclockwise

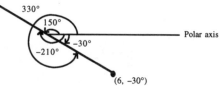

     (positive direction) and the point is located 6 units from the pole along the negative polar axis.
     (6, 330°): The polar axis is rotated 330° counterclockwise (positive direction) and the point is located
     6 units along the positive polar axis.

29. Use $r^2 = x^2 + y^2$ and $\tan \theta = \dfrac{y}{x}$

$r^2 = (6.913)^2 + (4.705)^2$ $r = \sqrt{(6.913)^2 + (4.705)^2} = 8.362$

$\tan \theta = \dfrac{4.705}{6.913}$ $\theta = 34.239°$ since the point is in the first quadrant

Polar coordinates: (8.362, 34.239°)

31. Use $r^2 = x^2 + y^2$ and $\tan \theta = \dfrac{y}{x}$

$r^2 = (-8.336)^2 + (4.291)^2$ $r = \sqrt{(-8.336)^2 + (4.291)^2} = 9.376$

$\tan \theta = \dfrac{4.291}{-8.336}$ $\theta = 180° + \tan^{-1}\left(\dfrac{4.291}{-8.336}\right) = 152.763°$ since the point is in the second
                                                                              quadrant

Polar coordinates: (9.376, 152.763°)

33. Use $r^2 = x^2 + y^2$ and $\tan \theta = \dfrac{y}{x}$

$r^2 = (-16.322)^2 + (-27.089)^2$ $r = \sqrt{(-16.322)^2 + (-27.089)^2} = 31.626$

$\tan \theta = \dfrac{-27.089}{-16.322}$ $\theta = \tan^{-1}\left(\dfrac{-27.089}{-16.322}\right) - 180° = -121.070°$ since the point is in the
                                                                                    third quadrant

Polar coordinates: (31.626, –121.070°)

35. Use $x = r\cos \theta,\ y = r\sin \theta$

          $x = 7.066 \cos (125.317°)$ $y = 7.066 \sin (125.317°)$

          $= -4.085$ $= 5.766$

     Rectangular coordinates: (–4.085, 5.766)

37. Use $x = r\cos \theta,\ y = r\sin \theta$

          $x = 3.768 \cos (-2.113)$ $y = 3.768 \sin (-2.113)$

          $= -1.944$ $= -3.228$

     Rectangular coordinates: (–1.944, –3.228)

39.   Use $x = r \cos \theta, \; y = r \sin \theta$

   $x = -9.028 \cos(-0.663)$ $y = -9.028 \sin(-0.663)$

      $= -7.115$ $= 5.557$

   Rectangular coordinates: $(-7.115, 5.557)$

41. $6x - x^2 = y^2$ The graph of $r = 0$ is the pole, and since the pole

   $6x = x^2 + y^2$ is included as a solution of $r - 6 \cos \theta = 0$

   Use $x = r \cos \theta$ and $x^2 + y^2 = r^2$ $\left(\text{let } \theta = \dfrac{\pi}{2} \right)$, we can discard $r = 0$ and keep only

   $6r \cos \theta = r^2$

      $0 = r^2 - 6r \cos \theta = r(r - 6 \cos \theta)$ $r - 6 \cos \theta = 0$ or $r = 6 \cos \theta$

      $r = 0$ or $r - 6 \cos \theta = 0$

43. $2x + 3y = 5$ 45. $x^2 + y^2 = 9$

   Use $x = r \cos \theta$ and $y = r \sin \theta$ Use $x^2 + y^2 = r^2$

      $2r \cos \theta + 3r \sin \theta = 5$ $r^2 = 9$ or $r = \pm 3$

      $r(2 \cos \theta + 3 \sin \theta) = 5$

47. $2xy = 1$ 49. $4x^2 - y^2 = 4$

   Use $x = r \cos \theta$ and $y = r \sin \theta$ Use $x = r \cos \theta$ and $y = r \sin \theta$

      $2r \cos \theta r \sin \theta = 1$ $4(r \cos \theta)^2 - (r \sin \theta)^2 = 4$

      $r^2 (2 \sin \theta \cos \theta) = 1$ $4r^2 \cos^2 \theta - r^2 \sin^2 \theta = 4$

      $r^2 = \dfrac{1}{2 \sin \theta \cos \theta}$ $r^2 (4 \cos^2 \theta - \sin^2 \theta) = 4$

        $= \dfrac{1}{\sin 2\theta}$ $r^2 [4(1 - \sin^2 \theta) - \sin^2 \theta] = 4$

                          $r^2 (4 - 4 \sin^2 \theta - \sin^2 \theta) = 4$

                          $r^2 (4 - 5 \sin^2 \theta) = 4$

                          $r^2 = \dfrac{4}{4 - 5 \sin^2 \theta}$

51. $r(2 \cos \theta + \sin \theta) = 4$

   $2r \cos \theta + r \sin \theta = 4$

   Use $x = r \cos \theta$ and $y = r \sin \theta$ $2x + y = 4$

53. $r = 8 \cos \theta$

We multiply both sides by r, which adds the pole to the graph. But the pole is already part of the graph $\left(\text{let } \theta = \dfrac{\pi}{2}\right)$, so we have changed nothing. $r^2 = 8r \cos \theta$

But $r^2 = x^2 + y^2$, $r \cos \theta = x$. Hence, $x^2 + y^2 = 8x$.

55. $r^2 \cos 2\theta = 4$

$r^2 (\cos^2 \theta - \sin^2 \theta) = 4$

$r^2 \cos^2 \theta - r^2 \sin^2 \theta = 4$

$(r \cos \theta)^2 - (r \sin \theta)^2 = 4$

Use $r \cos \theta = x$ and $r \sin \theta = y$

$x^2 - y^2 = 4$

57. $r = 4$, so $r^2 = 16$

Use $r^2 = x^2 + y^2$

$x^2 + y^2 = 16$

59. $\theta = 30°$ $\tan \theta = \tan 30°$

$\tan \theta = \dfrac{1}{\sqrt{3}}$ Use $\tan \theta = \dfrac{y}{x}$

$\dfrac{y}{x} = \dfrac{1}{\sqrt{3}}$ $y = \dfrac{1}{\sqrt{3}} x$

61. $r = \dfrac{3}{\sin \theta - 2}$

Multiply both sides by $\sin \theta - 2$, which is never 0 since $\sin \theta$ is never 2.

$r(\sin \theta - 2) = 3$ $r \sin \theta - 2r = 3$

Use $r \sin \theta = y$ and $r = -\sqrt{x^2 + y^2}$ $y + 2\sqrt{x^2 + y^2} = 3$

$y - 3 = -2\sqrt{x^2 + y^2}$ or $(y - 3)^2 = 4(x^2 + y^2)$

Note: The unusual choice of $r = -\sqrt{x^2 + y^2}$ is not a misprint. It is necessary to correspond with the fact that in the original polar equation, $\sin \theta < 2$; hence, $\sin \theta - 2$ is negative; hence, r is negative.

63. $d = \sqrt{r_1^2 + r_2^2 - 2r_1 r_2 \cos (\theta_2 - \theta_1)}$ $(r_1, \theta_1) = (2, 30°)$ $(r_2, \theta_2) = (3, 60°)$

$= \sqrt{2^2 + 3^2 - 2(2)(3) \cos (60° - 30°)}$

$= \sqrt{2.6076952\ldots} \approx 1.615$ units

EXERCISE 7.2   Sketching Polar Equations

1.

θ	0	$\dfrac{\pi}{6}$	$\dfrac{\pi}{4}$	$\dfrac{\pi}{3}$	$\dfrac{\pi}{2}$
Exact value r	10	$5\sqrt{3}$	$5\sqrt{2}$	5	0
Calculator value r	10	8.7	7.1	5	0

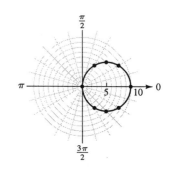

	θ	$\dfrac{2\pi}{3}$	$\dfrac{3\pi}{4}$	$\dfrac{5\pi}{6}$	π
Exact value	r	-5	$-5\sqrt{2}$	$-5\sqrt{3}$	-10
Calculator value	r	-5	-7.1	-8.7	-10

3.

	θ	$0°$	$30°$	$60°$	$90°$	$120°$	$150°$	$180°$	$210°$
Exact value	r	6	$\left(3 + \dfrac{3\sqrt{3}}{2}\right)$	$\dfrac{9}{2}$	3	$\dfrac{3}{2}$	$\left(3 - \dfrac{3\sqrt{3}}{2}\right)$	0	$\left(3 - \dfrac{3\sqrt{3}}{2}\right)$
Calculator value	r	6	5.6	4.5	3	1.5	0.4	0	0.4

	θ	$240°$	$270°$	$300°$	$330°$	$360°$
Exact value	r	$\dfrac{3}{2}$	3	$\dfrac{9}{2}$	$\left(3 + \dfrac{3\sqrt{3}}{2}\right)$	6
Calculator value	r	1.5	3	4.5	5.6	6

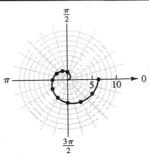

5.

θ	0	$\dfrac{\pi}{6}$	$\dfrac{\pi}{3}$	$\dfrac{\pi}{2}$	$\dfrac{2\pi}{3}$	$\dfrac{5\pi}{6}$	π	$\dfrac{7\pi}{6}$	$\dfrac{4\pi}{3}$	$\dfrac{3\pi}{2}$	$\dfrac{5\pi}{3}$	$\dfrac{11\pi}{6}$	2π
r	0	0.5	1.0	1.6	2.1	2.6	3.1	3.7	4.2	4.7	5.2	5.8	6.3

7. The graph consists of all points whose distance from the pole is 5, a circle with center at the pole, and radius 5.

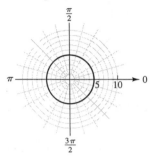

9. The graph consists of all points on a line that forms an angle of $\frac{\pi}{4}$ with the polar axis and passes through the pole.

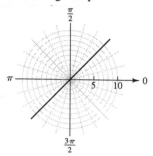

11. Set up a table that shows how r varies as θ varies through each set of quadrant values, then sketch the polar curve from the information in the table.

θ	$4 \cos \theta$	
0 to $\frac{\pi}{2}$	4 to 0	
$\frac{\pi}{2}$ to π	0 to -4	
π to $\frac{3\pi}{2}$	-4 to 0	Curve is traced out a second time in this region, although coordinate pairs are different
$\frac{3\pi}{2}$ to 2π	0 to 4	

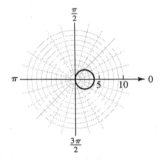

13. Set up a table that shows how r varies as 2θ varies through each set of quadrant values, then sketch the polar curve from the information in the table.

θ	2θ	$\cos 2\theta$	$8 \cos 2\theta$
0 to $\frac{\pi}{4}$	0 to $\frac{\pi}{2}$	1 to 0	8 to 0
$\frac{\pi}{4}$ to $\frac{\pi}{2}$	$\frac{\pi}{2}$ to π	0 to -1	0 to -8
$\frac{\pi}{2}$ to $\frac{3\pi}{4}$	π to $\frac{3\pi}{2}$	-1 to 0	-8 to 0
$\frac{3\pi}{4}$ to π	$\frac{3\pi}{2}$ to 2π	0 to 1	0 to 8
π to $\frac{5\pi}{4}$	2π to $\frac{5\pi}{2}$	1 to 0	8 to 0
$\frac{5\pi}{4}$ to $\frac{3\pi}{2}$	$\frac{5\pi}{2}$ to 3π	0 to -1	0 to -8
$\frac{3\pi}{2}$ to $\frac{7\pi}{4}$	3π to $\frac{7\pi}{2}$	-1 to 0	-8 to 0
$\frac{7\pi}{4}$ to 2π	$\frac{7\pi}{2}$ to 4π	0 to 1	0 to 8

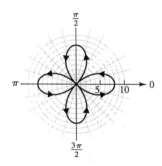

15. Set up a table that shows how r varies as 3θ varies through each set of quadrant values, then sketch the polar curve from the information in the table.

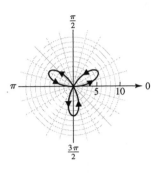

θ	3θ	$\sin 3\theta$	$6 \sin 3\theta$
0 to $\frac{\pi}{6}$	0 to $\frac{\pi}{2}$	0 to 1	0 to 6
$\frac{\pi}{6}$ to $\frac{\pi}{3}$	$\frac{\pi}{2}$ to π	1 to 0	6 to 0
$\frac{\pi}{3}$ to $\frac{\pi}{2}$	π to $\frac{3\pi}{2}$	0 to -1	0 to -6
$\frac{\pi}{2}$ to $\frac{2\pi}{3}$	$\frac{3\pi}{2}$ to 2π	-1 to 0	-6 to 0
$\frac{2\pi}{3}$ to $\frac{5\pi}{6}$	2π to $\frac{5\pi}{3}$	0 to 1	0 to 6
$\frac{5\pi}{6}$ to π	$\frac{5\pi}{3}$ to 3π	1 to 0	6 to 0

17. Set up a table that shows how r varies as θ varies through each set of quadrant values, then sketch the polar curve from the information in the table.

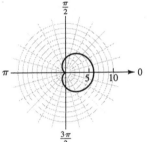

θ	$\cos \theta$	$3 \cos \theta$	$3 + 3 \cos \theta$
0 to $\frac{\pi}{2}$	1 to 0	3 to 0	6 to 3
$\frac{\pi}{2}$ to π	0 to -1	0 to -3	3 to 0
π to $\frac{3\pi}{2}$	-1 to 0	-3 to 0	0 to 3
$\frac{3\pi}{2}$ to π	0 to 1	0 to 3	3 to 6

19. Note that $r = 2 + 4 \cos \theta$ will equal 0 when $\cos \theta = -\frac{1}{2}$, that is, when $\theta = \frac{2\pi}{3}, \frac{4\pi}{3}$ (to list values between 0 and 2π only). Set up a table that shows how θ varies over particular invervals, then sketch the polar curve from the information in the table.

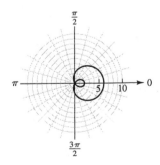

θ	$\cos \theta$	$4 \cos \theta$	$2 + 4 \cos \theta$
0 to $\frac{\pi}{2}$	1 to 0	4 to 0	6 to 2
$\frac{\pi}{2}$ to $\frac{2\pi}{3}$	0 to $-\frac{1}{2}$	0 to -2	2 to 0
$\frac{2\pi}{3}$ to π	$-\frac{1}{2}$ to -1	-2 to -4	0 to -2
π to $\frac{4\pi}{3}$	-1 to $-\frac{1}{2}$	-4 to -2	-2 to 0
$\frac{4\pi}{3}$ to $\frac{3\pi}{2}$	$-\frac{1}{2}$ to 0	-2 to 0	0 to 2
$\frac{3\pi}{2}$ to 2π	0 to 1	0 to 4	2 to 6

21. **(A)** $r = 5 + 5 \cos \theta$ **(B)** $r = 5 + 4 \cos \theta$

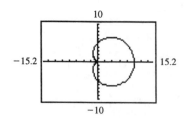 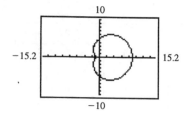

(C) $r = 4 + 5 \cos \theta$

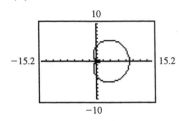

(D) If $a = b$, the graph will touch but not pass thorugh the origin. If $a > b$, the graph will not touch nor pass through the origin. If $a < b$, the graph will go through the origin and part of the graph will be inside the other part.

23. **(A)** $r = 9 \cos \theta$ $r = 9 \cos 3\theta$

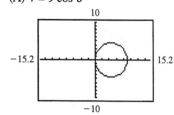 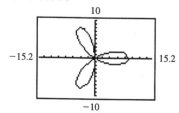

$r = 9 \cos 5\theta$

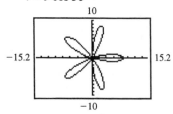

(B) and (C) Since $r = 9 \cos \theta$ has one leaf, $r = 9 \cos 3\theta$ has 3 leaves, and $r = 9 \cos 5\theta$ has 5 leaves, a reasonable guess for $r = 9 \cos 7\theta$ would be 7 leaves. A reasonable guess for $r = 9 \cos n\theta$ would be n leaves (n odd).

25. (A) $r = 9 \cos 2\theta$ $r = 9 \cos 4\theta$

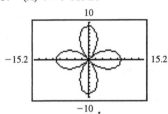

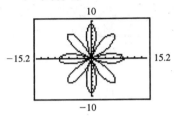

$r = 9 \cos 6\theta$

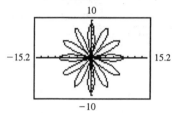

(B) and (C)   Since $r = 9 \cos 2\theta$ has 4 leaves, $r = 9 \cos 4\theta$ has 8 leaves, and $r = 9 \cos 6\theta$ has 12 leaves, a reasonable guess for $r = 9 \cos 8\theta$ would be 16 leaves. A reasonable guess for $r = 9 \cos n\theta$ would be $2n$ leaves (n even).

27. (A) $r = 9 \cos \dfrac{\theta}{2}$ (B) $r = 9 \cos \dfrac{\theta}{4}$

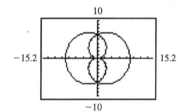

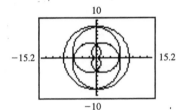

(C) n times.

29. The following table can be used to investigate how r varies as θ varies over particular intervals. We then sketch the polar curve from the information in the table.

θ	2θ	$\cos 2\theta$	$64 \cos 2\theta$	$r = \pm\sqrt{64 \cos 2\theta}$	
0 to $\dfrac{\pi}{4}$	0 to $\dfrac{\pi}{2}$	1 to 0	64 to 0	$\begin{cases} 8 \text{ to } 0 \\ -8 \text{ to } 0 \end{cases}$	The two branches of the curve are reflections of each other in the x axis
$\dfrac{\pi}{4}$ to $\dfrac{\pi}{2}$	$\dfrac{\pi}{2}$ to π	0 to -1	0 to -64	$\left.\begin{matrix}\\\\\\\end{matrix}\right\}$ No curve; no real square root of a negative no.	
$\dfrac{\pi}{2}$ to $\dfrac{3\pi}{4}$	π to $\dfrac{3\pi}{2}$	-1 to 0	-64 to 0		
$\dfrac{3\pi}{4}$ to π	$\dfrac{3\pi}{2}$ to 2π	0 to 1	0 to 64	$\begin{cases} 0 \text{ to } 8 \\ 0 \text{ to } -8 \end{cases}$	The two branches of the curve are reflections of each other in the x axis

θ	2θ	$\cos 2\theta$	$64 \cos 2\theta$	$r = \pm\sqrt{64 \cos 2\theta}$	
π to $\dfrac{5\pi}{4}$	2π to $\dfrac{5\pi}{2}$	1 to 0	64 to 0	$\begin{cases} 8 \text{ to } 0 \\ -8 \text{ to } 0 \end{cases}$	
$\dfrac{5\pi}{4}$ to $\dfrac{3\pi}{2}$	$\dfrac{5\pi}{2}$ to 3π	0 to -1	0 to -64	$\left.\begin{matrix}\\\\\\\end{matrix}\right\}$ No curve	This repeats the curve already traced out
$\dfrac{3\pi}{2}$ to $\dfrac{7\pi}{4}$	3π to $\dfrac{7\pi}{2}$	-1 to 0	-64 to 0		
$\dfrac{7\pi}{4}$ to 2π	$\dfrac{7\pi}{2}$ to 4π	0 to 1	0 to 64	$\begin{cases} 0 \text{ to } 8 \\ -8 \text{ to } 0 \end{cases}$	

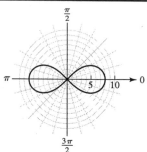

31. Here are graphs of $r = 1 + 2 \cos(n\theta)$ for $n = 1, 2, 3,$ and 4.

n = 1   n = 2

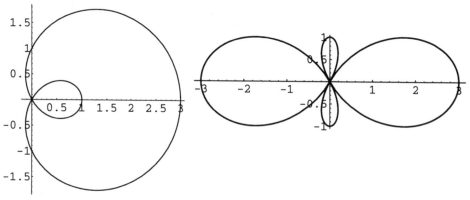

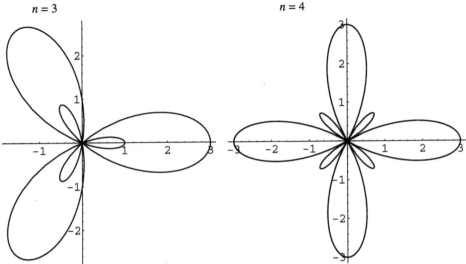

n = 3   n = 4

Generalizing from these graphs, we conclude that for each n, there are n large petals and n small petals. For n odd, the small petals are within the large petals, for n even the small petals are between the large petals.

33. We sketch the two graphs using rapid sketching techniques. Tables of how r varies for each curve as θ varies over particular intervals can be readily constructed.

θ	$2\cos\theta$	$2\sin\theta$
0 to $\frac{\pi}{2}$	2 to 0	0 to 2
$\frac{\pi}{2}$ to π	0 to –2	2 to 0
π to $\frac{3\pi}{2}$	–2 to 0	0 to –2
$\frac{3\pi}{2}$ to π	0 to 2	–2 to 0

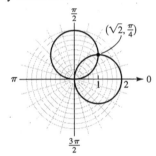

We solve the system, $r = 2\cos\theta$, $r = 2\sin\theta$, by equating the right sides: $2\cos\theta = 2\sin\theta$
$$\cos\theta = \sin\theta$$
$$1 = \tan\theta$$

The only solution of this equation, $0 \le \theta \le \pi$, is $\theta = \frac{\pi}{4}$. If we substitute this into either of the

original equations, we get $r = 2\cos\frac{\pi}{4} = 2\sin\frac{\pi}{4} = 2\left(\frac{\sqrt{2}}{2}\right) = \sqrt{2}$. Solution: $\left(\sqrt{2}, \frac{\pi}{4}\right)$

The sketch shows that the pole is on both graphs, however, the pole has no ordered pairs of

coordinates that simultaneously satisfy both equations. As $\left(0, \frac{\pi}{2}\right)$, it satisfies the first; as $(0, 0)$, it

satisfies the second; it is not a solution of the system.

35.   We graph the system using rapid sketching techniques. See Problem 13 for a table for $r = 8 \cos 2\theta$; a table for $r = 8 \sin \theta$ can be readily constructed.

θ	$8 \sin \theta$
0 to $\dfrac{\pi}{2}$	0 to 8
$\dfrac{\pi}{2}$ to π	8 to 0
π to $\dfrac{3\pi}{2}$	0 to -8
$\dfrac{3\pi}{2}$ to 2π	-8 to 0

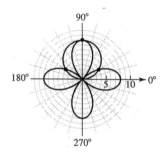

We solve the system: $r = 8 \sin \theta$, $r = 8 \cos 2\theta$, by equating the right-hand sides:

$$8 \sin \theta = 8 \cos 2\theta$$
$$\sin \theta = \cos 2\theta$$
$$\sin \theta = 1 - 2 \sin^2 \theta \qquad \text{Double-angle identity}$$
$$2 \sin^2 \theta + \sin \theta - 1 = 0$$
$$(2 \sin \theta - 1)(\sin \theta + 1) = 0 \qquad \text{Factoring}$$

Therefore, $2 \sin \theta - 1 = 0$ or $\sin \theta + 1 = 0$

$$
\begin{array}{ll}
2 \sin \theta - 1 = 0 & \sin \theta + 1 = 0 \\
\sin \theta = \dfrac{1}{2} & \sin \theta = -1 \\
\theta = 30^\circ, \, 150^\circ & \theta = 270^\circ
\end{array}
$$

If we substitute these values into either of the original equations, we get (choosing $r = 8 \sin \theta$ for ease of calculation):

$\theta = 30^\circ \qquad r = 8 \sin 30^\circ = 8\left(\dfrac{1}{2}\right) = 4 \qquad$ Solution: $(4, 30^\circ)$

$\theta = 150^\circ \qquad r = 8 \sin 150^\circ = 8\left(\dfrac{1}{2}\right) = 4 \qquad$ Solution: $(4, 150^\circ)$

$\theta = 270^\circ \qquad r = 8 \sin 270^\circ = 8(-1) = -8 \qquad$ Solution: $(-8, 270^\circ)$

The sketch shows that all three solutions are on both graphs; it also shows that the pole is on both graphs. However, the pole has no ordered pairs of coordinates that simultaneously satisfy both equations. As $(0, 0^\circ)$, it satisfies the first; as $(0, 45^\circ)$, it satisfies the second; it is not a solution of the system.

37.   (A) The America's Cup curve appears to pass through the point with polar coordinates $(9.5, 30^\circ)$. Speed = 9.5 knots.

(B) The America's Cup curve appears to pass through the point with polar coordinates $(12.0, 60^\circ)$. Speed = 12.0 knots.

(C) The America's Cup curve appears to pass through the point with polar coordinates $(13.5, 90^\circ)$. Speed = 13.5 knots.

(D) The America's Cup curve appears to pass through the point with polar coordinates $(12.0, 120^\circ)$. Speed = 12.0 knots.

39. (A) $r = \dfrac{8}{1 - 0.5 \cos \theta}$ (B) $r = \dfrac{8}{1 - \cos \theta}$

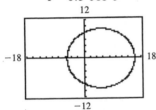

The graph is an ellipse.

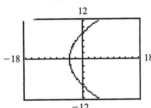

The graph is a parabola.

(C) $r = \dfrac{8}{1 - 2 \cos \theta}$

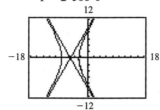

The graph is a hyperbola.

41. (A) At aphelion, $\theta = 0°$, hence
$$r = \frac{3.44 \times 10^7 \text{ mi}}{1 - 0.206 \cos 0°} = 4.33 \times 10^7 \text{ mi}$$

At perihelion, $\theta = 180°$, hence
$$r = \frac{3.44 \times 10^7 \text{ mi}}{1 - 0.206 \cos 180°} = 2.85 \times 10^7 \text{ mi}$$

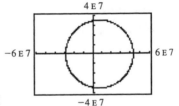

(B) The diagram in the text shows the areas swept out when the planet is near aphelion or perihelion as, approximately, triangles, whose areas would be proportional to their base (distance traveled by planet) and altitude (distance from planet to sum). Since at aphelion the distance from planet to sun is greater, the distance traveled by planet must be smaller in order to have equal area. Thus, the planet travels more slowly at aphelion, faster at perihelion.

EXERCISE 7.3   Complex Numbers in Rectangular and Polar Forms

1.

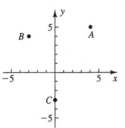

3.

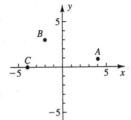

5.

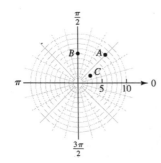

7.

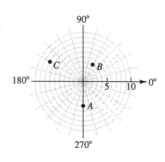

9. (A) A sketch shows that $\sqrt{3} - i$ is associated with a special
    30°–60° reference triangle in the fourth quadrant.

    Thus, $r = 2$ and $\theta = -\dfrac{\pi}{6}$ and the polar form for $\sqrt{3} - i$ is

    $$\sqrt{3} - i = 2\left[\cos\left(-\frac{\pi}{6}\right) + i \ \sin\left(-\frac{\pi}{6}\right)\right]$$
    $$= 2e^{(-\pi/6)i}$$

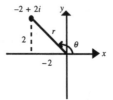

    (B) A sketch shows that $-2 + 2i$ is associated with a special 45°
    reference triangle in the second quadrant. Thus, $r = 2\sqrt{2}$ and
    $\theta = \dfrac{3\pi}{4}$ and the polar form for $-2 + 2i$ is

    $$-2 + 2i = 2\sqrt{2}\left[\cos\frac{3\pi}{4} + i \sin\frac{3\pi}{4}\right]$$
    $$= 2\sqrt{2}\ e^{(3\pi/4)i}$$

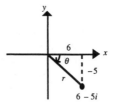

    (C) A sketch shows that $6 - 5i$ is not associated with a special
    reference triangle. We proceed as follows:

    $$r = \sqrt{6^2 + (-5)^2} = 7.81 \text{ to two decimal places}$$
    $$\theta = \tan^{-1}\left(-\frac{5}{6}\right) = -0.69 \text{ to two decimal places}$$

    Thus, the polar form for $6 - 5i$ is

    $$6 - 5i = 7.81\,[\cos(-0.69) + i \sin(-0.69)]$$
    $$= 7.81e^{(-0.69)i}$$

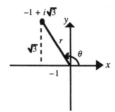

11. (A) A sketch shows that $-1 + i\sqrt{3}$ is associated with a special
    30°–60° reference triangle in the second quadrant. Thus, $r = 2$
    and $\theta = 120°$ and the polar form for $-1 + i\sqrt{3}$ is

    $$-1 + i\sqrt{3} = 2(\cos 120° + i \sin 120°)$$
    $$= 2e^{120°i}$$

(B) A sketch shows that $-3i$ is located on the negative y axis. Thus, $r = 3$ and $\theta = -90°$ and the polar form for $-3i$ is

$$-3i = 3[\cos(-90°) + i\sin(-90°)]$$
$$= 3e^{(-90°)i}$$

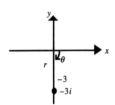

(C) A sketch shows that $-7 - 4i$ is not associated with a special reference triangle. We proceed as follows:

$$r = \sqrt{(-7)^2 + (-4)^2} = 8.06 \qquad \text{to two decimal places}$$
$$\theta = -180° + \tan^{-1}\frac{4}{7} = -150.26° \quad \text{to two decimal places}$$

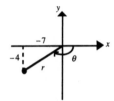

Thus, the polar form for $-7 - 4i$ is

$$-7 - 4i = 8.06[\cos(-150.26°) + i\sin(-150.26°)]$$
$$= 8.06e^{(-150.26°)i}$$

13. (A) $x + iy = 2e^{30°i}$

$$= 2(\cos 30° + i\sin 30°)$$
$$= 2\left(\frac{\sqrt{3}}{2} + i \cdot \frac{1}{2}\right)$$
$$= \sqrt{3} + i$$

(B) $x + iy = \sqrt{2}e^{(-3\pi/4)i}$

$$= \sqrt{2}\left[\cos\left(-\frac{3\pi}{4}\right) + i\sin\left(-\frac{3\pi}{4}\right)\right]$$
$$= \sqrt{2}\left[-\frac{1}{\sqrt{2}} + i\left(-\frac{1}{\sqrt{2}}\right)\right]$$
$$= -1 - i$$

(C) $x + iy = 5.71e^{(-0.48)i}$

$$= 5.71[\cos(-0.48) + i\sin(-0.48)]$$
$$= 5.06 - 2.64i$$

15. (A) $x + iy = \sqrt{3}e^{(-\pi/2)i}$

$$= \sqrt{3}\left[\cos\left(-\frac{\pi}{2}\right) + i\sin\left(-\frac{\pi}{2}\right)\right]$$
$$= \sqrt{3}\,[0 + i(-1)]$$
$$= -i\sqrt{3}$$

(B) $x + iy = \sqrt{2}\,e^{135°i}$

$$= \sqrt{2}\,(\cos 135° + i\sin 135°)$$
$$= \sqrt{2}\left(-\frac{1}{\sqrt{2}} + i \cdot \frac{1}{\sqrt{2}}\right)$$
$$= -1 + i$$

(C) $x + iy = 6.83e^{(-108.82°)i}$

$$= 6.83\,[\cos(-108.82°) + i\sin(-108.82°)]$$
$$= -2.20 - 6.46i$$

17. $z_1 z_2 = 6e^{132°i} \cdot 3e^{93°i}$

$$= 6 \cdot 3e^{i(132° + 93°)} = 18e^{225°i}$$

$$\frac{z_1}{z_2} = \frac{6e^{132°i}}{3e^{93°i}}$$
$$= \frac{6}{3}e^{i(132° - 93°)} = 2e^{39°i}$$

19. $z_1 z_2 = 3e^{67°i} \cdot 2e^{97°i}$

$$= 3 \cdot 2e^{i(67° + 97°)} = 6e^{164°i}$$

$$\frac{z_1}{z_2} = \frac{3e^{67°i}}{2e^{97°i}}$$
$$= \frac{3}{2}e^{i(67° - 97°)} = 1.5e^{(-30°)i}$$

21. $z_1 z_2 = 7.11e^{0.79i} \cdot 2.66e^{1.07i}$

$\dfrac{z_1}{z_2} = \dfrac{7.11e^{0.79i}}{2.66e^{1.07i}}$

$= 7.11 \cdot 2.66e^{\,i(0.79\,+\,1.07)} = 18.91e^{1.86i}$

$= \dfrac{7.11}{2.66}e^{\,i(0.79\,-\,1.07)} = 2.67e^{(-0.28)i}$

23. Directly, $(1 + i\sqrt{3})(\sqrt{3} + i) = \sqrt{3} + i + 3i + i^2\sqrt{3} = \sqrt{3} + 4i - \sqrt{3} = 4i$

Using polar forms,

$1 + i\sqrt{3} = 2e^{60°i}$

$\sqrt{3} + i = 2e^{30°i}$

Hence, $(1 + i\sqrt{3})(\sqrt{3} + i) = 2e^{60°i} \cdot 2e^{30°i} = 2 \cdot 2e^{\,i(60°\,+\,30°)} = 4e^{90°i}$

25. Directly, $(1 + i)^2 = 1 + 2i + i^2 = 1 + 2i + (-1) = 2i$

Using polar forms,

$1 + i = \sqrt{2}\,e^{45°i}$

Hence, $(1 + i)^2 = (1 + i)(1 + i) = \sqrt{2}\,e^{45°i} \cdot \sqrt{2}\,e^{45°i} = \sqrt{2} \cdot \sqrt{2}\,e^{\,i(45°\,+\,45°)} = 2e^{90°i}$

27. Directly, $(1 + i)^3 = (1 + i)(1 + i)^2 = (1 + i)\,2i$ (from the previous problem)

$= 2i + 2i^2 = 2i + 2(-1) = -2 + 2i$

Using polar forms,

$1 + i = \sqrt{2}\,e^{45°i}$

$(1 + i)^2 = 2e^{90°i}$ (from the previous problem)

Hence, $(1 + i)^3 = (1 + i)(1 + i)^2 = \sqrt{2}\,e^{45°i} \cdot 2e^{90°i} = \sqrt{2} \cdot 2e^{\,i(45°\,+\,90°)} = 2\sqrt{2}\,e^{135°i}$

29. $z^2 = zz = re^{i\theta} \cdot re^{i\theta} = rre^{\,i(\theta\,+\,\theta)} = r^2 e^{2i\theta}$

31. $(r^{1/2}\,e^{i\theta/2})^2 = r^{1/2}\,e^{i\theta/2}r^{1/2}\,e^{i\theta/2} = r^{1/2}\,r^{1/2}\,e^{\,i(\theta/2\,+\,\theta/2)} = re^{i\theta}$

Since $(r^{1/2}\,e^{i\theta/2})^2 = re^{i\theta}$, $r^{1/2}\,e^{i\theta/2}$ is a square root of $re^{i\theta}$.

33. Generalizing from the results of Problems 29 and 30, z^n appears to be $r^n e^{n\theta i}$.

35. (A) $8(\cos 0° + i \sin 0°) = 8(1 + 0i) = 8 + 0i$

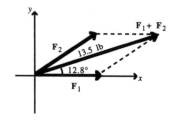

$6(\cos 30° + i \sin 30°) = 6\left(\dfrac{\sqrt{3}}{2} + i\dfrac{1}{2}\right) = 3\sqrt{3} + 3i$

$(8 + 0i) + (3\sqrt{3} + 3i) = (8 + 3\sqrt{3}) + 3i$

(B) $r = \sqrt{(8 + 3\sqrt{3})^2 + 3^2} \approx 13.5$

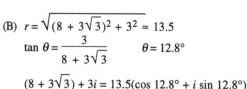

$\tan \theta = \dfrac{3}{8 + 3\sqrt{3}}$ $\theta = 12.8°$

$(8 + 3\sqrt{3}) + 3i = 13.5(\cos 12.8° + i \sin 12.8°)$

(C) We can interpret $13.5(\cos 12.8° + i \sin 12.8°)$ as a force of 13.5 lb at an angle of 12.8° with respect to the force $\mathbf{F}_1$.

EXERCISE 7.4 DeMoivre's Theorem and the *n*th Root Theorem

1. $(3e^{15°i})^3 = 3^3\,e^{3\cdot15°i} = 27e^{45°i}$

3. $(\sqrt2\,e^{45°i})^{10} = (\sqrt2)^{10}\,e^{(10\cdot45°)i} = (2^{1/2})^{10}\,e^{450°i} = 2^5e^{450°i} = 32e^{450°i}$ or $32e^{(450°-360°)i} = 32e^{90°i}$

5. $(\sqrt3 + i)^6 = (2e^{30°i})^6 = 2^6\,e^{(6\cdot30°)i} = 64e^{180°i}$

7. $(-1 + i)^4 = (\sqrt2e^{135°i})^4 = (\sqrt2)^4\,e^{(4\cdot135°)i} = (2^{1/2})^4\,e^{540°i} = 2^2\,e^{540°i}$
   $= 4e^{540°i} = 4(\cos540° + i\sin540°) = 4(-1 + 0i) = -4$

9. $(-\sqrt3 + i)^5 = (2e^{150°i})^5 = 2^5\,e^{(5\cdot150°)i} = 32e^{750°i}$
   $= 32(\cos750° + i\sin750°) = 32\left(\dfrac{\sqrt3}{2} + i\cdot\dfrac12\right) = 16\sqrt3 + 16i$

11. $\left(-\dfrac12 - \dfrac{\sqrt3}{2}i\right)^3 = (1e^{-120°i})^3 = 1^3\,e^{3(-120°)i} = e^{-360°i}$
    $= \cos(-360°) + i\sin(-360°) = 1 + 0i = 1$

13. From the *n*th root theorem, both roots are given by
    $$4^{1/2}\,e^{(30°/2 + k360°/2)i} \qquad k = 0, 1$$
    Thus,
    $$w_1 = 4^{1/2}\,e^{(15° + 0\cdot180°)i} = 2e^{15°i}$$
    $$w_2 = 4^{1/2}\,e^{(15° + 1\cdot180°)i} = 2e^{195°i}$$

15. From the *n*th root theorem, all three roots are given by
    $$8^{1/3}\,e^{(90°/3 + k360°/3)i} \qquad k = 0, 1, 2$$
    Thus,
    $$w_1 = 8^{1/3}\,e^{(30° + 0\cdot120°)i} = 2e^{30°i}$$
    $$w_2 = 8^{1/3}\,e^{(30° + 1\cdot120°)i} = 2e^{150°i}$$
    $$w_3 = 8^{1/3}\,e^{(30° + 2\cdot120°)i} = 2e^{270°i}$$

17. First, write $-1 + i$ in polar form.
    $$-1 + i = 2^{1/2}\,e^{135°i}$$
    From the *n*th root theorem, all five roots are given by
    $$(2^{1/2})^{1/5}\,e^{(135°/5 + k360°/5)i} \qquad k = 0, 1, 2, 3, 4$$
    Thus,
    $$w_1 = 2^{1/10}\,e^{(27° + 0\cdot72°)i} = 2^{1/10}\,e^{27°i}$$
    $$w_2 = 2^{1/10}\,e^{(27° + 1\cdot72°)i} = 2^{1/10}\,e^{99°i}$$
    $$w_3 = 2^{1/10}\,e^{(27° + 2\cdot72°)i} = 2^{1/10}\,e^{171°i}$$
    $$w_4 = 2^{1/10}\,e^{(27° + 3\cdot72°)i} = 2^{1/10}\,e^{243°i}$$
    $$w_5 = 2^{1/10}\,e^{(27° + 4\cdot72°)i} = 2^{1/10}\,e^{315°i}$$

19. First, write −8 in polar form.

$$-8 \doteq 8e^{180°i}$$

From the nth root theorem, all three roots are given by

$$8^{1/3}\, e^{(180°/3 + k360°/3)i} \quad k = 0, 1, 2$$

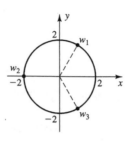

Thus,

$$w_1 = 8^{1/3}\, e^{(60° + 0 \cdot 120°)i} = 2e^{60°i}$$

$$w_2 = 8^{1/3}\, e^{(60° + 1 \cdot 120°)i} = 2e^{180°i}$$

$$w_3 = 8^{1/3}\, e^{(60° + 2 \cdot 120°)i} = 2e^{300°i}$$

The three roots are equally spaced around a circle with radius 2 at an angular increment of 120° from one root to the next.

21. First, write 1 in polar form.

$$1 = 1e^{0°i}$$

From the nth root theorem, all four roots are given by

$$1^{1/4}\, e^{(0°/4 + k360°/4)i} \quad k = 0, 1, 2, 3$$

Thus,

$$w_1 = 1^{1/4}\, e^{(0° + 0 \cdot 90°)i} = 1e^{0°i}$$

$$w_2 = 1^{1/4}\, e^{(0° + 1 \cdot 90°)i} = 1e^{90°i}$$

$$w_3 = 1^{1/4}\, e^{(0° + 2 \cdot 90°)i} = 1e^{180°i}$$

$$w_4 = 1^{1/4}\, e^{(0° + 3 \cdot 90°)i} = 1e^{270°i}$$

The four roots are equally spaced around a circle with radius 1 at an angular increment of 90° from one root to the next.

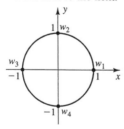

23. First, write $-i$ in polar form.

$$-i = 1e^{(-90°)i}$$

From the nth root theorem, all five roots are given by

$$1^{1/5}\, e^{(-90°/5 + k360°/5)i} \quad k = 0, 1, 2, 3, 4$$

Thus,

$$w_1 = 1^{1/5}\, e^{(-18° + 0 \cdot 72°)i} = 1e^{-18°i}$$

$$w_2 = 1^{1/5}\, e^{(-18° + 1 \cdot 72°)i} = 1e^{54°i}$$

$$w_3 = 1^{1/5}\, e^{(-18° + 2 \cdot 72°)i} = 1e^{126°i}$$

$$w_4 = 1^{1/5}\, e^{(-18° + 3 \cdot 72°)i} = 1e^{198°i}$$

$$w_5 = 1^{1/5}\, e^{(-18° + 4 \cdot 72°)i} = 1e^{270°i}$$

The five roots are equally spaced around a circle with radius 1 at an angular increment of 72° from one root to the next.

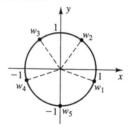

25. (A) $(-\sqrt{2} + i\sqrt{2})^4 + 16 = ((-\sqrt{2} + i\sqrt{2})^2)^2 + 16$

$= (2 - 4i + 2i^2)^2 + 16$

$= (-4i)^2 + 16$

$= 16i^2 + 16$

$= -16 + 16$

$= 0$

Thus, $-\sqrt{2} + i\sqrt{2}$ is a root of $x^4 + 16 = 0$.

(B) The four roots are equally spaced around the circle. Since there are 4 roots, the angle between successive roots on the circle is $\dfrac{360°}{4} = 90°$.

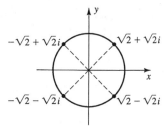

(C) $(-\sqrt{2} - i\sqrt{2})^4 + 16 = ((-\sqrt{2} - i\sqrt{2})^2)^2 + 16$

$= (2 + 4i + 2i^2)^2 + 16$

$= (4i)^2 + 16 = 16i^2 + 16 = -16 + 16 = 0$

$(\sqrt{2} - i\sqrt{2})^4 + 16 = ((\sqrt{2} - i\sqrt{2})^2)^2 + 16$

$= (2 - 4i + 2i^2)^2 + 16$

$= (-4i)^2 + 16 = 16i^2 + 16 = -16 + 16 = 0$

$(\sqrt{2} + i\sqrt{2})^4 + 16 = ((\sqrt{2} + i\sqrt{2})^2)^2 + 16$

$= (2 + 4i + 2i^2)^2 + 16$

$= (4i)^2 + 16 = 16i^2 + 16 = -16 + 16 = 0$

27. $x^3 + 27 = 0$

$x^3 = -27$

Therefore x is a cube root of -27, and there are three of them. First, write -27 in polar form and use the nth root theorem.

$-27 = 27e^{180°i}$

All three cube roots of -27 are given by

$27^{1/3} e^{(180°/3 + k360°/3)i}$ $k = 0, 1, 2$

Thus,

$$w_1 = 27^{1/3} e^{(60° + 0 \cdot 120°)i} = 3e^{60°i} = 3(\cos 60° + i \sin 60°) = 3\left(\frac{1}{2} + i\frac{\sqrt{3}}{2}\right) = \frac{3}{2} + \frac{3\sqrt{3}}{2}i$$

$$w_2 = 27^{1/3} e^{(60° + 1 \cdot 120°)i} = 3e^{180°i} = 3(\cos 180° + i \sin 180°) = 3[(-1) + i0] = -3$$

$$w_3 = 27^{1/3} e^{(60° + 2 \cdot 120°)i} = 3e^{300°i} = 3(\cos 300° + i \sin 300°) = 3\left[\frac{1}{2} + i\left(\frac{\sqrt{3}}{2}\right)\right] = \frac{3}{2} - \frac{3\sqrt{3}}{2}i$$

29. $x^3 - 64 = 0$
$$x^3 = 64$$

Therefore x is a cube root of 64, and there are three of them. First, write 64 in polar form and use the nth root theorem.

$$64 = 64e^{0°i}$$

All three cube roots of 64 are given by

$$64^{1/3}\, e^{(0°/3 + k360°/3)i} \qquad k = 0, 1, 2$$

Thus,

$w_1 = 64^{1/3}\, e^{(0° + 0\,\cdot\,120°)i} = 4e^{0°i} = 4(\cos 0° + i \sin 0°) = 4(1 + 0i) = 4$

$w_2 = 64^{1/3}\, e^{(0° + 1\,\cdot\,120°)i} = 4e^{120°i} = 4(\cos 120° + i \sin 120°) = 4\left[\left(-\dfrac{1}{2}\right) + i\,\dfrac{\sqrt{3}}{2}\right] = -2 + 2i\sqrt{3}$

$w_3 = 64^{1/3}\, e^{(0° + 2\,\cdot\,120°)i} = 4e^{240°i} = 4(\cos 240° + i \sin 240°) = 4\left[\left(-\dfrac{1}{2}\right) + i\left(-\dfrac{\sqrt{3}}{2}\right)\right] = -2 - 2i\sqrt{3}$

31.   For $k = 0$,

$$r^{1/n}\, e^{(\theta/n + k\,\cdot\,360°/n)i} = r^{1/n}\, e^{\theta/n\, i} = r^{1/n}\left(\cos \dfrac{\theta}{n} + i \sin \dfrac{\theta}{n}\right)$$

For $k = n$,

$$r^{1/n}\, e^{(\theta/n + k\,\cdot\,360°/n)i} = r^{1/n}\, e^{(\theta/n + n\,\cdot\,360°/n)i}$$
$$= r^{1/n}\, e^{(\theta/n + 360°)i}$$
$$= r^{1/n}\left[\cos\left(\dfrac{\theta}{n} + 360°\right) + i \sin\left(\dfrac{\theta}{n} + 360°\right)\right]$$
$$= r^{1/n}\left(\cos \dfrac{\theta}{n} + i \sin \dfrac{\theta}{n}\right)$$

Thus, the two are the same number.

33. $x^5 - 1 = 0$
$$x^5 = 1$$

Therefore x is a fifth root of 1, and there are five of them. First, write 1 in polar form and use the nth root theorem.

$$1 = 1e^{0°i}$$

All five fifth roots of 1 are given by

$$1^{1/5}\, e^{(0°/5 + k360°/5)i} \qquad k = 0, 1, 2, 3, 4$$

Thus,

$w_1 = 1^{1/5}\, e^{(0° + 0\,\cdot\,72°)i} = e^{0°i} = \cos 0° + i \sin 0° = 1 + i0 = 1$

$w_2 = 1^{1/5}\, e^{(0° + 1\,\cdot\,72°)i} = e^{72°i} = \cos 72° + i \sin 72° = 0.309 + 0.951i$

$w_3 = 1^{1/5}\, e^{(0° + 2\,\cdot\,72°)i} = e^{144°i} = \cos 144° + i \sin 144° = -0.809 + 0.588i$

$w_4 = 1^{1/5}\, e^{(0° + 3\,\cdot\,72°)i} = e^{216°i} = \cos 216° + i \sin 216° = -0.809 - 0.588i$

$w_5 = 1^{1/5}\, e^{(0° + 4\,\cdot\,72°)i} = e^{288°i} = \cos 288° + i \sin 288° = 0.309 - 0.951i$

35. $x^3 + 5 = 0$
      $x^3 = -5$

Therefore x is a cube root of -5, and there are three of them. First, write -5 in polar form and use the nth root theorem.

$$-5 = 5e^{180°i}$$

All three cube roots of -5 are given by

$$5^{1/3} \, e^{(180°/3 + k360°/3)i} \quad k = 0, 1, 2$$

Thus,

$w_1 = 5^{1/3} \, e^{(60° + 0 \cdot 120°)i} = 5^{1/3} \, e^{60°i} = 5^{1/3} (\cos 60° + i \sin 60°) = 0.855 + 1.481i$

$w_2 = 5^{1/3} \, e^{(60° + 1 \cdot 120°)i} = 5^{1/3} \, e^{180°i} = 5^{1/3} (\cos 180° + i \sin 180°) = -1.710$

$w_3 = 5^{1/3} \, e^{(60° + 2 \cdot 120°)i} = 5^{1/3} \, e^{300°i} = 5^{1/3} (\cos 300° + i \sin 300°) = 0.855 - 1.481i$

CHAPTER 7   REVIEW EXERCISE

1.

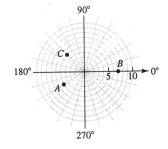

2.   Set up a table that shows how r varies as θ varies through each set of quadrant values, then sketch the polar curve from the information in the table.

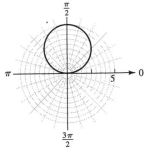

θ	$5 \sin \theta$	
0 to $\dfrac{\pi}{2}$	0 to 5	
$\dfrac{\pi}{2}$ to π	5 to 0	
π to $\dfrac{3\pi}{2}$	0 to -5	Curve is traced out a second time in this region, although coordinate pairs are different
$\dfrac{3\pi}{2}$ to 2π	-5 to 0	

3. Set up a table that shows how r varies as θ varies through each set of quadrant values, then sketch the polar curve from the information in the table.

θ	$\cos \theta$	$4 \cos \theta$	$4 + 4 \cos \theta$
0 to $\dfrac{\pi}{2}$	1 to 0	4 to 0	8 to 4
$\dfrac{\pi}{2}$ to π	0 to –1	0 to –4	4 to 0
π to $\dfrac{3\pi}{2}$	–1 to 0	–4 to 0	0 to 4
$\dfrac{3\pi}{2}$ to 2π	0 to 1	0 to 4	4 to 8

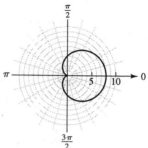

4. The graph consists of all points whose distance from the pole is 8, a circle with center at the pole, and radius 8.

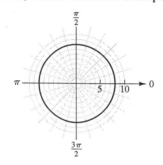

5.

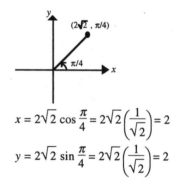

$$x = 2\sqrt{2}\,\cos\frac{\pi}{4} = 2\sqrt{2}\left(\frac{1}{\sqrt{2}}\right) = 2$$

$$y = 2\sqrt{2}\,\sin\frac{\pi}{4} = 2\sqrt{2}\left(\frac{1}{\sqrt{2}}\right) = 2$$

6. Use $r^2 = x^2 + y^2$ and $\tan \theta = \dfrac{y}{x}$

$r^2 = (-\sqrt{3})^2 + 1^2 = 4 \qquad r = 2 \qquad \tan \theta = \dfrac{1}{-\sqrt{3}} = -\dfrac{1}{\sqrt{3}} \qquad \theta = \dfrac{5\pi}{6}$ since the point is in the

second quadrant

Polar coordinates: $\left(2, \dfrac{5\pi}{6}\right)$

7.

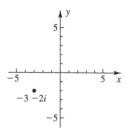

8.

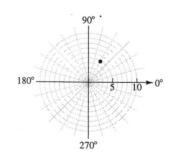

9. See figure at the right. The point with
    coordinates $(-8, 30°)$ can equally well be described
    as $(-8, -330°)$ or $(8, -150°)$ or $(8, 210°)$. Thus,
    $(-8, -330°)$: The polar axis is rotated 330°
    clockwise (negative direction) and the point is
    located 8 units from the pole along the negative
    polar axis. $(8, -150°)$: The polar axis is rotated
    150° clockwise (negative direction) and the point
    is located 8 units from the pole along the
    positive polar axis. $(8, 210°)$: The polar axis is
    rotated 210° counterclockwise (positive direction)
    and the point is located 6 units along the positive
    polar axis.

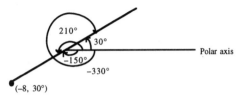

10. $z_1z_2 = 9e^{42°i} \cdot 3e^{37°i}$

    $\qquad = 9 \cdot 3e^{i(42° + 37°)} = 27e^{79°i}$

    $\dfrac{z_1}{z_2} = \dfrac{9e^{42°i}}{3e^{37°i}} = \dfrac{9}{3}e^{i(42° - 37°)} = 3e^{5°i}$

11. $(2e^{10°i})^4 = 2^4 e^{4 \cdot 10°i} = 16e^{40°i}$

12.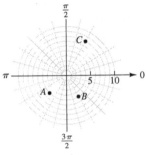

13. Set up a table that shows how r varies as 3θ varies through each set of quadrant values, then sketch
    the polar curve from the information in the table.

θ	3θ	$\sin 3\theta$	$8 \sin 3\theta$
0 to $\dfrac{\pi}{6}$	0 to $\dfrac{\pi}{2}$	0 to 1	0 to 8
$\dfrac{\pi}{6}$ to $\dfrac{\pi}{3}$	$\dfrac{\pi}{2}$ to π	1 to 0	8 to 0
$\dfrac{\pi}{3}$ to $\dfrac{\pi}{2}$	π to $\dfrac{3\pi}{2}$	0 to -1	0 to -8
$\dfrac{\pi}{2}$ to $\dfrac{2\pi}{3}$	$\dfrac{3\pi}{2}$ to 2π	-1 to 0	-8 to 0
$\dfrac{2\pi}{3}$ to $\dfrac{5\pi}{6}$	2π to $\dfrac{5\pi}{2}$	0 to 1	0 to 8
$\dfrac{5\pi}{6}$ to π	$\dfrac{5\pi}{2}$ to 3π	1 to 0	8 to 0

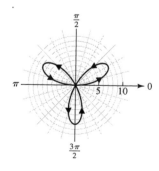

14. Set up a table that shows how r varies as 2θ varies through each set of quadrant values, then sketch the polar curve from the information in the table.

θ	2θ	$\sin 2\theta$	$4 \sin 2\theta$
0 to $\dfrac{\pi}{4}$	0 to $\dfrac{\pi}{2}$	0 to 1	0 to 4
$\dfrac{\pi}{4}$ to $\dfrac{\pi}{2}$	$\dfrac{\pi}{2}$ to π	1 to 0	4 to 0
$\dfrac{\pi}{2}$ to $\dfrac{3\pi}{4}$	π to $\dfrac{3\pi}{2}$	0 to -1	0 to -4
$\dfrac{3\pi}{4}$ to π	$\dfrac{3\pi}{2}$ to 2π	-1 to 0	-4 to 0
π to $\dfrac{5\pi}{4}$	2π to $\dfrac{5\pi}{2}$	0 to 1	0 to 4
$\dfrac{5\pi}{4}$ to $\dfrac{3\pi}{2}$	$\dfrac{5\pi}{2}$ to 3π	1 to 0	4 to 0
$\dfrac{3\pi}{2}$ to $\dfrac{7\pi}{4}$	3π to $\dfrac{7\pi}{2}$	0 to -1	0 to -4
$\dfrac{7\pi}{4}$ to 2π	$\dfrac{7\pi}{2}$ to 4π	-1 to 0	-4 to 0

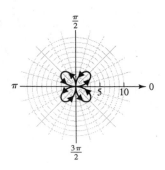

15. The graph consists of all points on a line that forms an angle of $\dfrac{\pi}{6}$ with the polar axis and passes through the pole.

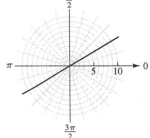

16. $8x - y^2 = x^2$
        $8x = x^2 + y^2$

Use $x = r \cos \theta$ and $x^2 + y^2 = r^2$

     $8r \cos \theta = r^2$

        $0 = r^2 - 8r \cos \theta = r(r - 8 \cos \theta)$

$r = 0$ or $r - 8 \cos \theta = 0$

The graph of $r = 0$ is the pole, and since the pole is included as a solution of $r - 8 \cos \theta = 0$

$\left(\text{let } \theta = \dfrac{\pi}{2} \right)$, we can discard $r = 0$ and keep only $r - 8 \cos \theta = 0$ or $r = 8 \cos \theta$

17. $r(3 \cos \theta - 2 \sin \theta) = -2$

$3r \cos \theta - 2r \sin \theta = -2$

Use $x = r \cos \theta$ and $y = r \sin \theta$ $\qquad$ $3x - 2y = -2$

18. $r = -3 \cos \theta$
We multiply both sides by r, which adds the pole to the graph. But the pole is already part of the graph $\left(\text{let } \theta = \dfrac{\pi}{2} \right)$, so we havė changed nothing. $r^2 = -3r \cos \theta$
But $r^2 = x^2 + y^2$, $r \cos \theta = x$. Hence, $x^2 + y^2 = -3x$

19.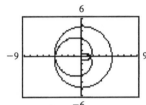

20.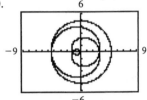

21.   A sketch shows that $-\sqrt{3} - i$ is associated with a special 30°–60° reference triangle in the third quadrant. Thus, $r = 2$ and $\theta = -150°$ and the polar form for $-\sqrt{3} - i$ is

$-\sqrt{3} - i = 2[\cos(-150°) + i \sin(-150°)]$

$\qquad = 2e^{(-150°)i}$

22. $x + iy = 3\sqrt{2}\, e^{(3\pi/4)i}$

$\qquad = 3\sqrt{2} \left(\cos \dfrac{3\pi}{4} + i \sin \dfrac{3\pi}{4} \right)$

$\qquad = 3\sqrt{2} \left(-\dfrac{1}{\sqrt{2}} + i \cdot \dfrac{1}{\sqrt{2}} \right)$

$\qquad = -3 + 3i$

23.   A sketch shows that $2 + 2i\sqrt{3}$ is associated with a special 30°–60° reference triangle in the first quadrant and $-\sqrt{2} + i\sqrt{2}$ is associated with a special 45° triangle in the second quadrant. Thus,

$2 + 2i\sqrt{3} = 4(\cos 60° + i \sin 60°)$

$\qquad = 4e^{60°i}$

$-\sqrt{2} + i\sqrt{2} = 2(\cos 135° + i \sin 135°)$

$\qquad = 2e^{135°i}$

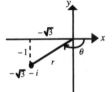

Therefore, $(2 + 2i\sqrt{3})(-\sqrt{2} + i\sqrt{2}) = 4e^{60°i} \cdot 2e^{135°i} = 4 \cdot 2e^{(60° + 135°)i} = 8e^{195°i}$

24.   Using the results of the previous problem,

$$\frac{-\sqrt{2} + i\sqrt{2}}{2 + 2i\sqrt{3}} = \frac{2e^{135°i}}{4e^{60°i}} = \frac{2}{4}\,e^{(135° - 60°)i} = 0.5e^{75°i}$$

25. $(-1 - i)^4 = (\sqrt{2}\,e^{225°i})^4 = (\sqrt{2})^4\,e^{(4 \cdot 225°)i} = (2^{1/2})^4\,e^{900°i} = 2^2\,e^{900°i}$
$$= 4e^{900°i} = 4(\cos 900° + i \sin 900°) = 4(-1 + 0i) = -4$$

26. $x^3 - 64 = 0$
$$x^3 = 64$$

Therefore, x is a cube root of 64, and there are three of them. First write 64 in polar form and use the nth root theorem.

$$64 = 64e^{0°i}$$

All three cube roots of 64 are given by

$$64^{1/3}\,e^{(0°/3 + k360°/3)i} \qquad k = 0, 1, 2$$

Thus,

$w_1 = 64^{1/3}\,e^{(0° + 0 \cdot 120°)i} = 4e^{0°i} = 4(\cos 0° + i \sin 0°) = 4(1 + 0i) = 4$

$w_2 = 64^{1/3}\,e^{(0° + 1 \cdot 120°)i} = 4e^{120°i} = 4(\cos 120° + i \sin 120°) = 4\left[\left(-\dfrac{1}{2}\right) + i\,\dfrac{\sqrt{3}}{2}\right] = -2 + 2i\sqrt{3}$

$w_3 = 64^{1/3}\,e^{(0° + 2 \cdot 120°)i} = 4e^{240°i} = 4(\cos 240° + i \sin 210°) = 4\left[\left(-\dfrac{1}{2}\right) + i\left(-\dfrac{\sqrt{3}}{2}\right)\right] = -2 - 2i\sqrt{3}$

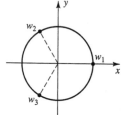

27.   First, write $-4\sqrt{3} - 4i$ in polar form.
$$-4\sqrt{3} - 4i = 8e^{210°i}$$

From the nth root theorem, all three roots are given by

$$8^{1/3}\,e^{(210°/3 + k360°/3)i} \qquad k = 0, 1, 2$$

Thus,

$w_1 = 8^{1/3}\,e^{(70° + 0 \cdot 120°)i} = 2e^{70°i}$

$w_2 = 8^{1/3}\,e^{(70° + 1 \cdot 120°)i} = 2e^{190°i}$

$w_3 = 8^{1/3}\,e^{(70° + 2 \cdot 120°)i} = 2e^{310°i}$

28.   To show that $2e^{30°i}$ is a square root of $2 + i2\sqrt{3}$, we need only show that $(2e^{30°i})^2 = 2 + i2\sqrt{3}$.

But, by DeMoivre's theorem, $(2e^{30°i})^2 = 2^2\, e^{2\,\cdot\,30°i} = 4e^{60°i} = 4(\cos 60° + i \sin 60°) = 4\left(\dfrac{1}{2} + i \cdot \dfrac{\sqrt{3}}{2}\right)$

$= 2 + i2\sqrt{3}$

29.   (A)

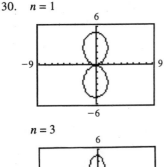

There are a total of three cube roots and they are spaced equally
around a circle of radius 2.

(B)   Since $w_3 = -2i = 2e^{-90°i}$, and each root is spaced 120° along the circle
from the last,

$$w_1 = 2e^{(-90° + 120°)i} = 2e^{30°i} = 2(\cos 30° + i \sin 30°) = 2\left(\frac{\sqrt{3}}{2} + i \cdot \frac{1}{2}\right) = \sqrt{3} + i$$

$$w_2 = 2e^{(30° + 120°)i} = 2e^{150°i} = 2(\cos 150° + i \sin 150°) = 2\left[\left(-\frac{\sqrt{3}}{2}\right) + i \cdot \frac{1}{2}\right] = -\sqrt{3} + i$$

(C) $(-2i)^3 = (2e^{-90°i})^3 = 2^3\, e^{3(-90°i)} = 8e^{-270°i} = 8[\cos(-270°) + i \sin(-270°)] = 8(0 + i) = 8i$

$(\sqrt{3} + i)^3 = (2e^{30°i})^3 = 2^3\, e^{3\,\cdot\,30°i} = 8e^{90°i} = 8(\cos 90° + i \sin 90°) = 8(0 + i) = 8i$

$(-\sqrt{3} + i)^3 = (2e^{150°i})^3 = 2^3\, e^{3\,\cdot\,150°i} = 8e^{450°i} = 8(\cos 450° + i \sin 450°) = 8(0 + i) = 8i$

30. $n = 1$ $n = 2$

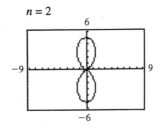

$n = 3$

Since each graph has 2 leaves, we expect 2 leaves for arbitrary n.

31.   (A) $r = \dfrac{2}{1 - 1.6 \sin \theta}$

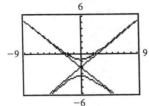

The graph is a hyperbola.

(B) $r = \dfrac{2}{1 - \sin \theta}$

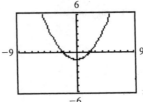

The graph is a parabola.

(C) $r = \dfrac{2}{1 - 0.4 \sin \theta}$

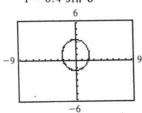

The graph is an ellipse.

32. $r(\sin \theta - 2) = 3$ $r \sin \theta - 2r = 3$

Use $r \sin \theta = y$ $r = -\sqrt{x^2 + y^2}$ $y + 2\sqrt{x^2 + y^2} = 3$

$y - 3 = -2\sqrt{x^2 + y^2}$ or $(y - 3)^2 = 4(x^2 + y^2)$

Note: See comment, Exercise 7.1, Problem 61.

33. $x^3 - 12 = 0$

$x^3 = 12$

Therefore, x is a cube root of 12, and there are three of them. First write 12 in polar form and use the nth root theorem.

$12 = 12e^{0°i}$

All three cube roots of 12 are given by

$12^{1/3} \, e^{(0°/3 + k360°/3)i}$ $k = 0, 1, 2$

Thus,

$w_1 = 12^{1/3} \, e^{(0° + 0 \cdot 120°)i} = 12^{1/3} \, e^{0°i} = 12^{1/3} (\cos 0° + i \sin 0°) = 2.289$

$w_2 = 12^{1/3} \, e^{(0° + 1 \cdot 120°)i} = 12^{1/3} \, e^{120°i} = 12^{1/3} (\cos 120° + i \sin 120°) = -1.145 + 1.983i$

$w_3 = 12^{1/3} \, e^{(0° + 2 \cdot 120°)i} = 12^{1/3} \, e^{240°i} = 12^{1/3} (\cos 240° + i \sin 240°) = -1.145 - 1.983i$

34. $[r^{1/3} e^{(\theta/3 + k \cdot 120°)i}]^3 = (r^{1/3})^3 e^{3(\theta/3 + k \cdot 120°)i}$ DeMoivre's theorem

$= re^{(\theta + k \cdot 360°)i}$ Algebra

$= r[\cos(\theta + k \cdot 360°) + i \sin(\theta + k \cdot 360°)]$ Definition of $e^{i\theta}$

$= r(\cos \theta + i \sin \theta)$ Periodic property of sine and
cosine functions

$= re^{i\theta}$ Definition of $e^{i\theta}$

35.   (A) The coordinates of P represent a simultaneous solution.

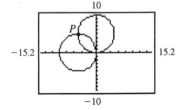

(B) We solve the system, $r = 10 \sin \theta$, $r = -10 \cos \theta$, by equating the right sides:

$10 \sin \theta = -10 \cos \theta$

$\sin \theta = -\cos \theta$

$\tan \theta = -1$

The only solution of this equation, $0 \le \theta \le \pi$, is $\theta = \dfrac{3\pi}{4}$. If we substitute this into either of

the original equations, we get $r = 10 \sin \dfrac{3\pi}{4} = -10 \cos \dfrac{3\pi}{4} = 10\left(\dfrac{\sqrt{2}}{2}\right) = 5\sqrt{2}$

Solution: $\left(5\sqrt{2}, \dfrac{3\pi}{4}\right)$

(C) The pole has no ordered pairs of coordinates that simultaneously satisfy both equations.

As $\left(0, \dfrac{\pi}{2}\right)$, it satisfies the first; as $(0, 0)$, it satisfies the second; it is not a solution of the
system.

319

CUMULATIVE REVIEW EXERCISE CHAPTERS 1—7

1. We compare α and β by changing to decimal degrees.

   Since $\theta_d = \dfrac{180}{\pi}\,\theta_r$, $\alpha_d = \dfrac{180°}{\pi}\,\alpha_r = \dfrac{180°}{\pi} \cdot \dfrac{2\pi}{7} = 51.42857°\ldots$

   Since $25' = \dfrac{25°}{60}$ and $40'' = \dfrac{40°}{3600}$, then, $\beta = 51°25'40'' = 51.427777°\ldots$ Thus, $\alpha > \beta$.

2. *Solve for the hypotenuse c:* $c^2 = a^2 + b^2$

   $$c = \sqrt{a^2 + b^2} = \sqrt{(1.27 \text{ cm})^2 + (4.65 \text{ cm})^2} = 4.82 \text{ cm}$$

   Solve for θ: We use the tangent. $\tan \theta = \dfrac{1.27}{4.65}$ $\theta = \tan^{-1}\dfrac{1.27}{4.65} = 15.3°$

   Solve for the complementary angle: $90° - \theta = 90° - 15.3° = 74.7°$

3. $P(a, b) = (7, -24)$

   $r = \sqrt{7^2 + (-24)^2} = 25$

   $\sec \theta = \dfrac{r}{a} = \dfrac{25}{7}$

   $\tan \theta = \dfrac{b}{a} = \dfrac{-24}{7}$

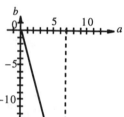

$P(7, -24)$

4. (A) $\cot x \sec x \sin x = \dfrac{\cos x}{\sin x} \cdot \dfrac{1}{\cos x} \cdot \sin x$ Quotient and reciprocal identities

   $= 1$ Algebra

   (B) $\tan \theta + \cot \theta = \dfrac{\sin \theta}{\cos \theta} + \dfrac{\cos \theta}{\sin \theta}$ Quotient identities

   $= \dfrac{\sin^2 \theta}{\sin \theta \cos \theta} + \dfrac{\cos^2 \theta}{\sin \theta \cos \theta}$ Algebra

   $= \dfrac{\sin^2 \theta + \cos^2 \theta}{\sin \theta \cos \theta}$ Algebra

   $= \dfrac{1}{\sin \theta \cos \theta}$ Pythagorean identity

   $= \dfrac{1}{\sin \theta} \cdot \dfrac{1}{\cos \theta}$ Algebra

   $= \csc \theta \sec \theta$ Reciprocal identity

   $= \sec \theta \csc \theta$ Algebra

5. Locate the 30°–60° reference triangle, determine (a, b) and r, then evaluate.

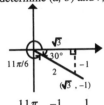

$$\sin \frac{11\pi}{6} = \frac{-1}{2} = -\frac{1}{2}$$

6. Locate the 30°–60° reference triangle, determine (a, b) and r then evaluate.

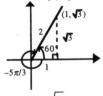

$$\tan \frac{-5\pi}{3} = \frac{\sqrt{3}}{1} = \sqrt{3}$$

7. $y = \cos^{-1}(-0.5)$ is equivalent to $\cos y = -0.5$. What y between 0 and π has cosine equal to -0.5? y must be associated with a reference triangle in the second quadrant. Reference triangle is a special 30°–60° triangle.

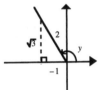

$$y = \frac{2\pi}{3} \qquad \cos^{-1}(-0.5) = \frac{2\pi}{3}$$

8. $y = \csc^{-1}(\sqrt{2})$ is equivalent to $\csc y = \sqrt{2}$ and $-\frac{\pi}{2} \le y \le \frac{\pi}{2}$, $y \ne 0$. What number between $-\frac{\pi}{2}$ and $\frac{\pi}{2}$ has cosecant equal to $\sqrt{2}$? y must be in the first quadrant.

$$\csc y = \sqrt{2} = \frac{\sqrt{2}}{1} \qquad y = \frac{\pi}{4}$$

Thus, $\csc^{-1}(\sqrt{2}) = \frac{\pi}{4}$

9. Calculator in degree mode: $\sin 43°22' = \sin(43.366...°)$ Convert to decimal degrees, if necessary

$$= 0.6867$$

10. Use the reciprocal relationship $\cot \theta = \frac{1}{\tan \theta}$

Calculator in radian mode: $\cot \frac{2\pi}{5} = \frac{1}{\tan \frac{2\pi}{5}} = 0.3249$

11. Calculator in radian mode: $\sin^{-1}(0.8) = 0.9273$

12. Calculator in radian mode: $\sec^{-1}(4.5) = \cos^{-1}\frac{1}{4.5} = 1.347$

13. $\sin x + \sin y = 2 \sin \frac{x + y}{2} \cos \frac{x - y}{2}$

$$\sin 3t + \sin t = 2 \sin \frac{3t + t}{2} \cos \frac{3t - t}{2} = 2 \sin 2t \cos t$$

14. An angle of radian measure 2.5 is the central angle of a circle subtended by an arc with measure 2.5 times that of the radius of the circle.

15. We are given two angles and a non-included side (*AAS*).
We use the law of sines.

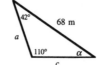

Solve for α: $\alpha + \beta + \gamma = 180°$

$\alpha = 180° - (42° + 110°) = 28°$

Solve for a: $\dfrac{\sin\ \alpha}{a} = \dfrac{\sin\ \beta}{b}$

$a = \dfrac{b\ \sin\ \alpha}{\sin\ \beta} = \dfrac{(68\ \text{m})\ \sin\ 28°}{\sin\ 110°} = 34\ \text{m}$

Solve for c: $\dfrac{\sin\ \gamma}{c} = \dfrac{\sin\ \beta}{b}$ $c = \dfrac{b\ \sin\ \gamma}{\sin\ \beta} = \dfrac{(68\ \text{m})\ \sin\ 42°}{\sin\ 110°} = 48\ \text{m}$

16. We are given two sides and the included angle (*SAS*). We use the law of cosines to find the third
side, then the law of sines to find a second angle.

Solve for b: $b^2 = a^2 + c^2 - 2ac\ \cos\ \beta$

$= 16^2 + 24^2 - 2(16)(24)\ \cos\ 34°$

$= 195.2991...$

$b = \sqrt{195.2991...} = 14\ \text{in}$

Since *a* is the shorter of the remaining sides, *α*, the angle opposite *a*, must be acute.

Solve for α: $\dfrac{\sin\ \alpha}{a} = \dfrac{\sin\ \beta}{b}$

$\sin\ \alpha = \dfrac{a\ \sin\ \beta}{b} = \dfrac{(16\ \text{in})(\sin\ 34°)}{14\ \text{in}} = 0.6402$

$\alpha = \sin^{-1}(0.6402) = 40°$

Solve for γ: $\alpha + \beta + \gamma = 180°$

$\gamma = 180° - (40° + 34°) = 106°$

17. We are given three sides (*SSS*). We solve for the largest angle, *γ* (largest because it is opposite the
largest side, *c*) using the law of cosines. We then solve for a second angle using the law of sines,
because it involves simpler calculations.

Solve for γ: $c^2 = a^2 + b^2 - 2ab\ \cos\ \gamma$

$\cos\ \gamma = \dfrac{a^2 + b^2 - c^2}{2ab}$

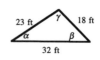

$\gamma = \cos^{-1}\dfrac{a^2 + b^2 - c^2}{2ab} = \cos^{-1}\dfrac{18^2 + 23^2 - 32^2}{2 \cdot 18 \cdot 23} = 102°$

Both *β* and *γ* must be acute, since they are smaller than *α*.

Solve for β: $\dfrac{\sin\ \beta}{b} = \dfrac{\sin\ \gamma}{c}$ $\sin\ \beta = \dfrac{b\ \sin\ \gamma}{c}$

$\beta = \sin^{-1}\dfrac{b\ \sin\ \gamma}{c}$ $\begin{cases} \beta \text{ is acute, because there is room for} \\ \text{only one obtuse angle in a triangle.} \end{cases}$

$= \sin^{-1}\dfrac{(23\ \text{ft})\ \sin\ 102°}{32\ \text{ft}} = 45°$

Solve for α: $\alpha + \beta + \gamma = 180°$

$\alpha = 180° - (\beta + \gamma) = 180° - (45° + 102°) = 33°$

18. The given information consists of two sides and the included angle; hence, we use the formula
$A = \dfrac{ab}{2} \sin \theta$ in the form $A = \dfrac{1}{2} ac \sin \beta = \dfrac{1}{2}$ (16 in)(24 in) sin 34° = 110 in²

19. The point with coordinates (–7, 30°) can equally well be described as (7, –150°). Thus: rotate the polar axis 150° clockwise (negative direction) and go 7 units along the positive polar axis.

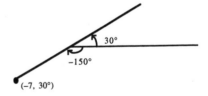

(–7, 30°)

20. Horizontal component |**H**|: $\cos 25° = \dfrac{|\mathbf{H}|}{13}$ |**H**| = 13 cos 25° = 12

Vertical component |**V**|: $\sin 25° = \dfrac{|\mathbf{V}|}{13}$ |**V**| = 13 sin 25° = 5.5

21. To find |**u** + **v**|: Apply the Pythagorean theorem to triangle OCB.

$|\mathbf{u} + \mathbf{v}|^2 = OB^2 = OC^2 + BC^2 = 6.4^2 + 3.9^2 = 56.17$

$|\mathbf{u} + \mathbf{v}| = \sqrt{56.17} = 7.5$.

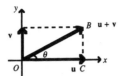

Solve triangle OCB for θ: $\tan \theta = \dfrac{BC}{OC} = \dfrac{|\mathbf{v}|}{|\mathbf{u}|}$ $\theta = \tan^{-1} \dfrac{|\mathbf{v}|}{|\mathbf{u}|}$ θ is acute

$\theta = \tan^{-1} \dfrac{3.9}{6.4} = 31°$

22. The algebraic vector ⟨a, b⟩ has coordinates given by
$a = x_b - x_a = (-3) - 4 = -7$ $b = y_b - y_a = 7 - (-2) = 9$
Hence, ⟨a, b⟩ = ⟨–7, 9⟩
Magnitude of $\langle a, b \rangle = |\langle -7, 9 \rangle| = \sqrt{a^2 + b^2} = \sqrt{(-7)^2 + 9^2} = \sqrt{130}$

23. $|\mathbf{u}| = \sqrt{2^2 + (-7)^2} = \sqrt{53}$ $|\mathbf{v}| = \sqrt{3^2 + 8^2} = \sqrt{73}$

$\cos \theta = \dfrac{\mathbf{u} \cdot \mathbf{v}}{|\mathbf{u}||\mathbf{v}|} = \dfrac{(2\mathbf{i} - 7\mathbf{j}) \cdot (3\mathbf{i} + 8\mathbf{j})}{\sqrt{53}\sqrt{73}} = \dfrac{2 \cdot 3 + (-7) \cdot 8}{\sqrt{53}\sqrt{73}}$ $\theta = \cos^{-1} \dfrac{-50}{\sqrt{53}\sqrt{73}} = 143.5°$

24.

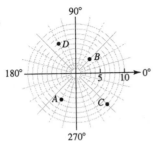

25. Set up a table that shows how r varies as θ varies through each set of quadrant values, then sketch the polar curve from the information in the table.

θ	$\sin \theta$	$5 \sin \theta$	$5 + 5 \sin \theta$
0 to $\dfrac{\pi}{2}$	0 to 1	0 to 5	5 to 10
$\dfrac{\pi}{2}$ to π	1 to 0	5 to 0	10 to 5
π to $\dfrac{3\pi}{2}$	0 to -1	0 to -5	5 to 0
$\dfrac{3\pi}{2}$ to 2π	-1 to 0	-5 to 0	0 to 5

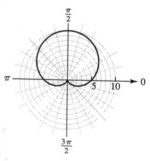

26.

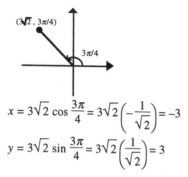

$(3\sqrt{2}, 3\pi/4)$

$3\pi/4$

$x = 3\sqrt{2} \cos \dfrac{3\pi}{4} = 3\sqrt{2}\left(-\dfrac{1}{\sqrt{2}}\right) = -3$

$y = 3\sqrt{2} \sin \dfrac{3\pi}{4} = 3\sqrt{2}\left(\dfrac{1}{\sqrt{2}}\right) = 3$

27. Use $r^2 = x^2 + y^2$ and $\tan \theta = \dfrac{y}{x}$

$r^2 = (-2\sqrt{3})^2 + 2^2 = 16$

$r = 4$

$\tan \theta = \dfrac{2}{-2\sqrt{3}} = -\dfrac{1}{\sqrt{3}}$

$\theta = \dfrac{5\pi}{6}$ since the point is in the second quadrant

Polar coordinates: $\left(4, \dfrac{5\pi}{6}\right)$

28. A sketch shows that $2 - 2i$ is associated with a special $45°$ reference triangle in the fourth quadrant. Thus, $r = 2\sqrt{2}$ and $\theta = -\dfrac{\pi}{4}$ and the polar form for $2 - 2i$ is

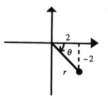

$2 - 2i = 2\sqrt{2}\left[\cos\left(-\dfrac{\pi}{4}\right) + i \sin\left(-\dfrac{\pi}{4}\right)\right]$

$\qquad = 2\sqrt{2}\, e^{(-\pi/4)i}$

29. $x + iy = 3e^{(3\pi/2)i}$

$\qquad = 3\left(\cos \dfrac{3\pi}{2} + i \sin \dfrac{3\pi}{2}\right)$

$\qquad = 3[0 + i(-1)]$

$\qquad = -3i$

30. $z_1 z_2 = 3e^{50°i} \cdot 5e^{15°i}$

$\qquad = 3 \cdot 5e^{(50° + 15°)i} = 15e^{65°i}$

$\dfrac{z_1}{z_2} = \dfrac{3e^{50°i}}{5e^{15°i}}$

$\qquad = \dfrac{3}{5} e^{i(50° - 15°)i} = 0.6e^{35°i}$

31. $(3e^{25°i})^4 = 3^4 e^{4 \cdot 25°i} = 81e^{100°i}$

32. Since the maximum value occurs at the end points of the interval, it would appear that A should be positive.

Since the maximum value of the function appears to be 3, and the minimum value appears to be -1,

$$A = \frac{3 - (-1)}{2} = 2 \quad \text{and} \quad k = \frac{3 + (-1)}{2} = 1$$

Since the maximum value is achieved at 0 and at 2, the period of the function is 2. Hence, $\frac{2\pi}{B} = 2$ and $B = \pi$.

Thus, the required function is $y = 1 + 2 \cos \pi x$.

33. Since $\sin \theta < 0$ and $\cot \theta > 0$, the terminal side of θ lies in quadrant III. We sketch a reference triangle and label what we know. Since

$\cot \theta = 4 = \frac{4}{1} = \frac{-4}{-1}$, we know that $a = -4$ and $b = -1$. We use the

Pythagorean theorem to find r:
$$(-4)^2 + (-1)^2 = r^2$$
$$r = \sqrt{17} \quad (r \text{ is never negative})$$

Therefore, $\csc \theta = \frac{r}{b} = \frac{\sqrt{17}}{-1} = -\sqrt{17} \qquad \cos \theta = \frac{a}{r} = \frac{-4}{\sqrt{17}}$

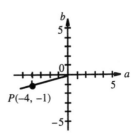

34. This graph is the graph of $y = 2 \sin(2x + \pi)$ moved up one unit. We first find the period and phase shift by solving

$$2x + \pi = 0 \qquad \text{and} \qquad 2x + \pi = 2\pi$$
$$x = -\frac{\pi}{2} \qquad\qquad\qquad x = -\frac{\pi}{2} + \pi$$

$$\text{Period} = \pi \qquad \text{Phase Shift} = -\frac{\pi}{2} \qquad \text{Frequency} = \frac{1}{\text{Period}} = \frac{1}{\pi}$$

We then sketch one period of the graph starting at $x = -\frac{\pi}{2}$ (the phase shift) and ending at

$x = -\frac{\pi}{2} + \pi = \frac{\pi}{2}$ (the phase shift plus one period). The graph is a basic sine curve relative to the horizontal line $y = 1$ (shown as a broken line) and the y axis. We then extend the graph from $-\pi$ to 2π.

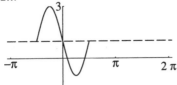
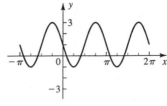

35.
$$\frac{\cos x}{1 - \sin x} + \frac{\cos x}{1 + \sin x} = \frac{\cos x (1 + \sin x)}{(1 - \sin x)(1 + \sin x)} + \frac{\cos x (1 - \sin x)}{(1 - \sin x)(1 + \sin x)} \qquad \text{Algebra}$$

$$= \frac{\cos x (1 + \sin x) + \cos x (1 - \sin x)}{(1 - \sin x)(1 + \sin x)} \qquad \text{Algebra}$$

$$= \frac{\cos x + \sin x \cos x + \cos x - \sin x \cos x}{1 - \sin^2 x} \qquad \text{Algebra}$$

$$= \frac{2 \cos x}{1 - \sin^2 x} \qquad\qquad \text{Algebra}$$

$$= \frac{2 \cos x}{\cos^2 x} \qquad\qquad \text{Pythagorean identity}$$

$$= \frac{2}{\cos x} \qquad\qquad \text{Algebra}$$

$$= 2 \sec x \qquad\qquad \text{Reciprocal identity}$$

36. $\tan \dfrac{\theta}{2} = \dfrac{\sin \theta}{1 + \cos \theta}$ Half-angle identity

$$= \frac{\dfrac{\sin \theta}{\sin \theta}}{\dfrac{1 + \cos \theta}{\sin \theta}} \qquad\qquad \text{Algebra}$$

$$= \frac{1}{\dfrac{1}{\sin \theta} + \dfrac{\cos \theta}{\sin \theta}} \qquad\qquad \text{Algebra}$$

$$= \frac{1}{\csc \theta + \cot \theta} \qquad\qquad \text{Reciprocal and quotient identities}$$

37. $\dfrac{\cos x - \sin x}{\cos x + \sin x} = \dfrac{(\cos x - \sin x)(\cos x - \sin x)}{(\cos x + \sin x)(\cos x - \sin x)}$ Algebra

$$= \frac{\cos^2 x - 2 \cos x \sin x + \sin^2 x}{\cos^2 x - \sin^2 x} \qquad \text{Algebra}$$

$$= \frac{1 - 2 \cos x \sin x}{\cos^2 x - \sin^2 x} \qquad \text{Pythagorean identity}$$

$$= \frac{1 - \sin 2x}{\cos 2x} \qquad \text{Double-angle identities}$$

$$= \frac{1}{\cos 2x} - \frac{\sin 2x}{\cos 2x} \qquad \text{Algebra}$$

$$= \sec 2x - \tan 2x \qquad \text{Reciprocal and quotient identities}$$

38. First draw a reference triangle in the first quadrant and find $\sin x$.

$$b = \sqrt{25^2 - 24^2} = 7 \qquad \sin x = \frac{b}{r} = \frac{7}{25}$$

We can now find $\tan \dfrac{x}{2}$ from the half-angle identity.

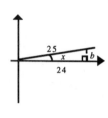

$$\tan \frac{x}{2} = \frac{1 - \cos x}{\sin x} = \frac{1 - \dfrac{24}{25}}{\dfrac{7}{25}} = \frac{25 - 24}{7} = \frac{1}{7}$$

To find $\sin 2x$, we use double-angle identity.

$$\sin 2x = 2 \sin x \cos x = 2 \cdot \frac{7}{25} \cdot \frac{24}{25} = \frac{336}{625}$$

39. (A) Graph both sides of the equation in the same viewing window.

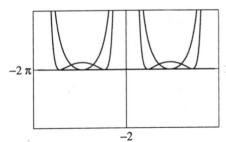

$$\frac{\cos^2 x}{(\cos x - 1)^2} = \frac{1 + \cos x}{1 - \cos x}$$ is not an identity, since the graphs do not match.

Try $x = \dfrac{\pi}{2}$.

Left side: $\dfrac{\cos^2 (\pi/2)}{[\cos(\pi/2) - 1]^2} = \dfrac{0}{(0 - 1)^2} = 0$

Right side: $\dfrac{1 + \cos(\pi/2)}{1 - \cos(\pi/2)} = \dfrac{1 + 0}{1 - 0} = 1$

This verifies that the equation is not an identity.

(B) Graph both sides of the equation in the same viewing window.

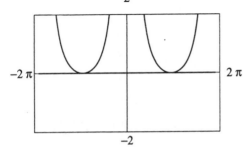

$$\frac{\sin^2 x}{(\cos x - 1)^2} = \frac{1 + \cos x}{1 - \cos x}$$ appears to be an identity, which we verify.

$$\frac{\sin^2 x}{(\cos x - 1)^2} = \frac{1 - \cos^2 x}{(\cos x - 1)^2}$$ Pythagorean identity

$$= \frac{1 - \cos^2 x}{(1 - \cos x)^2}$$ Algebra

$$= \frac{(1 - \cos x)(1 + \cos x)}{(1 - \cos x)(1 - \cos x)}$$ Algebra

$$= \frac{1 + \cos x}{1 - \cos x}$$ Algebra

40. Let $y = \sin^{-1}\frac{3}{4}$, then $\sin y = \frac{3}{4}$, $-\frac{\pi}{2} \le y \le \frac{\pi}{2}$.
Sketch the reference triangle associated with y,
then $\sec y = \sec\left(\sin^{-1}\frac{3}{4}\right)$, can be determined
directly from the triangle.

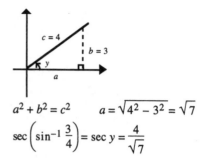

$$a^2 + b^2 = c^2 \qquad a = \sqrt{4^2 - 3^2} = \sqrt{7}$$

$$\sec\left(\sin^{-1}\frac{3}{4}\right) = \sec y = \frac{4}{\sqrt{7}}$$

41. (A)

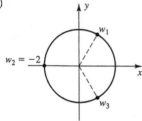

There are a total of three cube roots and they are spaced equally around a circle of radius 2.

(B) Since $w_2 = -2 = 2e^{180°i}$, and each root is spaced 120° along the circle from the last,

$$w_3 = 2e^{(180° + 120°)i} = 2e^{300°i} = 2(\cos 300° + i \sin 300°) = 2\left(\frac{1}{2} - i\frac{\sqrt{3}}{2}\right) = 1 - i\sqrt{3}$$

$$w_1 = 2e^{(180° - 120°)i} = 2e^{60°i} = 2(\cos 60° + i \sin 60°) = 2\left(\frac{1}{2} + i\frac{\sqrt{3}}{2}\right) = 1 + i\sqrt{3}$$

(C) $(1 - i\sqrt{3})^3 = (2e^{300°i})^3 = 2^3 e^{3 \cdot 300°i} = 8e^{900°i} = 8(\cos 900° + i \sin 900°) = 8(-1 + i0) = -8$

$(1 + i\sqrt{3})^3 = (2e^{60°i})^3 = 2^3 e^{3 \cdot 60°i} = 8e^{180°i} = 8(\cos 180° + i \sin 180°) = 8(-1 + i0) = -8$

42. First solve for x over one period, $0 \le x < 2\pi$. The add integer multiples of 2π to find all solutions.

$$\sin 2x + \sin x = 0 \qquad \text{Use double-angle identity}$$

$$2 \sin x \cos x + \sin x = 0$$

$$\sin x (2 \cos x + 1) = 0$$

$$\sin x = 0 \quad \text{or} \quad 2 \cos x + 1 = 0$$

$$x = 0, \pi \qquad\qquad \cos x = -\frac{1}{2}$$

$$x = \frac{2\pi}{3}, \frac{4\pi}{3}$$

Thus, the solutions over one period, $0 \le x < 2\pi$, are 0, π, $\dfrac{2\pi}{3}$, $\dfrac{4\pi}{3}$. Thus, if x can range over all real numbers,

$$x = \begin{cases} \left.\begin{array}{c} 0 + 2k\pi \\ \pi + 2k\pi \end{array}\right\} \text{ or } k\pi \\ \dfrac{2\pi}{3} + 2k\pi \\ \dfrac{4\pi}{3} + 2k\pi \end{cases} \qquad k \text{ any integer}$$

43. $2 \cos 2x = 5 \sin x - 4$

Solve for sin x and/or cos x: $2(1 - 2\sin^2 x) = 5 \sin x - 4$ $\qquad$ Use double-angle identity

$$2 - 4\sin^2 x = 5 \sin x - 4$$
$$0 = 4\sin^2 x + 5 \sin x - 6$$
$$0 = (4\sin x - 3)(\sin x + 2)$$

$4\sin x - 3 = 0$ $\qquad\qquad$ $\sin x + 2 = 0$

$\sin x = \dfrac{3}{4}$ $\qquad\qquad$ $\sin x = -2$

Solve over $0 \le x < 2\pi$: Sketch a graph of $y = \sin x$,
$y = \dfrac{3}{4}$, and $y = -2$ in the same coordinate system.

$\sin x = -2$ No solution (-2 is not in the domain of the sine function)

$\sin x = \dfrac{3}{4}$ From the graph, we see that the solutions are in the first and second quadrants.

$x = \sin^{-1}\dfrac{3}{4} = 0.8481$

$x = \pi - 0.8481 = 2.294$

Because the sine function is periodic with period 2π, all solutions are given by:
$x = 0.8481 + 2k\pi$, $x = 2.294 + 2k\pi$, k any integer.

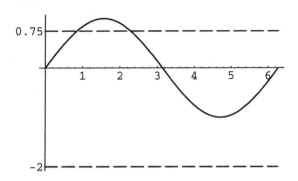

44. We are given two sides and a non-included angle (*SSA*).

(A) $a = 11.5$ cm

Solve for β: $\dfrac{\sin \beta}{b} = \dfrac{\sin \alpha}{a}$

$$\sin \beta = \dfrac{b \sin \alpha}{a} = \dfrac{17.4 \sin 49°30'}{11.5}$$

$$= 1.151$$

Since $\sin \beta = 1.151$ has no solution, no triangle exists with the given measurements. No solution.

(B) *Solve for β:* $\dfrac{\sin \beta}{b} = \dfrac{\sin \alpha}{a}$

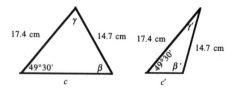

$$\sin \beta = \dfrac{b \sin \alpha}{a} = \dfrac{(17.4 \text{ cm}) \sin 49°30'}{14.7 \text{ cm}} = 0.9001$$

Two triangles are possible; angle β can be either acute or obtuse.

$\beta = \sin^{-1} 0.9001 = 64°10'$ ⟨⟩ $\beta' = 180° - \sin^{-1} 0.9001 = 115°50°$

Solve for γ and γ':

$\alpha + \beta + \gamma = 180°$ ⟨⟩ $\alpha' + \beta' + \gamma' = 180°$

$\gamma = 180° - (49°30' + 64°10')$ ⟨⟩ $\gamma' = 180° - (49°30' + 115°50')$

$= 66°20'$ ⟨⟩ $= 14°40'$

Solve for c and c':

$\dfrac{\sin \alpha}{a} = \dfrac{\sin \gamma}{c}$ ⟨⟩ $\dfrac{\sin \alpha}{a} = \dfrac{\sin \gamma'}{c'}$

$c = \dfrac{a \sin \gamma}{\sin \alpha} = \dfrac{(14.7 \text{ cm}) \sin 66°20'}{\sin 49°30'}$ ⟨⟩ $c = \dfrac{a \sin \gamma'}{\sin \alpha} = \dfrac{(14.7 \text{ cm}) \sin 14°40'}{\sin 49°30'}$

$= 17.7$ cm ⟨⟩ $= 4.89$ cm

(C) *Solve for β:* $\dfrac{\sin \beta}{b} = \dfrac{\sin \alpha}{a}$

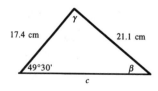

$$\sin \beta = \dfrac{b \sin \alpha}{a} = \dfrac{(17.4 \text{ cm}) \sin 49°30'}{21.1}$$

$$= 0.6271$$

$\beta = \sin^{-1} 0.6271 = 38°50'$

There is another solution of $\sin \beta = 0.6271$ that deserves brief consideration:

$\beta' = 180° - \sin^{-1} 0.6271 = 141°10'$.

However, there is not enough room in a triangle for an angle of 141°10' and an angle of 49°30', since their sum is greater than 180°.)

Solve for γ: $\alpha + \beta + \gamma = 180°$

$\gamma = 180° - (49°30' + 38°50') = 91°40'$

Solve for c: $\dfrac{\sin \alpha}{a} = \dfrac{\sin \gamma}{c}$

$c = \dfrac{a \sin \gamma}{\sin \alpha} = \dfrac{(21.1 \text{ cm}) \sin 91°40'}{\sin 49°30'} = 27.7$ cm

45. $\angle BCD = 40.0°$. Hence, $\angle OCB = 180° - 40.0° = 140.0°$.

    We can find $|\mathbf{u} + \mathbf{v}|$ using the law of cosines:

    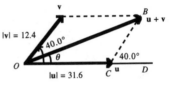

$$|\mathbf{u} + \mathbf{v}|^2 = |\mathbf{u}|^2 + |\mathbf{v}|^2 - 2|\mathbf{u}||\mathbf{v}|\cos(OCB)$$
$$= 31.6^2 + 12.4^2 - 2(31.6)(12.4)\cos 140.0°$$
$$= 1752.65370\ldots$$
$$|\mathbf{u} + \mathbf{v}| = \sqrt{1752.65370\ldots} = 41.9$$

To find θ, we use the law of sines:
$$\frac{\sin\theta}{|\mathbf{v}|} = \frac{\sin OCB}{|\mathbf{u} + \mathbf{v}|}$$
$$\frac{\sin\theta}{12.4} = \frac{\sin 140.0°}{41.9}$$
$$\sin\theta = \frac{12.4}{41.9}\sin 140.0°$$
$$\theta = \sin^{-1}\left(\frac{12.4}{41.9}\sin 140.0°\right) = 11.0°$$

46. (A) $3\mathbf{u} - 4\mathbf{v} = 3\langle 1, -2\rangle - 4\langle 0, 3\rangle = \langle 3, -6\rangle + \langle 0, -12\rangle = \langle 3, -18\rangle$

    (B) $3\mathbf{u} - 4\mathbf{v} = 3(2\mathbf{i} + 3\mathbf{j}) - 4(-\mathbf{i} + 5\mathbf{j}) = 6\mathbf{i} + 9\mathbf{j} + 4\mathbf{i} - 20\mathbf{j} = 10\mathbf{i} - 11\mathbf{j}$

47. $|\mathbf{v}| = \sqrt{7^2 + (-24)^2} = 25 \qquad \mathbf{u} = \frac{1}{|\mathbf{v}|}\mathbf{v} = \frac{1}{25}\langle 7, -24\rangle = \left\langle \frac{7}{25}, -\frac{24}{25}\right\rangle$ or $\langle 0.28, -0.96\rangle$

48. The algebraic vector $\langle a, b\rangle$ has coordinates given by
$$a = x_b - x_a = (-1) - (-3) = 2 \qquad b = y_b - y_a = 5 - 2 = 3$$
    Hence, $\langle a, b\rangle = \langle 2, 3\rangle = \langle 2, 0\rangle + \langle 0, 3\rangle = 2\langle 1, 0\rangle + 3\langle 0, 1\rangle = 2\mathbf{i} + 3\mathbf{j}$

49. (A) $\mathbf{u} \cdot \mathbf{v} = \langle 4, 0\rangle \cdot \langle 0, -5\rangle = 0 + 0 = 0$ Thus, $\mathbf{u}$ and $\mathbf{v}$ are orthogonal.

    (B) $\mathbf{u} \cdot \mathbf{v} = \langle 3, 2\rangle \cdot \langle -3, 4\rangle = -9 + 8 = -1 \neq 0$ Thus, $\mathbf{u}$ and $\mathbf{v}$ are not orthogonal.

    (C) $\mathbf{u} \cdot \mathbf{v} = (\mathbf{i} - 2\mathbf{j}) \cdot (6\mathbf{i} + 3\mathbf{j}) = 6 - 6 = 0$ Thus, $\mathbf{u}$ and $\mathbf{v}$ are orthogonal.

Cumulative Review Exercise Chapters 1—7

50. Set up a table that shows how r varies as 2θ varies through each set of quadrant values, then sketch the polar curve from the information in the table.

θ	2θ	$\cos\theta$	$8\cos 2\theta$
0 to $\frac{\pi}{4}$	0 to $\frac{\pi}{2}$	1 to 0	8 to 0
$\frac{\pi}{4}$ to $\frac{\pi}{2}$	$\frac{\pi}{2}$ to π	0 to -1	0 to -8
$\frac{\pi}{2}$ to $\frac{3\pi}{4}$	π to $\frac{3\pi}{2}$	-1 to 0	-8 to 0
$\frac{3\pi}{4}$ to π	$\frac{3\pi}{2}$ to 2π	0 to 1	0 to 8
π to $\frac{5\pi}{4}$	2π to $\frac{5\pi}{2}$	1 to 0	8 to 0
$\frac{5\pi}{4}$ to $\frac{3\pi}{2}$	$\frac{5\pi}{2}$ to 3π	0 to -1	0 to -8
$\frac{3\pi}{2}$ to $\frac{7\pi}{4}$	3π to $\frac{7\pi}{2}$	-1 to 0	-8 to 0
$\frac{7\pi}{4}$ to 2π	$\frac{7\pi}{2}$ to 4π	0 to 1	0 to 8

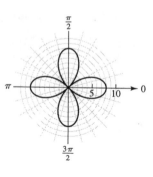

51. $x^2 = 6y$

Use $x = r\cos\theta$ and $y = r\sin\theta$

$$(r\cos\theta)^2 = 6(r\sin\theta)$$
$$r^2\cos^2\theta = 6r\sin\theta$$
$$r^2\cos^2\theta - 6r\sin\theta = 0$$
$$r(r\cos^2\theta - 6\sin\theta) = 0$$
$$r = 0 \text{ or } r\cos^2\theta - 6\sin\theta = 0$$
$$r\cos^2\theta = 6\sin\theta$$
$$r = \frac{6\sin\theta}{\cos^2\theta} = 6\frac{\sin\theta}{\cos\theta}\frac{1}{\cos\theta} = 6\tan\theta\sec\theta$$

The graph of $r = 0$ is the pole, and since the pole is included as a solution of $r = 6\tan\theta\sec\theta$ (let $\theta = 0$), we can discard $r = 0$ and keep only $r = 6\tan\theta\sec\theta$.

52. $r = 4\sin\theta$

We multiply both sides by r, which adds the pole to the graph. But the pole is already part of the graph (let $\theta = 0$), so we have changed nothing. $r^2 = 4r\sin\theta$. But $r^2 = x^2 + y^2$, $r\sin\theta = y$. Hence, $x^2 + y^2 = 4y$.

53. A sketch shows that $3 + 3i$ is associated with a special 45° reference triangle in the first quadrant and $-1 + i\sqrt{3}$ is associated with a special 30°–60° reference triangle in the second quadrant. Thus,

$$3 + 3i = 3\sqrt{2}(\cos 45° + i\sin 45°)$$
$$= 3\sqrt{2}\,e^{45°i}$$
$$-1 + i\sqrt{3} = 2(\cos 120° + i\sin 120°)$$
$$= 2e^{120°i}$$

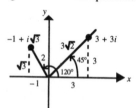

Therefore, $(3 + 3i)(-1 + i\sqrt{3}) = 3\sqrt{2}\,e^{45°i} \cdot 2e^{120°i} = 3\sqrt{2}\cdot 2e^{(45° + 120°)i} = 6\sqrt{2}\,e^{165°i}$

54. Using the results of the previous problem,

$$\frac{-1 + i\sqrt{3}}{3 + 3i} = \frac{2e^{120°i}}{3\sqrt{2}\,e^{45°i}} = \frac{2}{3\sqrt{2}}\,e^{(120° - 45°)i} = \frac{\sqrt{2}}{3}\,e^{75°i}$$

55. $(1 - i)^6 = (\sqrt{2}\,e^{-45°i})^6 = (\sqrt{2})^6\,e^{6(-45°i)} = (2^{1/2})^6\,e^{-270°i} = 2^3\,e^{-270°i}$
$$= 8e^{-270°i} = 8[\cos(-270°) + i\sin(-270°)] = 8[0 + i \cdot 1] = 8i$$

56. First, write $8i$ in polar form.
$$8i = 8e^{90°i}$$

From the nth root theorem, all three roots are given by
$$8^{1/3}\,e^{(90°/3 + k360°/3)i} \qquad k = 0, 1, 2$$

Thus,

$$w_1 = 8^{1/3}\,e^{(30° + 0 \cdot 120°)i} = 2e^{30°i} = 2(\cos 30° + i\sin 30°) = 2\left(\frac{\sqrt{3}}{2} + i\,\frac{1}{2}\right) = \sqrt{3} + i$$

$$w_2 = 8^{1/3}\,e^{(30° + 1 \cdot 120°)i} = 2e^{150°i} = 2(\cos 150° + i\sin 150°) = 2\left(-\frac{\sqrt{3}}{2} + i\,\frac{1}{2}\right) = -\sqrt{3} + i$$

$$w_3 = 8^{1/3}\,e^{(30° + 2 \cdot 120°)i} = 2e^{270°i} = 2(\cos 270° + i\sin 270°) = 2\,(0 + i(-1)) = -2i$$

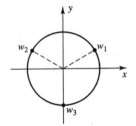

57.

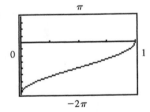

58.

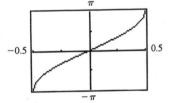

59.

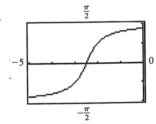

60. Graph $y1 = \tan x$ and $y2 = 5$ in the same viewing window in a graphing utility and find the points of intersection using an automatic intersection routine. The intersection points are found to be -1.768 and 1.373.

    Check: $\tan(-1.768) = 5.005$
    $\tan(1.373) = 4.990$

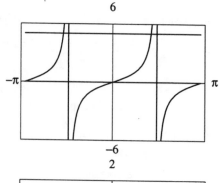

61. Graph $y1 = \cos x$ and $y2 = \sqrt[3]{x}$ in the same viewing window in a graphing utility and find the points of intersection using an automatic intersection routine. The intersection point is found to be 0.582.

    Check: $\cos(0.582) = 0.835$

    $\sqrt[3]{0.582} = 0.835$

    Since $|\cos x| \le 1$, while $|\sqrt[3]{x}| > 1$ for real x not shown, there can be no other solutions.

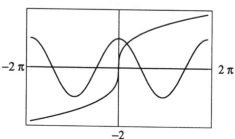

62. Graph $y1 = 3 \sin 2x \cos 3x$ and $y2 = 2$ in the same viewing window in a graphing utility and find the points of intersection using an automatic intersection routine. The intersection points are found to be 3.909, 4.313, 5.111, and 5.516.

    Check: $3 \sin[2(3.909)] \cos[3(3.909)] = 2.002$

    (The remaining checking is left to the student.)

63.

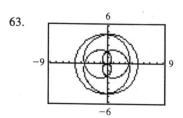

64. (A) $r = \dfrac{2}{1 - 0.7 \sin(\theta + 0.6)}$

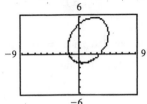

The graph is an ellipse.

(B) $r = \dfrac{2}{1 - \sin(\theta + 0.6)}$

The graph is a parabola.

(C) $r = \dfrac{2}{1 - 1.5 \sin(\theta + 0.6)}$

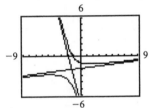

The graph is a hyperbola.

65. Since $\theta = \dfrac{s}{r}$, and $s = 8$, and $r = 2$, we have $\theta = \dfrac{8}{2} = 4$ rad.

Since $\cos \theta = \dfrac{a}{r}$ and $\sin \theta = \dfrac{b}{r}$, we have

$a = r \cos \theta = 2 \cos 4$ and $b = r \sin \theta = 2 \sin 4$

Thus, $(a, b) = (2 \cos 4, 2 \sin 4) = (-1.307, -1.514)$.

66. Since $\tan \theta = \dfrac{b}{a}$, we have $\tan \theta = \dfrac{-1.2}{-1.6} = 0.75$. Since (a, b) is in Quadrant III,

$\theta = \tan^{-1} \theta + \pi = \tan^{-1} 0.75 + \pi = 3.785$ rad

Since $\theta = \dfrac{s}{r}$, we can write $3.785 = \dfrac{s}{2}$, $s = 2(3.785) = 7.570$ units

67. We first find the period and phase shift by solving

$$\pi x + \frac{\pi}{4} = 0 \qquad \text{and} \qquad \pi x + \frac{\pi}{4} = 2\pi$$

$$x = -\frac{1}{4} \qquad\qquad\qquad x = -\frac{1}{4} + 2$$

$$\text{Period} = 2 \qquad\qquad \text{Phase Shift} = -\frac{1}{4}$$

Now, since $2 \sec\left(\pi x + \dfrac{\pi}{4}\right) = \dfrac{1}{\dfrac{1}{2}\cos\left(\pi x + \dfrac{\pi}{4}\right)}$, we graph $y = \dfrac{1}{2}\cos\left(\pi x + \dfrac{\pi}{4}\right)$ for one cycle from

$-\dfrac{1}{4}$ to $-\dfrac{1}{4} + 2 = \dfrac{7}{4}$ with a broken line graph, then take reciprocals. We also place asymptotes through the x intercepts of the cosine graph to guide us when we sketch the secant function.

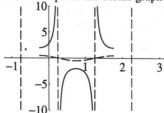

We then extend the one cycle over the required interval from -1 to 3.

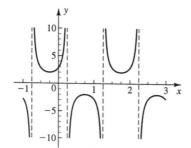

68. Let $y = \tan^{-1} x \quad -\dfrac{\pi}{2} < y < \dfrac{\pi}{2}$ or, equivalently, $x = \tan y \quad -\dfrac{\pi}{2} < y < \dfrac{\pi}{2}$

Geometrically,

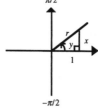   or

In either case, $r = \sqrt{x^2 + 1}$

$$\sec(2\tan^{-1} x) = \sec(2y) = \dfrac{1}{\cos 2y} = \dfrac{1}{\cos^2 y - \sin^2 y} = 1 \div (\cos^2 y - \sin^2 y)$$

$$= 1 \div \left[\left(\dfrac{1}{\sqrt{x^2 + 1}}\right)^2 - \left(\dfrac{x}{\sqrt{x^2 + 1}}\right)^2\right]$$

$$= 1 \div \left[\dfrac{1}{x^2 + 1} - \dfrac{x^2}{x^2 + 1}\right]$$

$$= 1 \div \dfrac{1 - x^2}{x^2 + 1} = \dfrac{x^2 + 1}{1 - x^2} \text{ or } \dfrac{1 + x^2}{1 - x^2}$$

69. $\tan 3x = \tan(x + 2x)$ Algebra

$$= \frac{\tan x + \tan 2x}{1 - \tan x \tan 2x}$$ Sum identity

$$= \frac{\tan x + \dfrac{2 \tan x}{1 - \tan^2 x}}{1 - \tan x \cdot \dfrac{2 \tan x}{1 - \tan^2 x}}$$ Double-angle identity

$$= \frac{(1 - \tan^2 x)}{(1 - \tan^2 x)} \cdot \frac{\tan x + \dfrac{2 \tan x}{1 - \tan^2 x}}{1 - \dfrac{2 \tan^2 x}{1 - \tan^2 x}}$$ Algebra

$$= \frac{(1 - \tan^2 x) \tan x + 2 \tan x}{1 - \tan^2 x - 2 \tan^2 x}$$ Algebra

$$= \frac{\tan x - \tan^3 x + 2 \tan x}{1 - 3 \tan^2 x}$$ Algebra

$$= \frac{3 \tan x - \tan^3 x}{1 - 3 \tan^2 x}$$ Algebra

$$= \frac{\tan x (3 - \tan^2 x)}{1 - 3 \tan^2 x}$$ Algebra

$$= \frac{1}{\cot x} \frac{3 - \tan^2 x}{1 - 3 \tan^2 x}$$ Reciprocal identity

$$= \frac{3 - \tan^2 x}{\cot x - 3 \tan^2 x \cot x}$$ Algebra

$$= \frac{3 - \tan^2 x}{\cot x - 3 \tan x \cdot (\tan x \cot x)}$$ Algebra

$$= \frac{3 - \tan^2 x}{\cot x - 3 \tan x \cdot 1}$$ Reciprocal identity

$$= \frac{3 - \tan^2 x}{\cot x - 3 \tan x}$$ Algebra

70. Use $r = \sqrt{x^2 + y^2}$ and $r \cos \theta = x$

$$r(\cos \theta + 1) = 1$$

$$r \cos \theta + r = 1$$

$$x + \sqrt{x^2 + y^2} = 1$$

$$x = 1 - \sqrt{x^2 + y^2}$$

$$x - 1 = -\sqrt{x^2 + y^2} \text{ or, squaring both sides}$$

$$(x - 1)^2 = (-\sqrt{x^2 + y^2})^2 = x^2 + y^2$$

71. (A)

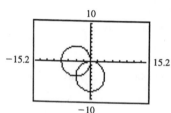

(B) The pole has no ordered pairs of solutions that simultaneously satisfy both equations.

As $(0, 0)$, it satisfies the first; as $\left(0, \dfrac{\pi}{2}\right)$ it satisfies the second; it is not a solution of the system.

72. $x^3 - 4 = 0$
$x^3 = 4$

Therefore x is a cube root of 4, and there are three of them. First, write 4 in polar form and use the nth root theorem.

$$4 = 4e^{0°i}$$

All three cube roots of 4 are given by

$$4^{1/3}\, e^{(0°/3 + k360°/3)i} \qquad k = 0, 1, 2$$

Thus,

$$w_1 = 4^{1/3}\, e^{(0° + 0\,\cdot\,120°)i} = 4^{1/3}\, e^{0°i} = 4^{1/3}\,(\cos 0° + i \sin 0°) = 1.587$$

$$w_2 = 4^{1/3}\, e^{(0° + 1\,\cdot\,120°)i} = 4^{1/3}\, e^{120°i} = 4^{1/3}\,(\cos 120° + i \sin 120°) = -0.794 + 1.375i$$

$$w_3 = 4^{1/3}\, e^{(0° + 2\,\cdot\,120°)i} = 4^{1/3}\, e^{240°i} = 4^{1/3}\,(\cos 240° + i \sin 240°) = -0.794 - 1.375i$$

73. (A) By DeMoivre's theorem, $(\cos \theta + i \sin \theta)^3 = (e^{\theta i})^3 = e^{3\theta i} = \cos 3\theta + i \sin 3\theta$

By the binomial theorem,

$$(\cos \theta + i \sin \theta)^3 = \cos^3 \theta + 3 \cos^2 \theta (i \sin \theta) + 3 \cos \theta (i \sin \theta)^2 + (i \sin \theta)^3$$
$$= \cos^3 \theta + 3i \cos^2 \theta \sin \theta - 3 \cos \theta \sin^2 \theta - i \sin^3 \theta$$

Thus,

$$\cos 3\theta + i \sin 3\theta = \cos^3 \theta - 3 \cos \theta \sin^2 \theta + i\,(3 \cos^2 \theta \sin \theta - \sin^3 \theta)$$

Equating the real and imaginary parts of the left and right sides, we obtain

$$\cos 3\theta = \cos^3 \theta - 3 \cos \theta \sin^2 \theta \quad \text{and} \quad \sin 3\theta = 3 \cos^2 \theta \sin \theta - \sin^3 \theta$$

(B) $\cos 3\theta = \cos (\theta + 2\theta)$ Algebra

$\qquad = \cos \theta \cos 2\theta - \sin \theta \sin 2\theta$ Sum identity

$\qquad = \cos \theta (\cos^2 \theta - \sin^2 \theta) - \sin \theta (2 \sin \theta \cos \theta)$ Double-angle identity

$\qquad = \cos^3 \theta - \cos \theta \sin^2 \theta - 2 \cos \theta \sin^2 \theta$ Algebra

$\qquad = \cos^3 \theta - 3 \cos \theta \sin^2 \theta$ Algebra

$\sin 3\theta = \sin (\theta + 2\theta)$ Algebra

$\qquad = \sin \theta \cos 2\theta + \cos \theta \sin 2\theta$ Sum identity

$\qquad = \sin \theta (\cos^2 \theta - \sin^2 \theta) + \cos \theta (2 \sin \theta \cos \theta)$ Double-angle identities

$\qquad = \sin \theta \cos^2 \theta - \sin^3 \theta + 2 \sin \theta \cos^2 \theta$ Algebra

$\qquad = 3 \sin \theta \cos^2 \theta - \sin^3 \theta$ Algebra

74. The graph of $f(x)$ is shown in the figure.
The graph appears to be a basic cosine curve with period π,

amplitude $= \dfrac{1}{2}(y_{max} - y_{min}) = \dfrac{1}{2}[2 - (-4)] = 3$, displaced

downward by $k = 1$ unit.
It appears that $g(x) = -1 + 3\cos 2x$ would be an
appropriate choice.

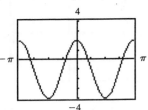

We verify $f(x) = g(x)$ as follows:

$$\begin{aligned}
f(x) &= 2\cos^2 x - 4\sin^2 x \\
&= 2\cos^2 x - 4(1 - \cos^2 x) && \text{Pythagorean identity} \\
&= 2\cos^2 x - 4 + 4\cos^2 x && \text{Algebra} \\
&= 6\cos^2 x - 4 && \text{Algebra} \\
&= 3(2\cos^2 x - 1) - 1 && \text{Algebra} \\
&= 3\cos 2x - 1 && \text{Double-angle identity} \\
&= -1 + 3\cos 2x = g(x) && \text{Algebra}
\end{aligned}$$

75. The graph of $f(x)$ is shown in the figure.
The graph appears to be have vertical asymptotes

$x = -\dfrac{3\pi}{4}, -\dfrac{\pi}{4}, \dfrac{\pi}{4}$, and $\dfrac{3\pi}{4}$ and period π. It appears to have

high and low points with y coordinates -4 and -2,
respectively. It appears that $g(x) = \sec 2x - 3$ would be an
appropriate choice.

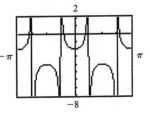

We verify $f(x) = g(x)$ as follows:

$$\begin{aligned}
f(x) &= \frac{6\sin^2 x - 2}{2\cos^2 x - 1} \\[2mm]
&= \frac{3(2\sin^2 x - 1) + 1}{2\cos^2 x - 1} && \text{Algebra} \\[2mm]
&= \frac{1 - 3(1 - 2\sin^2 x)}{2\cos^2 x - 1} && \text{Algebra} \\[2mm]
&= \frac{1 - 3\cos 2x}{\cos 2x} && \text{Double-angle identities} \\[2mm]
&= \frac{1}{\cos 2x} - 3 && \text{Algebra} \\[2mm]
&= \sec 2x - 3 = g(x) && \text{Reciprocal identity}
\end{aligned}$$

76. The graph of $f(x)$ is shown in the figure.

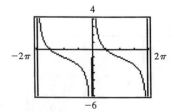

The graph appears to be have vertical asymptotes $x = -2\pi$, $x = 0$, and $x = -2\pi$, and period 2π. It appears to have x intercepts $-\dfrac{3\pi}{2}$ and $\dfrac{\pi}{2}$, and symmetry with respect to points where the curve crosses the line $y = -1$. It appears to be a cotangent curve displaced downward by $|k| = 1$ unit. It appears that $g(x) = -1 + \cot\dfrac{x}{2}$ would be an appropriate choice.

We verify $f(x) = g(x)$ as follows:

$$f(x) = \frac{\sin x + \cos x - 1}{1 - \cos x}$$

$$= \frac{\sin x - (1 - \cos x)}{1 - \cos x} \qquad \text{Algebra}$$

$$= \frac{\sin x}{1 - \cos x} - \frac{1 - \cos x}{1 - \cos x} \qquad \text{Algebra}$$

$$= 1 + \frac{1 - \cos x}{\sin x} - 1 \qquad \text{Algebra}$$

$$= 1 + \tan\frac{x}{2} - 1 \qquad \text{Half-angle identity}$$

$$= \cot\frac{x}{2} - 1 = g(x) \qquad \text{Reciprocal identity}$$

77. We are given two angles and the included side (ASA). We find the third angle, then apply the law of sines to find side BC.

$$\angle ABC + \angle BCA + \angle CAB = 180°$$

$$\angle ABC = 180° - (52° + 77°) = 51°$$

$$\frac{\sin CAB}{BC} = \frac{\sin ABC}{AC}$$

$$BC = \frac{AC \sin CAB}{\sin ABC} = \frac{(520 \text{ ft}) \sin 77°}{\sin 51°} = 650 \text{ ft}$$

78. Here we are given two sides and the included angle, hence we can use the law of cosines to find side BC.

$$BC^2 = AB^2 + AC^2 - 2(AB)(AC) \cos CAB$$

$$= (580)^2 + (430)^2 - 2(580)(530) \cos 64° = 302{,}640.4\ldots$$

$$BC = 550 \text{ ft}$$

79. (A) Triangle ABC is a right triangle.

$$\tan BAC = \frac{BC}{AC}$$

$$BC = AC \tan BAC$$

$$= (35 \text{ ft}) \tan 54° = 48 \text{ ft.}$$

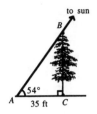

(B) Here triangle ABC is an oblique triangle.

$$\angle BAC = 54° - 11° = 43°$$

$$\angle BCA = 90° + 11° = 101°.$$

We are given two angles and the included side.
We find the third angle, then apply the law of sines to find side BC.

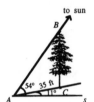

$$\angle ABC + \angle ACB + \angle BAC = 180°$$

$$\angle ABC = 180° - (43° + 101°) = 36°$$

$$\frac{\sin ABC}{AC} = \frac{\sin BAC}{BC}$$

$$BC = \frac{AC \sin BAC}{\sin ABC} = \frac{(35 \text{ ft}) \sin 43°}{\sin 36°} = 41 \text{ ft.}$$

80. In previous exercises, we have solved similar problems using right triangle methods. (See Chapter 1, Review Exercise, Problem 37, for example.)

For comparison, we solve this problem using oblique triangle methods. We are given two angles, $\angle ABC = 180° - 67° = 113°$ and $\angle CAB = 42°$, and the included side, hence, we can find the third angle, then use the law of sines to find the other two sides.

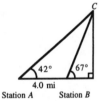

$$\angle ABC + \angle CAB + \angle BCA = 180°$$

$$\angle BCA = 180° - (113° + 42°) = 25°$$

Solve for BC:

$$\frac{\sin CAB}{BC} = \frac{\sin BCA}{AB}$$

$$BC = \frac{AB \sin CAB}{\sin BCA}$$

$$= \frac{(4.0 \text{ mi}) \sin 42°}{\sin 25°}$$

$$= 6.3 \text{ mi from Station } B$$

Solve for AC:

$$\frac{\sin ABC}{AC} = \frac{\sin BCA}{AB}$$

$$AC = \frac{AB \sin ABC}{\sin BCA}$$

$$= \frac{(4.0 \text{ mi}) \sin 113°}{\sin 25°}$$

$$= 8.7 \text{ mi from Station } A$$

81. (A) Period $= \dfrac{1}{\text{Frequency}} = \dfrac{1}{70 \text{ Hz}} = \dfrac{1}{70}$ sec. Since Period $= \dfrac{2\pi}{B}$, $B = \dfrac{2\pi}{\text{Period}} = \dfrac{2\pi}{1/70} = 140\pi$

(B) Frequency $= \dfrac{1}{\text{Period}} = \dfrac{1}{0.0125 \text{ sec}} = 80$ Hz. Since Period $= \dfrac{2\pi}{B}$, $B = \dfrac{2\pi}{\text{Period}} = \dfrac{2\pi}{0.0125} = 160\pi$

(C) Period $= \dfrac{2\pi}{B} = \dfrac{2\pi}{100\pi} = \dfrac{1}{50}$ sec. Frequency $= \dfrac{1}{\text{Period}} = \dfrac{1}{(1/50)\ \text{sec}} = 50$ Hz

82. The height of the wave from trough to crest is the difference in height between the crest (height A) and the trough (height $-A$). In this case, $A = 2$ ft.

$$A - (-A) = 2A = 2(2\ \text{ft}) = 4\ \text{ft.}$$

To find the wavelength λ, we note, $\quad \lambda = 5.12T^2 \quad\quad T = 4$ sec $\quad\quad \lambda = 5.12(4)^2 \approx 82$ ft

To find the speed S, we use

$$S = \sqrt{\dfrac{g\lambda}{2\pi}} \quad\quad g = 32\ \text{ft/sec}^2 \quad\quad S = \sqrt{\dfrac{32(82)}{2\pi}} \approx 20\ \text{ft/sec}$$

83. Area $OCBA$ = Area of Sector OCB + Area of triangle OAB. Area of Sector $OCB = \dfrac{1}{2}r^2\theta = \dfrac{1}{2}\cdot 1^2 \cdot \theta$

$$= \dfrac{1}{2}\theta$$

Area of right triangle $OAB = \dfrac{1}{2}$ (base)(height) $= \dfrac{1}{2}xy$

Since OAB is a right triangle, we have

$$\sin\theta = \dfrac{x}{1} \quad\quad x = \sin\theta \quad\quad\quad\quad \cos\theta = \dfrac{y}{1} \quad\quad y = \cos\theta$$

Hence, area of triangle $OAB = \dfrac{1}{2}xy = \dfrac{1}{2}\sin\theta\cos\theta.$

Thus, Area of $OCBA = \dfrac{1}{2}\theta + \dfrac{1}{2}\sin\theta\cos\theta$

84. Since $x = \sin\theta$, $\theta = \sin^{-1}x$ (θ is acute)

Since OAB is a right triangle, applying the Pythagorean theorem, we have

$$x^2 + y^2 = 1^2$$
$$y^2 = 1 - x^2$$
$$y = \sqrt{1 - x^2}$$

Thus, Area of $OCBA = \dfrac{1}{2}\theta + \dfrac{1}{2}xy$ (see previous problem)

$$= \dfrac{1}{2}\sin^{-1}x + \dfrac{1}{2}x\sqrt{1 - x^2}$$

85. We are to solve $\dfrac{1}{2}\theta + \dfrac{1}{2}\sin\theta\cos\theta = 0.5$. We graph

$\text{y1} = \dfrac{1}{2}\theta + \dfrac{1}{2}\sin\theta\cos\theta$ and y2 = 0.5 on the interval

from 0 to $\dfrac{\pi}{2}$. From the figure, we see that y1 and y2

intersect once on the interval. Using an automatic intersection routine the solution is found to be $\theta = 0.553$.

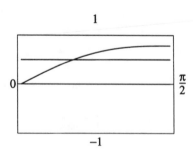

86. We are to solve $\dfrac{1}{2}\sin^{-1} x + \dfrac{1}{2}x\sqrt{1-x^2} = 0.4$.

We graph $y1 = \dfrac{1}{2}\sin^{-1} x + \dfrac{1}{2}x\sqrt{1-x^2}$ and $y2 = 0.4$
on the interval from 0 to 1. From the figure, we see
that y1 and y2 intersect once on the interval. Using an
automatic intersection routine the solution is found to
be $x = 0.412$.

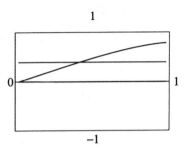

87. (A) Since the right triangles in the figure are similar, we can write $\dfrac{r}{h} = \dfrac{R}{H}$.

Since $\tan \alpha = \dfrac{r}{h} = \dfrac{R}{H}$, we can write $\dfrac{r}{h} = \tan \alpha$ $r = h \tan \alpha$

$\dfrac{R}{H} = \tan \alpha$ $R = H \tan \alpha$

Then, $R = H \tan \alpha = (H - h + h) \tan \alpha = (H - h) \tan \alpha + h \tan \alpha$
$R = (H - h) \tan \alpha + r$

(B) Solving the previous equation for $\tan \alpha$, we can write

$R - r = (H - h) \tan \alpha$ $\tan \alpha = \dfrac{R - r}{H - h}$

Since α and β are complementary angles, we can write

$\tan \beta = \cot \alpha = \dfrac{1}{\tan \alpha} = 1 + \tan \alpha = 1 + \dfrac{R - r}{H - h} = \dfrac{H - h}{R - r}$

Thus, $\beta = \tan^{-1}\left(\dfrac{H - h}{R - r}\right)$.

88. We require θ such that the actual velocity $\mathbf{R}$ will be the resultant of the
apparent velocity $\mathbf{v}$ and the wind velocity $\mathbf{w}$. From the diagram it should be
clear that

$\sin \theta = \dfrac{|\mathbf{w}|}{|\mathbf{v}|} = \dfrac{81.5}{265}$ $\theta = \sin^{-1}\dfrac{81.5}{265} = 18°$

The ground speed for this course will be the magnitude $|\mathbf{R}|$ of the actual
velocity. In the right triangle, ABC, we have

$\cos \theta = \dfrac{|\mathbf{R}|}{|\mathbf{v}|}$ $|\mathbf{R}| = |\mathbf{v}| \cos \theta = 265 \cos\left(\sin^{-1}\dfrac{81.5}{265}\right)$

$= 265 \sqrt{1 - \left(\dfrac{81.5}{265}\right)^2} = 252 \text{ mph}$

89. (A) First, form a force diagram with all force vectors in standard position at the origin.

Let $\mathbf{F}_1 =$ the tension in the left side
$\mathbf{F}_2 =$ the tension in the right side

$$|\mathbf{F}_1| = T_L \qquad |\mathbf{F}_2| = T_R \qquad |\mathbf{W}| = w$$

Write each force vector in terms of $\mathbf{i}$ and $\mathbf{j}$ unit vectors.

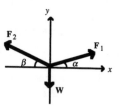

$$\mathbf{F}_1 = T_R \cos \alpha \mathbf{i} + T_R \sin \alpha \mathbf{j} \qquad \mathbf{F}_2 = T_L (-\cos \beta)\mathbf{i} + T_L \sin \beta \mathbf{j} \qquad \mathbf{W} = -w\mathbf{j}$$

For the system to be in static equilibrium, we must have $\mathbf{F}_1 + \mathbf{F}_2 + \mathbf{W} = \mathbf{0}$ which becomes, on addition,

$$(T_R \cos \alpha - T_L \cos \beta)\mathbf{i} + (T_R \sin \alpha + T_L \sin \beta - w)\mathbf{j} = 0\mathbf{i} + 0\mathbf{j}$$

Since two vectors are equal if and only if their corresponding components are equal, we are led to the following system of equations in T_L and T_R :

$$T_R \cos \alpha - T_L \cos \beta = 0 \qquad T_R \sin \alpha + T_L \sin \beta - w = 0$$

Solving, $\quad T_R = T_L \dfrac{\cos \beta}{\cos \alpha} \qquad\qquad T_L \sin \alpha \dfrac{\cos \beta}{\cos \alpha} + T_L \sin \beta = w$

$$T_L \left(\frac{\sin \alpha \cos \beta}{\cos \alpha} + \sin \beta \right) = w$$

$$T_L \left(\frac{\sin \alpha \cos \beta + \cos \alpha \sin \beta}{\cos \alpha} \right) = w$$

$$T_L \frac{\sin (\alpha + \beta)}{\cos \alpha} = w$$

Thus, $T_L = \dfrac{w \cos \alpha}{\sin (\alpha + \beta)}$. Hence, $T_R = \dfrac{w \cos \alpha}{\sin (\alpha + \beta)} \dfrac{\cos \beta}{\cos \alpha} = \dfrac{w \cos \beta}{\sin (\alpha + \beta)}$

(B) If $\alpha = \beta$, then

$$T_L = T_R = \frac{w \cos \alpha}{\sin (\alpha + \alpha)} = \frac{w \cos \alpha}{\sin 2\alpha} = \frac{w \cos \alpha}{2 \sin \alpha \cos \alpha} = \frac{w}{2 \sin \alpha} = \frac{w}{2} \frac{1}{\sin \alpha}$$

$$= \frac{1}{2} w \csc \alpha$$

90. We can apply the law of cosines to the triangle shown in the figure. Then,

$$100^2 = r^2 + r^2 - 2r \cdot r \cdot \cos \theta$$

$$10{,}000 = 2r^2 - 2r^2 \cos \theta = 2r^2 (1 - \cos \theta)$$

$$5000 = r^2 (1 - \cos \theta)$$

(A) Given $\theta = 10°$, then $5000 = r^2 (1 - \cos 10°)$

$$r^2 = \frac{5000}{1 - \cos 10°}$$

$$r = \sqrt{\frac{5000}{1 - \cos 10°}} = 574 \text{ ft}$$

(B) Given $r = 2000$, then $5000 = (2000)^2 (1 - \cos \theta)$

$$\frac{5000}{(2000)^2} = 1 - \cos \theta$$

$$\cos \theta = 1 - \frac{5000}{(2000)^2}$$

$$\theta = \cos^{-1}\left[1 - \frac{5000}{(2000)^2}\right] = 2.9°$$

91. (A) Since $s = r\theta$, we can write $r\theta = 50$. To determine θ, we note that triangle ABC is a right triangle, with side $AC = r - 1$. Then,

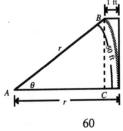

$$\cos \theta = \frac{AB}{AC} = \frac{r - 1}{r}$$

$$\theta = \cos^{-1}\left(\frac{r - 1}{r}\right)$$

Thus, $r \cos^{-1}\left(\frac{r - 1}{r}\right) = 50$. To solve this, we

graph $y1 = r \cos^{-1}\left(\frac{r - 1}{r}\right)$ and $y2 = 50$ on the interval from 1000 to 2000. From the figure, we see that y1 and y2 intersect once on the interval.

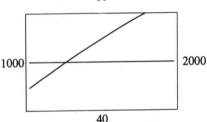

Using an automatic intersection routine, the solution is found to be $r = 1{,}250$ ft.

(B) From Problem 90, we have $5000 = r^2 (1 - \cos \theta)$. If $r = 1250$, then

$$5000 = (1250)^2 (1 - \cos \theta)$$

$$\frac{5000}{(1250)^2} = 1 - \cos \theta$$

$$\cos \theta = 1 - \frac{5000}{(1250)^2}$$

$$\theta = \cos^{-1}\left[1 - \frac{5000}{(1250)^2}\right] = 4.6°$$

92. The areas of the end pieces are given by the formula from Section 5.4, Problem 69 to be $\frac{1}{2}r^2 (\theta - \sin \theta)$. Subtracting this twice from the area of the total cross-section, πr^2, we obtain

$$A = \pi r^2 - \frac{1}{2}r^2 (\theta - \sin \theta) - \frac{1}{2}r^2 (\theta - \sin \theta) = \pi r^2 - r^2 (\theta - \sin \theta)$$

$$= \pi r^2 - r^2\theta + r^2 \sin \theta = r^2 (\pi - \theta + \sin \theta)$$

93. If all three pieces of the log have the same cross-sectional area, then each of these areas is one-third of the entire area, that is,

$$r^2(\pi - \theta + \sin \theta) = \frac{1}{3}\pi r^2$$

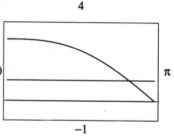

Thus, $\pi - \theta + \sin \theta = \frac{1}{3}\pi$. To solve this, we graph

$y1 = \pi - \theta + \sin \theta$ and $y2 = \frac{1}{3}\pi$ on the interval from 0 to π.

From the figure, we see that y1 and y2 intersect once on the interval.

Using an automatic intersection routine, the solution is found to be $\theta = 2.6053$ rad.

94. (A)

x months	1, 13	2, 14	3, 15	4, 16	5, 17	6, 18	7, 19	8, 20
$y \left(\begin{array}{c}\text{twilight} \\ \text{duration}\end{array}\right)$	1.62	1.82	2.35	2.98	3.55	4.12	4.05	3.50

x months	9, 21	10, 22	11, 23	12, 24
$y \left(\begin{array}{c}\text{twilight} \\ \text{duration}\end{array}\right)$	2.80	2.22	1.80	1.57

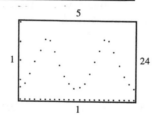

(B) From the table, Max $y = 4.12$ and Min $y = 1.57$. Then,

$$A = \frac{\text{Max } y - \text{Min } y}{2} = \frac{4.12 - 1.57}{2} = 1.275$$

$$B = \frac{2\pi}{\text{Period}} = \frac{2\pi}{12} = \frac{\pi}{6}$$

$$k = \text{Min } y + A = 1.57 + 1.275 = 2.845$$

From the plot in (A) or the table, we estimate the smallest value of x for which $y = k = 2.845$ to be approximately 3.4. Then, this is the phase shift for the graph. Substitute $B = \frac{\pi}{6}$ and $x = 3.4$ into the phase-shift equation.

$$x = -\frac{C}{B} \qquad 3.4 = -\frac{C}{\pi/6} \qquad C = \frac{-3.4\pi}{6} = -1.8$$

Thus, the equation required is $y = 2.845 + 1.275 \sin\left(\frac{\pi x}{6} - 1.8\right)$.

(C)

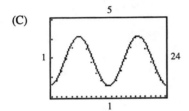

APPENDICES

APPENDIX A   Comments on Numbers

EXERCISE A.1   Real Numbers

1. There are infinitely many negative integers. Examples include –3. The only integer that is neither positive nor negative is 0. There are infinitely many positive integers. Examples include 5.

3. There are infinitely many rational numbers that are not integers. Examples include $\frac{2}{3}$.

5. (A) True.   (B) False (6 is a real number, but is not irrational.)   (C) True.

7. (A) $0.36\overline{36}$, rational   (B) $0.7\overline{77}$, rational   (C) $2.64575131...$, irrational
   (D) $1.625\overline{00}$, rational

9. (A) Since $\frac{26}{9} = 2.8\overline{88}$, it lies between 2 and 3.

   (B) Since $-\frac{19}{5} = -3.8$, it lies between –4 and –3.

   (C) Since $-\sqrt{23} = -4.7958...$, it lies between –5 and –4.

11. $y + 3$ 13. $(3 \cdot 2) x$ 15. $7x$ 17. $(5 + 7) + x$

19. $3m$ 21. $u + v$ 23. $(2 + 3) x$

25. (A) True. (This is the commutative property for addition of real numbers.)
    (B) False. For example, $4 - 2 \neq 2 - 4$.
    (C) True. (This is the commutative property for multiplication of real numbers.)
    (D) False. For example, $8 \div 2 \neq 2 \div 8$

EXERCISE A.2   Complex Numbers

1. $(3 - 2i) + (4 - 7i) = 3 - 2i + 4 + 7i = 3 + 4 - 2i + 7i = 7 + 5i$

3. $(3 - 2i) - (4 + 7i) = 3 - 2i - 4 - 7i = 3 - 4 - 2i + 7i = -1 - 9i$

5. $(6i)(3i) = 18i^2 = 18(-1) = -18$ 7. $(2i)(3 - 4i) = 6i - 8i^2 = 6i - 8(-1) = 8 + 6i$

9. $(3 - 4i)(1 - 2i) = 3 - 10i + 18i^2 = 3 - 10i + 8(-1) = 3 - 10i - 8 = -5 - 10i$

11. $(3 + 5i)(3 - 5i) = 9 - 25i^2 = 9 - 25(-1) = 9 + 25 = 34$

13. $\dfrac{1}{2 + i} = \dfrac{1}{2 + i} \dfrac{2 - i}{2 - i} = \dfrac{2 - i}{4 - i^2} = \dfrac{2 - i}{4 + 1} = \dfrac{2 - i}{5} = \dfrac{2}{5} - \dfrac{1}{5} i$

15. $\dfrac{2 - i}{3 + 2i} = \dfrac{2 - i}{3 + 2i} \dfrac{3 - 2i}{3 - 2i} = \dfrac{6 - 7i + 2i^2}{9 - 4i^2} = \dfrac{6 - 7i - 2}{9 + 4} = \dfrac{4 - 7i}{13} = \dfrac{4}{13} - \dfrac{7}{13} i$

17. $\dfrac{-1+2i}{4+3i} = \dfrac{-1+2i}{4+3i}\,\dfrac{4-3i}{4-3i} = \dfrac{-4+11i-6i^2}{16-9i^2} = \dfrac{-4+11i+6}{16+9} = \dfrac{2+11i}{25} = \dfrac{2}{25} + \dfrac{11}{25}\,i$

19. $(3+\sqrt{-4}) + (2-\sqrt{-16}) = (3+i\sqrt{4}) + (2-i\sqrt{16}) = 3+2i+2-4i = 3+2+2i-4i = 5-2i$

21. $(5-\sqrt{-1}) - (2-\sqrt{-36}) = (5-i) - (2-i\sqrt{36}) = (5-i) - (2-6i) = 5-i-2+6i = 5-2-i$
$+\,6i = 3+5i$

23. $(-3-\sqrt{-1})(-2+\sqrt{-49}) = (-3-i)(-2+i\sqrt{49}) = (-3-i)(-2+7i) = 6-19i-7i^2 = 6-19i+7$
$= 13-19i$

25. $\dfrac{5-\sqrt{-1}}{2+\sqrt{-4}} = \dfrac{5-i}{2+i\sqrt{4}} = \dfrac{5-i}{2+2i} = \dfrac{5-i}{2+2i}\,\dfrac{2-2i}{2-2i} = \dfrac{10-12i+2i^2}{4-4i^2} = \dfrac{10-12i-2}{4+4} = \dfrac{8-12i}{8}$
$= 1 - \dfrac{3}{2}\,i$

27. $(1-i)^2 = 2(1-i) + 2 = 1-2i+i^2 - 2+2i+2 = 1-2i-1-2+2i+2 = 0+0i = 0$

29. $\left(\dfrac{-1}{2}+\dfrac{\sqrt{3}}{2}\,i\right)^3 = \left(-\dfrac{1}{2}+\dfrac{\sqrt{3}}{2}\,i\right)\left(-\dfrac{1}{2}+\dfrac{\sqrt{3}}{2}\,i\right)^2$

$= \left(-\dfrac{1}{2}+\dfrac{\sqrt{3}}{2}\,i\right)\left[\left(-\dfrac{1}{2}\right)^2 + 2\left(-\dfrac{1}{2}\right)\left(\dfrac{\sqrt{3}}{2}\,i\right) + \left(\dfrac{\sqrt{3}}{2}\,i\right)^2\right]$

$= \left(-\dfrac{1}{2}+\dfrac{\sqrt{3}}{2}\,i\right)\left[\dfrac{1}{4} - \dfrac{2\sqrt{3}}{4}\,i + \dfrac{3}{4}\,i^2\right] = \left(-\dfrac{1}{2}+\dfrac{\sqrt{3}}{2}\,i\right)\left(\dfrac{1}{4} - \dfrac{2\sqrt{3}}{4}\,i - \dfrac{3}{4}\right)$

$= \left(-\dfrac{1}{2}+\dfrac{\sqrt{3}}{2}\,i\right)\left(-\dfrac{1}{2} - \dfrac{2\sqrt{3}}{4}\,i\right)$

$= \left(-\dfrac{1}{2}\right)\left(-\dfrac{1}{2}\right) + \left(-\dfrac{1}{2}\right)\left(-\dfrac{2\sqrt{3}}{4}\,i\right) + \left(\dfrac{\sqrt{3}}{2}\,i\right)\left(-\dfrac{1}{2}\right) + \left(\dfrac{\sqrt{3}}{2}\,i\right)\left(-\dfrac{2\sqrt{3}}{4}\,i\right)$

$= \dfrac{1}{4} + \dfrac{\sqrt{3}}{4}\,i - \dfrac{\sqrt{3}}{4}\,i - \dfrac{3}{4}\,i^2 = \dfrac{1}{4} + \dfrac{3}{4} = 1$

EXERCISE A.3   Significant Digits

1. $640 = 6.40. \times 10^2 = 6.4 \times 10^2$
2 places left
|
positive exponent

3. $5{,}460{,}000{,}000 = 5.460{,}000{,}000. \times 10^9 = 5.46 \times 10^9$
9 places left
|
positive exponent

5. $0.73 = 0.7.3 \times 10^{-1} = 7.3 \times 10^{-1}$
1 place right
|
negative exponent

7. $0.00000032 = 0.0000003.2 \times 10^{-7} = 3.2 \times 10^{-7}$
7 places right
|
negative exponent

9. $0.0000491 = 0.00004.91 \times 10^{-5} = 4.91 \times 10^{-5}$ 11. $67{,}000{,}000{,}000 = 6.7{,}000{,}000{,}000. \times 10^{10}$

         5 places right   10 places left

         negative exponent   positive exponent

13. $5.6 \times 10^4 = 5.6 \times 10{,}000 = 56{,}000$ 15. $9.7 \times 10^{-3} = 9.7 \times 0.001 = 0.0097$

17. $4.61 \times 10^{12} = 4.61 \times 1{,}000{,}000{,}000{,}000 = 4{,}610{,}000{,}000{,}000$

19. $1.08 \times 10^{-1} = 1.08 \times 0.1 = 0.108$

21. 12.3 has a decimal point. From the first nonzero digit (1) to the last digit (3), there are 3 digits. 3 significant digits.

23. 12.300 has a decimal point. From the first nonzero digit (1) to the last digit (0), there are 5 digits. 5 significant digits.

25. 0.01230 has a decimal point. From the first nonzero digit (1) to the last digit (0), there are 4 digits. 4 significant digits.

27. 6.7×10^{-1} is in scientific notation. There are 2 digits in 6.7. 2 significant digits.

29. 6.700×10^{-1} is in scientific notation. There are 4 digits in 6.700. 4 significant digits.

31. 7.090×10^5 is in scientific notation. There are 4 digits in 7.090. 4 significant digits.

33. 635,000   35. 86.8 (convention of leaving the digit before the 5 alone, if it is even)

37. 0.00465   39. $734 = 7.34 \times 10^2 \approx 7.3 \times 10^2$ 41. $0.040 = 4.0 \times 10^{-2}$

43. $0.000435 = 4.35 \times 10^{-4} \approx 4.4 \times 20^{-4}$ (convention of rounding the digit before the 5 up, if it is odd)

45. 3, since there are three significant digits in the number (32.8) with the least number of significant digits in the calculation.

47. 2, since there are two significant digits in the numbers (360 and 1,200) with the least number of significant digits in the calculation.

49. 1, since there is one significant digit in the number (6×10^4) with the least number of significant digits in the calculation.

51. $\dfrac{6.07}{0.5057}$ 6.07 has the least number of significant digits (3).

       $= 12.0$ Answer must have 3 significant digits.

53. $(6.14 \times 10^9)(3.154 \times 10^{-1})$ 6.14×10^9 has the least number of significant digits (3).

       $= 1.94 \times 10^9$ Answer must have 3 significant digits.

55. $\dfrac{6{,}730}{(2.30)(0.0551)}$ All numbers in the calculation have 3 significant digits.

       $= 53{,}100$ Answer must have 3 significant digits.

57. $C = 2\pi(25.31 \text{ cm})$

 $\quad = 159.0 \text{ cm}$

There are 4 significant digits in 25.31 cm.

Answer must have 4 significant digits.

59. $A = \dfrac{1}{2}(22.4 \text{ ft})(8.6 \text{ ft})$

 $\quad = 96 \text{ ft}^2$

8.6 has the least number of significant digits (2)

Answer must have 2 significant digits

61. $s = 4\pi(1.5 \text{ mm})^2$

 $\quad = 28 \text{ mm}^2$

There are 2 significant digits in 1.5 mm

Answer must have 2 significant digits

63. $V = \ell w h$

 $h = \dfrac{V}{\ell w}$

 $h = \dfrac{24.2 \text{ cm}^3}{(3.25 \text{ cm})(4.50 \text{ cm})}$

 $\quad = 1.65 \text{ cm}$

All numbers in the calculation have 3 significant digits

Answer must have 3 significant digits

65. $V = \dfrac{1}{3}\pi r^2 h$

 $3V = \pi r^2 h$

 $r^2 = \dfrac{3V}{\pi h}$

 $r = \sqrt{\dfrac{3V}{\pi h}} = \sqrt{\dfrac{3(1200 \text{ in}^3)}{\pi(6.55 \text{ in})}}$

 $\quad = 13 \text{ in}$

1200 in^3 has the least number of significant digits (2)

Answer must have 2 significant digits

APPENDICES

APPENDIX B   Functions and Inverse Functions

EXERCISE B.1   Functions

1. $f(x) = 4x - 1$
    $f(1) = 4(1) - 1$
    $= 3$

3. $f(x) = 4x - 1$
    $f(-1) = 4(-1) - 1$
    $= -5$

5. $f(x) = 4x - 1$
    $f(0) = 4(0) - 1$
    $= -1$

7. $g(x) = x - x^2$
    $g(1) = 1 - 1^2$
    $= 0$

9. $g(x) = x - x^2$
    $g(5) = 5 - 5^2$
    $= -20$

11. $g(x) = x - x^2$
    $g(-2) = -2 - (-2)^2$
    $= -2 - 4 = -6$

13. $f(0) + g(0) = 1 - 2 \cdot 0 + 4 - 0^2 = 1 + 4 = 5$

15. $\dfrac{f(3)}{g(1)} = \dfrac{1 - 2 \cdot 3}{4 - 1^2} = \dfrac{-5}{3} = -\dfrac{5}{3}$

17. $2f(-1) = 2[1 - 2(-1)] = 2(1 + 2) = 6$

19. $f(2 + h) = 1 - 2(2 + h) = 1 - 4 - 2h = -3 - 2h$

21. $\dfrac{f(2 + h) - f(2)}{h} = \dfrac{[1 - 2(2 + h)] - (1 - 2 \cdot 2)}{h} = \dfrac{-3 - 2h - (-3)}{h} = \dfrac{-2h}{h} = -2$

23. $g[f(2)] = g(1 - 2 \cdot 2) = g(-3) = 4 - (-3)^2 = 4 - 9 = -5$

25. $x^2 + y^2 = 25$ does not specify a function, since both $(3, 4)$ and $(3, -4)$ are solutions, in which the same domain value corresponds to more than one range value.

27. $2x - 3y = 6$ specifies a function, since each domain value x corresponds to exactly one range value $\left(\dfrac{2x - 6}{3} \right)$.

29. $y^2 = x$ does not specify a function, since both $(9, 3)$ and $(9, -3)$ are solutions, in which the same domain value corresponds to more than one range value.

31. $y = |x|$ specifies a function, since each domain value x corresponds to exactly one range value $(|x|)$.

33. $f(x) = x^2 - x + 1$ Domain $X = \{-2, -1, 0, 1, 2\}$
    $f(-2) = (-2)^2 - (-2) + 1 = 7$
    $f(-1) = (-1)^2 - (-1) + 1 = 3$
    $f(0) = 0^2 - 0 + 1 = 1$
    $f(1) = 1^2 - 1 + 1 = 1$
    $f(2) = 2^2 - 2 + 1 = 3$ Range $Y = \{1, 3, 7\}$

35. G does not specify a function since the domain value -4 corresponds to more than one range value (3 and 0). F specifies a function.
    Domain of $F =$ Set of all first components $= \{-2, -1, 0\} = X$
    Range of $F =$ Set of all second components $= \{0, 1\} = Y$

37. $s(t) = 4.88t^2$

$s(0) = 4.88(0)^2 = 0$

$s(1) = 4.88(1)^2 = 4.88$ m

$s(2) = 4.88(2)^2 = 19.52$ m

$s(3) = 4.88(3)^2 = 43.92$ m

39. $\dfrac{s(2 + h) - s(2)}{h} = \dfrac{4.88(2 + h)^2 - 4.88(2)^2}{h} = \dfrac{4.88(4 + 4h + h^2) - 19.52}{h}$

$= \dfrac{19.52 + 19.52h + 4.88h^2 - 19.52}{h} = \dfrac{19.52h + 4.88h^2}{h} = 19.52 + 4.88h$

As h gets closer to 0, this gets closer to 19.52; the average speed $\dfrac{s(2 + h) - s(2)}{h}$ tends to a quantity called the speed at $t = 2$.

EXERCISE B.2   Inverse Functions

1. This function is one-to-one since each range element corresponds to exactly one domain element.

3. This function is not one-to-one since the range element 9 corresponds to several domain elements.

5. This function passes the horizontal line test; it is one-to-one.

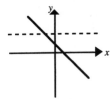

7. This function fails the horizontal line test; it is not one-to-one.

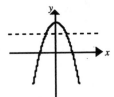

9. This function passes the horizontal line test; it is one-to-one.

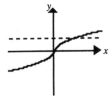

11. The function fails the horizontal line test; it is not one-to-one.

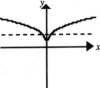

13. First, we note that f is not one-to-one, since the range element 0 corresponds to two domain elements, -1 and 2. The function g is one-to-one, since each range element corresponds to exactly one domain element. Reversing the ordered pairs in the function g produces the inverse function.

$g^{-1} = \{(-8, -2), (1, 1), (8, 2)\}$

Its domain is $\{-8, 1, 8\}$. Its range is $\{-2, 1, 2\}$.

15.

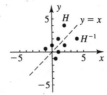

17. Replace $f(x)$ with y:

    Interchange the variables x and y to form f^{-1}:

                        f: $y = 2x - 7$

                        f^{-1}: $x = 2y - 7$

    Solve for y in terms of x: $x + 7 = 2y$

$$y = \frac{x + 7}{2}$$

    Replace y with $f^{-1}(x)$: $f^{-1}(x) = \dfrac{x + 7}{2}$

19. Replace $h(x)$ with y:

    Interchange the variables x and y to form h^{-1}:

                        h: $y = \dfrac{x + 3}{3}$

                        h^{-1}: $x = \dfrac{y + 3}{3}$

    Solve for y in terms of x: $3x = y + 3$

                              $y = 3x - 3$

    Replace y with $h^{-1}(x)$: $h^{-1}(x) = 3x - 3$

21.

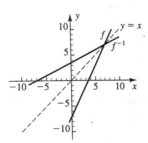

23. Replace $f(x)$ with y:

    Interchange the variables x and y to form f^{-1}:

                        f: $y = 2x - 7$

                        f^{-1}: $x = 2y - 7$

    Solve for y in terms of x: $x + 7 = 2y$

$$y = \frac{x + 7}{2}$$

    Replace y with $f^{-1}(x)$: $f^{-1}(x) = \dfrac{x + 7}{2}$

    Thus, $f^{-1}(3) = \dfrac{3 + 7}{2}$ $f^{-1}(3) = 5$

25. Replace $h(x)$ with y:

    Interchange the variables x and y to form h^{-1}:

                        h: $y = \dfrac{x}{3} + 1$

                        h^{-1}: $x = \dfrac{y}{3} + 1$

Solve for y in terms of x: $\quad x - 1 = \dfrac{y}{3}$

$$y = 3(x - 1) = 3x - 3$$

Replace y with $h^{-1}(x)$: $h^{-1}(x) = 3x - 3$

Thus, $h^{-1}(2) = 3 \cdot 2 - 3 \quad h^{-1}(2) = 3$

27. $f[f^{-1}(4)] = 2[f^{-1}(4)] - 7 = 2\left(\dfrac{4 + 7}{2}\right) - 7 = 4 + 7 - 7 = 4$

29. $h^{-1}[h(x)] = 3h(x) - 3 = 3\left(\dfrac{x}{3} + 1\right) - 3 = x + 3 - 3 = x$

31. $h[h^{-1}(x)] = \dfrac{h^{-1}(x)}{3} + 1 = \dfrac{3x - 3}{3} + 1 = x - 1 + 1 = x$